Bryophytes and Lichens in a Changing Environment

Edited by

Jeffrey W. Bates

Department of Biology
Imperial College of Science, Technology and Medicine
London

and

Andrew M. Farmer

English Nature
Peterborough

CLARENDON PRESS · OXFORD
1992

Oxford University Press, Walton Street, Oxford OX2 6DP
Oxford New York Toronto
Delhi Bombay Calcutta Madras Karachi
Petaling Jaya Singapore Hong Kong Tokyo
Nairobi Dar es Salaam Cape Town
Melbourne Auckland
and associated companies in
Berlin Ibadan

Oxford is a trade mark of Oxford University Press

Published in the United States
by Oxford University Press, New York

British Library Cataloguing in Publication Data
Data available

Library of Congress Cataloging in Publication Data
Bryophytes and lichens in a changing environment / edited by Jeffrey
W. Bates and Andrew M . Farmer.
1. Bryophytes—Ecology. 2. Lichens—Ecology. I. Bates, Jeffrey.
II. Farmer, Andrew.
QK533.B72 1992 588'.045—dc20 91–43550
ISBN 0–19–854291–7

Typeset by Cambridge Composing (UK) Ltd
Printed in Great Britain by
Biddles Ltd, Guildford & King's Lynn

Preface

This volume is a synthesis of new information about the responses of bryophytes and lichens to changing environmental conditions, particularly those brought about by human activities around the world. The subjects covered and the geographical scope have necessarily been dictated by the availability of information on organisms which are still incorrectly regarded as obscure or unimportant by many biologists, and consequently receive limited research attention. Bryophytes and lichens are often treated independently. This is exemplified by the separation of national societies (British Bryological Society, British Lichen Society) in the UK, in contrast to the situation in the USA (American Bryological and Lichenological Society). This separation has obvious advantages in considering taxonomy, evolutionary biology, cell biology, and some aspects of physiology, where there are few overlaps between the two groups. However, from an ecological viewpoint we would argue that there are important similarities between bryophytes and lichens. Ecologists are concerned more with functional than with phylogenetic similarities and distinctions between organisms (see Chapter 1) and there are many communities that are composed of complex associations of bryophytes and lichens. To restrict a consideration of the effects of environmental change to one sub-section of these communities would provide a very incomplete understanding. Eight of the fourteen chapters in this volume treat both groups of organisms together.

Bryophytes and lichens have long been recognized as sensitive indicators of environmental conditions. At various times the use of one group or the other has been recommended to indicate the presence of particular mineral deposits (geological prospecting), soil and site conditions (forestry), presence and levels of atmospheric and aquatic pollutants (biomonitoring), ages of rock surfaces and prehistoric megaliths (lichenometry), and historical continuity of forest cover (bioindication). This indicator power follows from certain fundamental characteristics which are shared by these two taxonomically unrelated groups of organisms. First, both groups of organisms are poikilohydric—there is very limited control of the uptake and loss of water. Consequently, periods of metabolic activity are determined much more immediately by the pattern of rainfall and conditions for evaporation than in vascular plants. The absence of a root system frees both groups from a reliance on soil, and bryophytes and lichens often grow together on hard substrates, like rock and bark, which are impermeable to roots. The lack of an effective cuticle in most bryophytes and lichens allows relatively free entry of solutions and gases to the surfaces of the majority of the living cells. It is probably for this reason that these organisms are more sensitive to their atmospheric chemical environment than terrestrial vascular plants, where water and solutes are obtained principally from well-buffered soil reservoirs and entry is controlled via the roots.

A second factor linking bryophytes and lichens is their small size and keen exploitation of boundary layer niches, where, during the periods important for metabolic activity, the microclimate, including carbon dioxide concentrations, may differ profoundly from conditions experienced else-where in the vicinity. In general, lichens appear to tolerate higher irradi-ances and more frequent desiccation than bryophytes, but there is wide overlap in their responses to these environmental factors. In spite of poikilohydrism, both groups contain desiccation-intolerant species (rarer amongst lichens) and probably also many which are susceptible to photo-inhibition at quite low irradiances.

A third common feature of many bryophytes and lichens is low growth rate, partly imposed by the interruptions characteristic of a poikilohydric lifestyle, compared to vascular plants. Bryophytes and lichens also often have a poor ability to disperse to or colonize new localities. Although not particularly different from tracheophytes in this respect, their poikilohydric nature means that they are sometimes hypersensitive to increasing micro-climatic dryness or pollutants and often more readily adopt a relict status in man-altered landscapes.

In our choice of material we have attempted to emphasize areas where there is current interest and new information and have omitted topics which have been well reviewed elsewhere. Naturally, the emphasis is on man-made changes because this is where environmental change is most dramatic and has resulted in most concern. The environmental changes considered include climate change and rising CO_2 concentrations, changes in intensities and types of atmospheric and aquatic pollutants, habitat destruction, the consequences of intensive agriculture, and the introduction of exotic species. The chapters fall loosely into five sections.

The first section (Chapters 1–3) is intended as an introduction to the organisms, and an appraisal of their ecological function and importance in ecosystems, and how environmental change may affect this function. Chapter 3 introduces the physiological characteristics of bryophytes and explores their responses to climatic variables, light, and CO_2 concentration.

Next follow four chapters (4–7) devoted to the analysis of species distribution patterns. Changes which have been recorded in historic times or deduced from studies of microfossils and macroscopic fragments are used to interpret past environmental change. The mechanisms by which populations spread or retreat are covered in two chapters (5 and 6) describing exotic species invasions, range alterations in established species in different situations, and reinvasions with declining pollution levels. Here we chose to omit direct coverage of the effects of SO_2 on bryophytes and lichens because this topic has been extensively reviewed elsewhere, most recently in Nash and Wirth's *Lichens, bryophytes and air quality*. Useful references to this subject can be found, however, in Chapters 6, 11, and 12 of the current work. We very much regret that a commissioned chapter on lichen distribution patterns failed to materialize.

Our next section (Chapters 8–10) covers the effects of specific land management regimes on forest species, which are in decline in most parts of the world, and on the rather more commonplace bryophytes and lichens which manage to coexist with modern intensive agriculture. The authors of the chapters (8 and 9) on forest management make practical suggestions for the conservation of threatened bryophytes and lichens, which it is hoped will be heeded.

The effects of rapidly changing pollutant regimes on species found in a number of different habitat types are described in a fourth group (Chapters 11–13). The book concludes (Chapter 14) with a consideration of the peculiar genetic systems of bryophytes and lichens and an examination of the evolutionary potential of both groups to accommodate future environmental changes.

We greatly appreciate the hard work put in by all authors and their adherence to a tight schedule. Lastly, we express thanks to the staff of Oxford University Press for constructive advice and encouragement.

London   J.B.
Peterborough   A.F.
August 1991

Contents

Contributors

Jeffrey W. Bates
Department of Biology, Imperial College at Silwood Park, Ascot, Berkshire SL5 7PY, UK.

J. Nigel B. Bell
As above, *and* Imperial College Centre for Environmental Technology, 48 Prince's Gardens, London SW7 2PE, UK.

Dennis H. Brown
Department of Botany, The University, Bristol BS8 1UG, UK.

Heinjo J. During
Department of Plant Ecology and Evolutionary Biology, University of Utrecht, P.O. Box 800.84, 3508 TB Utrecht, The Netherlands.

Andrew M. Farmer
English Nature, Northminster House, Peterborough PE1 1UA, UK.

Oliver L. Gilbert
Department of Landscape, The University of Sheffield, Box No. 595, Arts Tower, Sheffield S10 2UJ, UK.

Janice M. Glime
Department of Biological Sciences, Michigan Technological University, Houghton, Michigan 49931, USA.

S. Rob Gradstein
Institute of Systematic Botany, Heidelberglaan 2, 3584 CS Utrecht, The Netherlands.

Peter Kuhry
Department of Botany, The University of Alberta, Edmonton, Alberta, Canada T6G 2E9.

John A. Lee
Department of Environmental Biology, Williamson Building, Oxford Road, Manchester M13 9PL, UK.

Royce E. Longton
Department of Botany, School of Plant Sciences, University of Reading, Reading RG6 2AS, UK.

Walter C. Oechel
Department of Biology, San Diego State University, San Diego, California 92182–0057, USA.

Francis Rose
'Rotherhurst', 36 St Mary's Road, Liss, Hampshire GU33 7AH, UK.

Wilfred B. Schofield
Department of Botany, University of British Columbia, Vancouver, British Columbia, Canada V6T 127.

A. Jonathan Shaw
Department of Biology, Ithaca College, Ithaca, New York 14850, USA.

Lars Söderström
Department of Ecological Botany, University of Umeå, S–901 87 Umeå, Sweden.

Colin J. Studholme
Department of Environmental Biology, Williamson Building, Oxford Road, Manchester M13 9PL, UK.

Bjartmar Sveinbjörnsson
Department of Biological Sciences, University of Alaska, Anchorage, 3211 Providence Drive, Anchorage, Alaska 99508, USA.

Dale H. Vitt
Department of Botany, The University of Alberta, Edmonton, Alberta, Canada T6G 2E9.

1

Ecological classifications of bryophytes and lichens

Heinjo J. During

1.1 Introduction

Attempts to classify plants using common characters or responses to the environment are as old as botany (cf. Barkman 1988). Apart from taxonomical classifications, plants have been grouped into chorological types, sociological and ecological species groups, growth-forms, life-forms, and life-history strategies. Such classifications are useful for making comparisons of the ecology of species and communities in different geographical regions (Smith 1982).

Due to the peculiar life cycles and physiologies of bryophytes and lichens, classifications made for phanerogams are often inapplicable. Therefore, separate classifications, notably of growth forms, were published rather early (Giesenhagen 1910; Frey 1924). Classifications of the life-forms of cryptogams, however, have been less elaborate (Barkman 1958). More recently, the position of both bryophytes and lichens in the triangular strategy model of Grime (1979) has been analysed (Grime *et al.* 1990; Rogers 1990), and some general concepts of life-history theory have been applied to bryophytes (During 1979) which may also be useful in lichenology (Sipman 1983). Sociological species groups (species that tend to occur together) and ecological groups (species with similar ecological amplitudes) are generally confined to restricted regions or particular habitat types and will not be further discussed here.

Bryophytes and lichens show some striking similarities in ecology, although they are taxonomically fundamentally different. Both groups are poikilohydric and of small stature, and asexual propagation predominates. This leads to similarities in the constraints and trade-offs determining life-history variation in both groups. After introducing the peculiarities of bryophyte and lichen life-histories I will briefly discuss current growth-

form and life-form classifications and review recent attempts to apply the
plant strategy types of Grime (1979) and life-history theory to these groups.

1.2 The biology of bryophytes and lichens

1.2.1 Bryophytes

The bryophytes comprise some rather distantly related groups of Cormo-
phytes, the most important of which are the mosses (Bryopsida), the
liverworts (Hepaticopsida), and the hornworts (Anthocerotopsida) (Schuster 1984; Vitt 1984). All groups are characterized by a diplohaplont life
cycle: from a germinating (haploid) spore a mostly filamentous protonema
develops on which one or, in most mosses, several buds arise. These buds
grow into green plants, the gametophytes, which may be thallose or consist
of stems and leaves.

Sexual organs on these gametophytes produce gametes, and after
fertilization a (diploid) sporophyte develops that remains connected with
and dependent on the gametophyte. In the sporophyte, meiosis occurs and
spores are produced. Transport of spermatozoids is only possible in water;
therefore, gamete dispersal distances are very restricted.

Some 50 per cent of the bryophytes are dioecious (Mischler 1988). In
such species, male and female plants have to be in close proximity for
fertilization and sporophyte formation to occur. In several monoecious
species, distribution in space and/or time of ripe male and female organs
(antheridia and archegonia, respectively) is such that outcrossing is promoted (Longton and Schuster 1983), but since self-incompatibility is rare
or absent (Wyatt and Anderson 1984), sporophytes on monoecious plants
will often be the result of self-fertilization. Since the gametophytes are
haploid, this will result in replication of the same genotype. In combination
with the high frequency of asexual propagation, this would suggest low
levels of genetic variation in bryophytes. In the few examples studied to
date, however, genetic variation seems to be as high as it is in phanerogams
(Wyatt *et al.* 1989; Ennos 1990). Low risks of genet mortality due to a
combination of asexual propagation and a fine-grained scale of mortality
events may be important in this respect (Cook 1979; During and Van
Tooren 1987).

1.2.2 Lichens

Lichens constitute a unique group of 'plants' that consist of two (occasionally three or even four) unrelated components, fungi (the mycobiont)
and algae or cyanobacteria (the photobiont), living in a close symbiotic
association (Hale 1983). By convention, the fungus is the name-giving
component of the association. Most lichens belong to the Ascomycetes, but

in a few cases the mycobiont is a Basidiomycete or Deuteromycete (Hawksworth and Hill 1984).

Morphologically, ascolichens often show little resemblance to their non-lichenized relatives, as they form distinct, long-lived thalli. Sexual repro-ductive processes are assumed to be similar to those in related, non-lichenized Ascomycetes; they result in the formation of asci with ascospores in cup-shaped apothecia or more or less closed, flask-shaped perithecia. Most of the few (10–20 species) lichenized Basidiomycetes are much more similar to their non-lichenized relatives (Hale 1983).

In most species, the photobiont is a unicellular green alga (often *Trebouxia* spp.), but in some lichen groups cyanobacteria (e.g. *Nostoc* spp.) form the 'algal' component (Hale 1983). Reproduction is usually by simple mitotic divisions, and sexual reproduction of algae in the lichenized state is unknown, although it perhaps occurs in free-living colonies (Longton 1988*a*). Asexual propagation is very common (although less so in crustose lichens), either by specialized propagules or simply by broken-off frag-ments. An important aspect of asexual propagation in lichens is that in this way propagules will contain both the mycobiont and the photobiont. This is usually not the case with spore dispersal, although in some species with algae in the perithecia one or two algal cells may be dispersed with the spore (Pyatt 1973).

1.2.3 Similarities and differences in ecology

The structure and life of bryophytes and lichens differs in several ecologi-cally important aspects from those of vascular plants. Both bryophytes and lichens are poikilohydric; in many species, the plants frequently desiccate completely, but rapidly resume photosynthesis and growth upon rewetting. Even though bryophytes do not have real roots, in some so-called endohy-dric species transport of water is partially internal; in most other (ectohy-dric) species, water is taken up over the whole surface of the plant (Buch 1947). Nutrients are acquired along the same route, which may partially explain the high sensitivity of many species to air and water pollution (Brown 1984). Lichens generally behave as ectohydric bryophytes in this respect (Longton 1988*a*), although specialized structures may be important sites of water uptake in some species (Larson 1981).

Net carbon gain during periods of physiological activity is strongly determined by the length of the active period, which is related again to the water content of the plants and the resistance to evaporation. Since stomata are absent and a reasonably effective cuticle is found in only a few endohydric bryophytes, evaporation rates are usually strongly affected by the boundary layer resistance, which in turn largely depends on the interplay between wind speed and morphology of the plants or colonies

(Larson and Kershaw 1976; Proctor 1980). In both bryophytes and lichens the relationship between evaporation rate and water potential is such that the plants desiccate rapidly once the water potential has dropped below a certain threshold value, thus reducing the time during which respiration exceeds photosynthesis (Proctor 1982; Jahns 1984).

Depending on morphological and anatomical features such as the presence or absence of internal conducting tissues, direction and form of leaves and leaf bases, and papillosity of the cells, water is distributed more or less evenly and rapidly over the whole shoot. Such structures are particularly important for plants growing in drought-prone habitats, allowing photosynthesis to be carried on as long as possible. Thus, the variation in anatomical characters, growth form of individual shoots and of colonies may often be interpreted in terms of their water economy.

Productivity of bryophytes in the field ranges from very low values in extreme environments such as exposed rocks and outer branches of trees to remarkably high rates in some continuously moist communities of *Sphagnum* spp. in mire hollows and of several mosses in subantarctic regions (Longton 1984). Productivity of lichens generally tends to be lower, perhaps due to the lower chlorophyll content per unit of thallus (Robinson *et al.* 1989). Yet also in this group considerable variation occurs between species (Topham 1977; Rogers 1990).

Sexual reproduction of bryophytes and of the mycobionts of lichens results in the production of numerous spores. Spore size is commonly in the order of 10–20 μm, but in some groups (much) larger spores (25–200 μm) are produced (cf. Sipman 1983). Partly due to the low point of release in at least the terricolous species, most of the spores are deposited very close to the source (Miles and Longton 1987). Theoretically, at least some of the small spores should easily be dispersed by wind over thousands of kilometres (Van Zanten and Pocs 1981), but the potential dispersal distance of larger spores rapidly decreases with size.

Several lines of evidence suggest that successful sexual reproduction is rare in bryophytes. Thus, many taxa rarely or never produce sporophytes or do so only in a small part of their range (Longton and Schuster 1983), and chances of successful establishment of spores after dispersal appear to be extremely small in most of the species studied to date (Miles and Longton 1990). Although the reproductive processes of both the mycobionts and the photobionts of lichens are poorly understood (Hale 1983; Longton 1988*a*), gamete dispersal distances may be assumed to be very low also in this group, and successful sexual reproduction may be equally rare. In some species it seems to be effective, however (Ott 1987*a*, *b*), and for many crustose lichens it seems to be the only way of propagation.

1.3 Growth-forms

Bryophytes and lichens are small plants—from some mm to a few dm tall, with only very few species reaching 1 m in height—only species that grow hanging from branches may reach larger dimensions. Yet they show an astonishing variety of forms and shapes. Among the hepatics we find thallose Marchantiales with their variety of reproductive structures, but also tiny, but complexly built epiphyllous Lejeuneaceae. Mosses may show up as tiny, simple leaf rosettes enclosing a round, simply structured sporophyte such as in *Ephemerum*, but also as the erect, several dm tall stems of *Dawsonia* with its complex leaf and stem anatomy and its asymmetrical capsules with numerous peristome teeth. The morphology of lichens ranges from the loose hyphal mats with interspersed algal cells of *Lepraria* to the richly branched, anatomically much more differentiated thalli of *Cladonia* and *Usnea*.

In both groups, several attempts have been made to summarize this variation in a restricted number of growth-form types, i.e. groups of plants that are similar in architecture or general physiognomy. Although the bewildering variation in the tropics may not yet have been treated adequately (Richards 1984; Sipman and Harris 1989), for both bryophytes and lichens fairly generally accepted classifications are available now.

1.3.1 Bryophyte growth-forms

Giesenhagen (1910) was the first to describe growth-form types of bryophytes. Strongly impressed by the richness of the epiphytic bryophyte vegetation of the tropics of SE Asia, he recognized seven growth-form types based on the general physiognomy of the plants. In a rather different approach, Meusel (1935) defined a number of growth-form types on the basis of inherent branching patterns and architecture of the shoots. However, since bryophyte colonies generally function as integrated units, later authors returned to a more physiognomical basis to distinguish growth forms. The classification elaborated by Gimingham and co-workers (Gimingham and Robertson 1950; Gimingham and Birse 1957) has been very successful, and with some minor changes and additions this is in common usage (Table 1.1).

1.3.2 Lichen growth-forms

In many studies of the ecology and distribution of lichens, a simple system of growth-forms is used that has become more or less classical; it comprises crustose, foliose, and fruticose lichens (Hale 1983). Crustose lichens grow in or closely appressed to the substrate; they may consist of an indetermi-

Table 1.1 Growth form types of bryophytes after Gimingham and Birse (1957) and Mägdefrau (1982), schematically

Vertical extension of shoots (cm)	Direction of main shoots		
	Erect	Radiating from central point	Various, often horizontal or ascending
	(Acrocarps)		(Pleurocarps)
0.1–1	*Turfs* open turfs	*Cushions*	*Mats* thalloid smooth
0.5–3	short turfs	small cushions	thread-like rough
3 –> 30	tall turfs sphagnoids	large cushions	Carpets, Wefts, Tails, Dendroids, Pendants

nate hyphal mat enclosing algal colonies, but usually a distinct thallus is formed with an upper cortex, an algal layer and a medulla. Characteristically, a lower cortex is lacking, however. Foliose lichens also form flattened thalli that grow more or less appressed to the substrate, but usually a lower cortex is present and the lower side of the thallus is not connate with the substrate. Fruticose lichens have more or less intensely branched thalli with an outer cortex, a thin algal layer, a medulla, and a more or less hollow centre or a dense central cord (Hale 1983) that extend in three dimensions.

More elaborate systems of lichen growth-forms have been proposed by, e.g. Frey (1924), Hilitzer (1925), and Ochsner (1928); Barkman (1958) presented an overview of the various classifications. The system presented here (Table 1.2) is mainly based on Barkman (1958), with some additions derived from Creveld (1981) and Barkman (1988).

1.3.3 Ecological relevance of growth-forms

Growth-forms are defined on the basis of plant architecture only, without direct reference to ecological 'adaptations'. Yet, the distribution of each growth-form type over different habitats is to some extent constrained by competition and abiotic environmental conditions, notably water relations. Thus, during post-fire succession of lichens in lichen–conifer stands near Lake Abatibi (Quebec, Canada) there was a clear shift in dominant growth-form from crustose to simple fruticose (*Cladonia* sect. *Cladonia*) and finally to shrubby fruticose (*Cladonia* sect. *Cladina*) forms, which was considered to be due to the interplay between ecophysiological differences in drought tolerance and changing microclimatic conditions (Clayden and Bouchard 1983). Similarly, along an altitudinal gradient in the Andes crustose lichens were most abundant from sea level to 2000 m altitude, while foliose lichens dominated from 2000 to 4000 m, and fruticose lichens were abundant only

Table 1.2 Growth form types of lichens, schematically (mainly after Barkman 1958, 1988)

Connection with substrate	Growth form	
Very intimate or thallus immersed	Crustose	leprose type non-leprose crustose type crustose with stalked apothecia
	Placodioid	
Closely appressed	Foliose	*Parmelia* type *Lobaria* type
	Filamentous	
	Squamose	without podetia podetia with cups podetia cylindrical, unbranched podetia sparingly branched
Attached at one point only	Gelatinous	
	Umbilicaria	
	Fruticose	*Eucetraria* type *Ramalina* type

above 4000 m (Sipman 1989). The sequence from crustose through foliose to fruticose lichens has often been assumed to reflect successional stage, but at least in epiphytic lichens this does not seem to be generally applicable; lichens of 'climax' epiphyte communities generally are large foliose species (e.g. *Lobaria* spp.), whereas the first colonizing species may also be foliose (Topham 1977).

The ecological 'indicator value' of bryophyte growth-forms is well known since the classical studies of Birse (Gimingham and Birse 1957; Birse 1957, 1958*a*, *b*). Water relations appear largely to determine the distribution of growth-form types (e.g. Gimingham and Smith 1971; Mägdefrau 1982). Thus, the growth-form patterns in the epiphytic vegetation of a forest in Colombia were strongly dependent on the location within the canopy, but phorophyte species also influenced the growth-forms found (Van Leerdam *et al.* 1990). Here and in a dry evergreen forest in Guyana, some species and growth-forms were restricted to one or a few height zones (specialists), whereas others had a wider vertical distribution in the canopy (generalists; cf. Cornelissen and Ter Steege 1989).

1.4 Life-forms and plant strategies

The recognition of life-forms and plant strategies is based in principle on the view that (populations of) organisms are in some way adapted to their environment. Gould and Lewontin (1979; Mischler 1988) strongly argued that this may be a very dangerous enterprise; the potential adaptive value of characters should be taken as a hypothesis and investigated rigorously, not taken for granted based on adaptive 'story-telling'. The following account of attempts to classify bryophytes and lichens into life-forms and strategies will be considered by some as precisely that. Yet in my opinion it is a valid approach to use generalized hypotheses about the ecological importance of differences in morphological, physiological, and life-history characters as a basis for the grouping of plants into broad categories. If cautiously used, such classifications may be useful in providing a tentative structure to the 'web of life' and in guiding further research regarding key factors in specific environments or regions.

The essential difference from growth-form systems is that with life-forms the emphasis is not on architecture or general morphology of the plants, but on the interaction between plant and environment. Thus, the well-known system of Raunkiaer (1934) is based on the position of meristematic tissues relative to the soil surface. Of course, this has clear implications for morphology (at least size), but geophytic, hemicryptophytic, and chamaephytic grasses may have very similar growth-forms in summer, to take just one example.

The concepts of life-form and plant (c.q. life-history) strategy represent two different approaches to generalize about plant/environment interactions; life-forms usually focus on specific plant characteristics, whereas plant strategies refer to co-adapted traits fitting generalized environmental constellations. The distinction is somewhat arbitrary, however; most schemes take a position between these two extremes.

1.4.1 Life-forms

Ideally, a life-form system should reflect the adaptation of the plants to the crucial limiting factors in the environment (Barkman 1958). Since at the present state of knowledge of the ecology of cryptogams this is obviously impossible, published life-form classifications are based on a more or less arbitrary choice of parameters. Some authors circumvent the problem by advocating so-called 'dynamical systems' (e.g. Du Rietz 1931) in which several separate classifications, each starting from a different viewpoint, are proposed (Barkman 1958; Segal 1966). Two such classifications will be presented here.

The first approach focuses on water relations, since water is one of the most important environmental factors for poikilohydric plants like bryophytes and lichens. A provisional system based on the predominant source of water was presented for epiphytes by Barkman (1958), who distinguished three groups:

1. Atmophyta, plants with a relatively low water capacity (50–250 per cent of dry weight) for which the main source of water is atmospheric

humidity (water vapour, mist). This life-form includes the leprose and most fruticose lichens, especially the strongly sorediose ones like *Lepraria incana*. The most typical representatives of this life form are even unable to absorb liquid water and often live in places where this is never available.

2. Amphiphyta, plants with a medium water capacity (300–500 per cent) and very high osmotic potentials (80–130 MPa), which profit both from atmospheric humidity and from liquid water. According to Barkman (1958) this group primarily consists of non-leprose crustose and foliose lichens.

3. Ombrophyta, plants with a high water capacity (650–1700 per cent) and relatively low osmotic potentials (1–10 MPa), for which the main (often the only) water source is liquid water. This life-form comprises the lichens with cyanobacteria as photobiont (Lange *et al.* 1986) and all bryophytes.

A useful subdivision of the Ombrophyta might be based on the degree to which internal water transport is possible, as reflected in Buch's (1947) division of bryophytes into endo-, mixo- and ectohydric forms (cf. Watson 1971). In endohydric bryophytes, water transport is to a fairly large extent internal via more or less elaborate conducting systems, and consequently the soil is an important source of water for this group. Characteristically, the thalli or leaves are more or less water-repellent (Proctor 1984), and full hydration of desiccated plants may take half an hour or more. Well-known examples are *Polytrichum* spp. and a number of Marchantiales. In ectohydric bryophytes, on the other hand, internal water transport is negligible, and water is taken up over the whole plant surface. Rain and run-off are the main sources of water for these plants. Water-repellent layers are absent or restricted to very specific locations such as apices of papillae (Proctor 1982), and soaking of desiccated plants results in full saturation within minutes. Most foliose hepatics and pleurocarpous mosses belong to this category, but also a substantial number of large acrocarpous species. Finally, mixohydric bryophytes have some rudimentary internal conducting system and take up water both over the plant surface and via the rhizoids from the soil. A major part of this category is formed by a group of small, acrocarpous mosses which grow on loamy or clayey soils that dry out frequently but, due to their fine texture, maintain a moist top layer for some time after a rainfall event.

The second approach, due to Mattick (1951) and Barkman (1958), is based on the small stature of bryophytes and lichens and starts from the observation that there is a steep microclimatological gradient in the upper few mm of the substrate and the lower few cm of the adjacent air layer. The plants are classified according to the layer in which they live. Barkman

(1958) partially used Raunkiaer's terminology, although in a different sense; the terms do not refer to the position of the hibernating buds, but to the level where most photosynthetic tissue is found. Although there may be confusion with Raunkiaer's system, I will refrain from introducing new terms here. The following categories were distinguished:

1. Endophyta, growing mainly in the upper few mm of the substrate (endolithic and endophloeic lichens; comprises the Cryptolichenes of Mattick 1951).

2. Hemicryptophyta, growing closely appressed to the substrate, plant height up to 2 mm. Barkman distinguished forms that cannot easily be torn from the substrate (Hemicryptophyta connata, notably crustose lichens) from those that may be detached more easily (Hemicryptophyta appressa; foliose lichens and smooth and thallose mats of bryophytes).

3. Chamaephyta, growing loosely adherent to the substrate, plant height c. 0.5–1 cm.

4. Phanerophyta, shrubby and dendroid lichens and mosses with the main part of their photosynthetic tissues at a distance of more than 1 cm from the substrate.

Additionally, Barkman recognized the epibryophytic life-form of small cryptogams growing epiphytically on large bryophytes. Hyperepiphytes on epiphytic lichens are also quite common. This classification was extensively used by Barkman (1958) to characterize epiphytic cryptogam communities, but has not received much attention since. Since it is very easy to use and is based on a characteristic that has obvious ecological relevance, however, it may yet prove to be useful.

This is perhaps the best place to draw attention to the peculiar 'vagant life-form' (Weber 1977): plants that grow wholly without attachment to the substrate. Most bryophytes producing 'moss balls' (e.g. *Grimmia ovalis*) may also be found in 'normal', attached forms; they belong to, or are closely related to, species that characteristically form dense cushions (Beck *et al.* 1986). However, a number of lichens (notably desert species, e.g. *Aspicilia esculenta*) are only known in this form. This life-form seems to be connected with areas with low vegetation and a high frequency of strong winds.

1.4.2 Strategies

In recent years, the focus of ecological classifications has shifted from the modalities of single traits to integrated sets of characters that are predicted to occur predominantly in a restricted number of combinations and in response to particular sets of ecological conditions. Such recurrent combinations of characteristics have been termed 'strategies' (Grime 1979); an alternative definition with a slightly different emphasis is provided by the circumscription of life-history tactics by Stearns (1976): 'sets of co-adapted traits designed, by natural selection, to solve particular ecological problems'. Many different classifications of plant strategies have been proposed, only a few of which can be discussed in this chapter.

The C–S–R model

The coenotypes described by Ramensky (1938) represent one of the earliest attempts to categorize plants (originally only dominant plants) into strategy types. The main categories are:

(1) violents, species that grow aggressively and by their high competitive ability, both below and above ground, are able to outcompete other species;

(2) patients, species that survive and even gain dominance in harsh environments by their ability to tolerate extreme environmental conditions;

(3) explerents, species of low competitive ability that yet may become temporarily very abundant after disturbances.

Later, Rabotnov (1975) applied this concept to all plants, and drew attention to a fourth group, the pioneers—species able to colonize a new, for other species yet unsuitable, substrate and thus making 'initially lifeless territories suitable for patients or violents'.

Starting from a different viewpoint, Grime (1974, 1977) proposed a somewhat similar model for plant strategies. This model is based on the assumption that growth and survival of plants and plant populations are mainly determined by two complexes of environmental factors, stress and disturbance. In this context, stress consists of the phenomena which restrict photosynthetic production such as shortages of light, water, and mineral nutrients, whereas disturbance refers to the partial or total destruction of plant biomass due to herbivores, pathogens, and human influences, or natural catastrophes such as hurricanes, frost, drought, soil erosion, and fire (Grime 1979). Furthermore, it is considered highly unlikely that plants are able to tolerate extreme levels of both stress and disturbance. In environments with low levels of both factors, competition becomes the dominant factor.

Generally, plant species will tend to occur in a limited range of combinations of these restricting factors. Graphically, the model may be viewed as a triangle, in which the sides of the triangle represent gradients from high to low disturbance, competition, and stress, respectively (Fig. 1.1). The strategies of species occurring near the vertices of the triangle have been designated as Competitors, Stress tolerators, and Ruderals (from which the term C–S–R model is derived); intermediate strategies may be named accordingly (e.g. stress-tolerant ruderals).

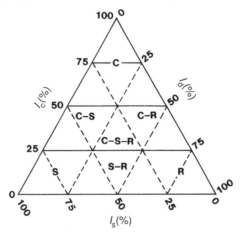

Fig. 1.1 Model describing the various equilibria between competition, stress, and disturbance in vegetation and the location of primary and secondary strategies. I_c: relative importance of competition; I_s: relative importance of stress; I_d: relative importance of disturbance; C: competitor; S: stress tolerator; R: ruderal; C–S, C–R, S–R, C–S–R: secondary strategies (after Grime 1977, reproduced from *American Naturalist* by permission of University of Chicago Press).

Competitors are characterized by, *inter alia*, a moderate to long life span, a relatively low reproductive effort, high potential relative growth rates, high and dense canopies of leaves, copious litter, and high morphological plasticity. Stress tolerators usually have a long to very long life span, a low reproductive effort, low potential relative growth rates, sparse and often persistent litter, and low morphological plasticity. Ruderals resemble Competitors in some respects, but their life span is very short and reproductive effort is high. With respect to the coenotypes of Ramensky (1938) and Rabotnov (1975), the three categories seem to be more or less equivalent to the violents, patients, and explerents mentioned before.

According to Grime (1979), lichens are confined to the stress-tolerant corner of the model, whereas bryophytes are more wide-ranging with the centre of the distribution in the stress-tolerant ruderals. While confirming that all lichens are subject to some degree of stress, Topham (1977) showed that lichen species and communities can be ordinated in a triangular subspace with the same axes of competition and stress. She observed that information about growth rates of lichens is too scanty to be very useful in this respect; yet she provided methods to determine the relative position of lichen species in this triangle using an index of capacity for overgrowth and some related characteristics as well as an index for propagule size. Along similar lines, Kiss (1985) used a combination of propagation mechanisms and growth form types to define 'life strategies' of lichens which clearly differ in tolerance to air pollution (Kiss 1988).

Rogers (1988, 1989, 1990) was able to derive relative growth rate (RGR) measures for 34 lichen species growing under field conditions. Due to the severe problems involved in cultivating lichens it is not yet possible to obtain maximum RGR values under more controlled conditions, although there is rapid progress in this field (Hawksworth and Hill 1984). RGR ranged from 0.5 mg g^{-1} week^{-1} in extremely stress-tolerant epilithic crustose species such as *Rhizocarpon obscurata* to 70 mg g^{-1} week^{-1} in some epiphyllous ruderals (*c.* 0.1–10 mg g^{-1} day^{-1}). In view of the range reported for phanerogams grown under optimal conditions (4–400 mg g^{-1} day^{-1}, Poorter 1990) this means, that lichens indeed are restricted to the stress-tolerant part of the full triangular space, but there is still considerable variation as regards RGR. It has proved difficult to apply the characteristics given by Grime (1974) for estimating competitive ability of lichens, but using a morphological index based on thallus thickness (c.q. height in the case of fruticose species) and thallus diameter in combination with the RGR values found, Rogers (1990) displayed the 34 species in a meaningful way in a triangular frame (Fig. 1.2). The morphological index may indeed represent a suitable measure of competitive ability for crustose and perhaps also foliose species (Rogers 1988), but whether it is also useful in comparisons with fruticose species remains to be seen.

A more detailed discussion of the position of bryophytes in this context is provided by Grime *et al.* (1990). Bryophytes also seem to occupy a triangular subspace of the full triangle. In an important contribution, Furness and Grime (1982) showed large variations among bryophytes in potential RGR under laboratory conditions, which could be correlated with the degree of stress in the environment. High maximum RGR values were found for the short-lived ruderal *Funaria hygrometrica* (*c.* 50 mg g^{-1} day^{-1}) and for *Brachythecium rutabulum* (70 mg g^{-1} day^{-1}), a species with a more competitive strategy, whereas species from continuously unproductive habitats, especially epiliths, characteristically reached much lower values (5–20 mg g^{-1} day^{-1}). These values are nearly an order of magnitude higher than those found in lichens (Rogers 1990). However, actual growth rates under field conditions are more in line with the lichen data. Thus,

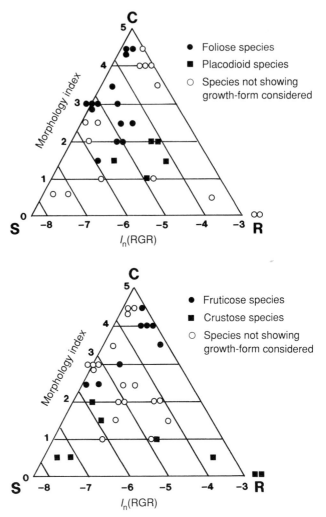

Fig. 1.2 The relationship between growth form and plant strategy of lichens. C, S, and R are as defined in Fig. 1.1. The morphology index is based on thickness and diameter of the thalli. (Reproduced from Rogers (1990) by courtesy of the Editor, *The Lichenologist*.)

maximum RGR values of *B. rutabulum* and *Thuidium tamariscinum* in the field were only 10 and 7 mg g^{-1} day^{-1}, respectively (Rincon and Grime 1989*b*).

According to Grime *et al.* (1990) the category of Competitors s.s. is not represented among bryophytes, but they suggested that if the plasticity needed to 'forage' for resources is used as a measure of competitive ability, stress-tolerators stand clearly apart from more competitive species such as *B. rutabulum*, *Eurhynchium praelongum*, and *T. tamariscinum*, since a high morphological plasticity tended to be associated with high relative growth rates (Rincon and Grime 1989*a*).

Life-history strategies

The traits that play an important role in discussions about life-history strategies are mostly connected with reproduction and mortality, and with the effects of constant or fluctuating environments on these (Stearns 1976). Examples are the size and number of offspring, the reproductive effort and its age distribution, age of first reproduction, and the variation in these traits among the progeny (Stearns 1976). With regard to fluctuating environments, an important distinction is that between organisms that pass through periods of low environmental quality in a highly resistant stage (seeds, spores; avoidance) and those that survive as (more or less) fully developed mature individuals (tolerance).

It is obviously impossible to maximize all traits simultaneously. For example, larger offspring often have better chances of survival to maturity; however, size of the offspring has direct consequences for the numbers that can be produced. Therefore, an essential concept in this context is that of 'trade-offs', balances in the allocation to various functions or structures in response to specific habitat conditions (Southwood 1988). Although the trade-offs between size and number of offspring and those between reproductive effort and risk of adult mortality are perhaps the most obvious ones, Stearns (1989) shows that in the analysis of life-histories many potential trade-offs may be recognized.

Much of the theory developed in this context refers to animals. In the case of plants, and bryophytes and lichens in particular, there is an added complication: the size of offspring (seeds, spores) is not only related to juvenile mortality risk, but also to dispersal capacity. Spores that are smaller than 20 μm have the potential for very wide dispersal, whereas the dispersal capacity rapidly declines with size (e.g. Schmidt 1918).

The division into r-selected and K-selected species based on the work of McArthur and Wilson (1967) represents the best known example of life-history strategies. Originally, these concepts only referred to optimal strategies under selection regimes of low versus high density dependence; r-selected species are characterized by high population growth rates, whereas in K-selected species utilization of resources leads to a high population density at carrying capacity. Through the incorporation of other elements, such as predictability of the environment, successional stage, and reproductive effort, by later authors (e.g. Pianka 1970; Gadgil and Solbrig 1972) the concepts of r- and K-strategies gradually widened to the contrast between rapidly growing ephemeral populations with early reproduction and many, small-sized offspring in early successional environments (r-species) and slow-growing, long-lasting populations of highly competitive plants or animals with late reproduction and a few, large offspring in dense 'climax' environments (K-species).

Application of these concepts to lichens has been rather limited, perhaps due to the fact that most lichens have been considered as K-strategists, but the group as a whole does cover a large part of the r-K continuum (Topham 1977). A discussion of r- and K-strategies with respect to bryophytes may be found in Slack (1977). Obviously, short-lived 'fugitive' bryophytes may be separated from longer-lived perennial ones, but the lack of information on mortality patterns, causes of mortality, and relative importance of density-dependent interactions in bryophytes precludes a more detailed discussion of this topic.

In a related, more detailed approach, Noble and Slatyer (1979) showed that the actual successional sequence during secondary succession after disturbance can be predicted fairly well from three 'vital attributes' of the species involved: (1) the method of arrival or persistence of the species at the site during and after disturbance; (2) the ability to establish and grow to maturity in the developing community; and (3) the time taken for the species to reach critical life stages. Different combinations of specific 'values' of these attributes were used to define 15 frequently encountered 'species types'. Noble and Slatyer (1979) stress that their model only refers to recurrent but fairly rare disturbances and not to catastrophic events or to continuous 'disturbance' such as permanent grazing.

Application of this approach to bryophytes and lichens is hampered by the fundamental differences with phanerogams as regards life-history and physiology. Criterion (3) especially causes problems, because the sequence propagule → juvenile → immature individual → reproductive individual is much less straightforward in these groups. Notably, the production of asexual propagules and the capability of regrowth from dispersed plant fragments are hardly linked to a specific life stage. Perhaps due to such complications, the analysis of the successional development of bryophyte communities on Dutch forest earth banks was not very illuminating (Van Tooren and During 1988).

For bryophytes, During (1979) proposed a provisional system of six life strategies (Table 1.3), that was based on three major trade-offs: (1) few, large spores versus many, small spores; (2) survival of the difficult season as spores only, discarding the gametophyte (avoidance) versus survival of the gametophyte (tolerance); and (3) for the tolerance group, potential life span of the gametophyte, which is negatively correlated with reproductive effort (RE). Large spores have a low dispersal capacity, but probably better

Table 1.3 Preliminary system of life strategies of bryophytes after During (1979)

Potential life span (yr)	Spores Numerous, very light (<20 µm)	Few, large (> 20 µm)	Reproductive effort
<1	Fugitives	Annual shuttle	High
Few	Colonists	Short-lived shuttle	Medium
Many	Perennial stayers	Long-lived shuttle	Low

chances of successful establishment and a longer life span in the diaspore bank. Small spores, on the other hand, are produced in such large numbers that many, in spite of the highly leptokurtic dispersal pattern will reach distant sites. Investment in asexual reproduction is usually low in species of the avoidance type (protonemal gemmae only), whereas it may be considerable in some groups with longer potential life span. The negative correlation between investment in asexual and in sexual reproduction that is often observed (e.g. Schofield 1981) may constitute an important additional trade-off.

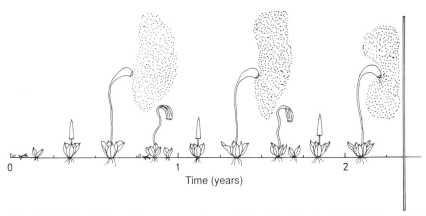

Fig. 1.3 The fugitive life strategy, schematically (after During 1979). The vertical bar symbolizes the end of the period during which the habitat is suitable for the species.

Ephemeral species with high RE and small spores (fugitives, Fig. 1.3) are expected to occur preferentially in habitats that occur unpredictably and are suitable for a very short time only. The prime example of this category is *Funaria hygrometrica*. Colonists (Fig. 1.4) such as *Bryum bicolor*, with a potential life span of several years and a fairly high RE and small spores, will be found in habitats that are suitable for some years but

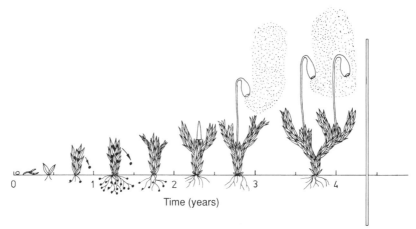

Fig. 1.4   The colonist life strategy, schematically (after During 1979). Meaning of vertical bar as in Fig. 1.3.

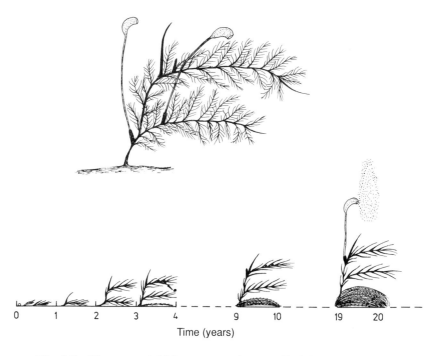

Fig. 1.5   The perennial life strategy, schematically (after During 1979).

disappear thereafter. Asexual reproduction is often prominently present in this group. Perennials (Fig. 1.5) are presumed to occur in more permanent habitats; the small spores produced seem to serve the occasional establishment of new populations. Many pleurocarps (e.g. *Brachythecium rutabulum*) and a number of tall acrocarpous mosses belong to this type. Production of specialized asexual propagules is not rare in this group, but investment in it is generally low. The three categories of shuttle species, in which spore size is larger, are thought to be adapted to microhabitats that disappear predictably at varying rates but reappear frequently within the same community. Annual shuttle species (e.g. *Ephemerum* spp., Fig. 1.6) are

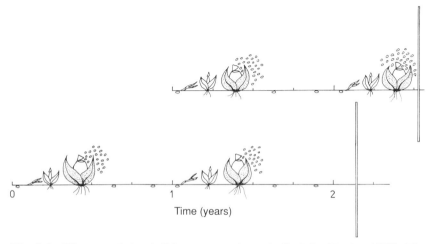

Time (years)

Fig. 1.6   The annual shuttle life strategy, schematically (after During 1979). The upper series symbolizes the possibility of shifting to another gap in the same community.

found in seasonally suitable microsites within a community, whereas short-lived shuttle species (e.g. *Bryum warneum*, Fig. 1.7) are found in somewhat longer-lasting microsites. *Splachnum* species occurring on dung patches in mires also belong to this group; their spores are rather small, but they are dispersed in small clumps by flies over rather short distances. In annual and short-lived shuttle species production of asexual propagules is rare and often restricted to protonemal gemmae. However, some 'superfertile' (Longton and Schuster 1983) short-lived shuttle species produce both spores and gemmae in large amounts (e.g. *Cololejeunea* spp.). Long-lived shuttle species (Fig. 1.8) are characteristic of long-lasting microsites such as tree branches; examples include *Leucodon sciuroides* and *Garovaglia* spp. Asexual propagules are often produced in large amounts by species of this category.

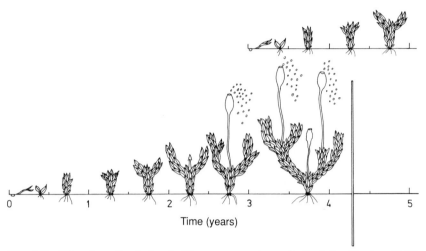

Fig. 1.7 The short-lived shuttle life strategy, schematically (after During 1979). For further explanation see Fig. 1.6.

Longton and Schuster (1983) drew attention to a tendency towards monoecism with increasing importance of spores for population mainten-ance. While perennials, the most 'primitive' strategy in their view, are generally dioecious, in the other categories monoecism is rather more common. Dioecious colonists and long-lived shuttle species often show a tendency towards reduced sporophyte production and increased reliance on asexual propagules for population maintenance.

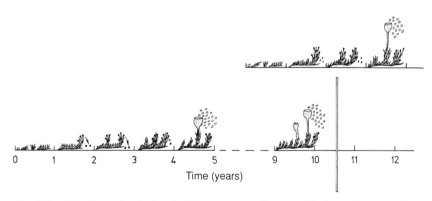

Fig. 1.8 The long-lived shuttle life strategy, schematically (after During 1979). For further explanation see Fig. 1.6.

1.5 An improved life-history classification

My original paper (During 1979) was called 'a preliminary review', and indeed, much of this system is tentative or hypothetical (cf. Longton and Schuster 1983; Whitehouse 1985). Thus, very few data on life span, reproductive effort, etc. were available. Gradually, this situation is changing.

A first estimate of reproductive effort of a few species has recently been published by Longton (1988*b*). In a coastal sand dune system in Spain, he determined the ratio of gametophyte to sporophyte biomass in patches of the fugitive *Funaria hygrometrica* and of the colonist *Tortella flavovirens*; the results, a ratio of 6.4:1 for *F. hygrometrica* and 356:1 for *T. flavovirens*, clearly show how different species within one habitat may be in this respect.

Another important aspect, that of spore dispersal and survival in high air currents, has been much studied by Van Zanten (e.g. 1978). Thus, within a large sample of neotropical liverworts, species with small spores seem to be somewhat more resistant to the inhospitable conditions found at such high altitudes than species with larger spores (Van Zanten and Gradstein 1988). After dispersal, spores that land in a suitable site must germinate and produce gametophytes. We still know very little about the importance of spore type and size for establishment chances, but recent studies (Longton and Miles 1982; Miles and Longton 1987, 1990) indicate that successful establishment from spores may be much rarer than anticipated earlier, except in fugitives and some weedy colonists.

There is still very little quantitative information about the fate of the plants after establishment—how mortality is distributed over age or size classes, at what age the production of spores or asexual propagules begins, and what proportion of gametophytes contribute to these modes of reproduction. There are some data on the population dynamics of a few, mainly long-lived, species (e.g. Collins 1976; Watson 1979), but insufficient for a more general picture. A complication is that usually only the fate of shoots or ramets is followed, not that of genets or even connected ramet systems. For ecological purposes, the fate of colonies (cushions, clumps) is probably most relevant.

From our studies on permanent quadrats in bryophyte communities (During and Ter Horst 1985; During and Van Tooren 1988), an indication of the dynamics of species of various life-strategies can be gained. Thus, in a study on earth banks in a forest where the vegetation had been charted repeatedly, species dynamics were quite variable, depending on both the life-history of the species and the size of the colonies (During and Ter Horst 1985). There was a very gradual transition in mean life span from supposedly short- to long-lived species, but with tremendous variation within each species.

From recent permanent grid studies in a Dutch chalk grassland and in various habitats around Barcelona, Spain, it became clear that the groups of colonists and perennials are still very heterogeneous (During and Van Tooren 1988; During *et al.* in press). An earlier attempt to divide the colonist strategy type into two groups based on the frequency of sporophyte production (During 1980, 1982) was not very successful.

In order to arrive at a classification into categories of species that are ecologically more similar, a combination with the ideas of Rabotnov, Ramensky, and Grime may prove more fruitful (Table 1.4). In particular,

Table 1.4 Revised system of bryophyte life strategies

	Spores		
Potential life span (yr)	Numerous, very light (<20 μm)	Few, large (> 20 μm)	Reproductive effort
<1	Fugitives	Annual shuttle	High
Few	Colonists Ephemeral colonists Colonists s.s. Pioneers	Medium Short-lived shuttle Long-lived shuttle	
Many	Perennial stayers Competitive per. Stress-tolerant per.	Dominants	Low

Rabotnov's (1975) concept of pioneers as species that 'make initially lifeless territories suitable for patients or violents' might be useful. Thus, we see pioneers as species that are able to colonize in very harsh environments and usually occur in the earliest stages of primary successions, in contrast to colonists s.s. that colonize open (usually as a result of disturbance) but potentially productive habitats, thus occurring more commonly in the earlier stages of secondary successions (cf. Grubb 1986, 1987). Potential life span of pioneers tends to be moderately long, their growth rate is low, but RE is often fairly high; population maintenance is mainly by asexual reproduction, however (During 1990). Potential life span of colonists s.s. is rather short, their growth rate is usually high, and investment in both sexual and asexual reproduction is often quite high. Gemma production may serve as a rapid way of population expansion during colonizing episodes (Joenje and During 1977).

The old, heterogeneous colonist strategy type contained yet another group of species differing fundamentally from both pioneers and colonists s.s.: the 'gap-dependent species' (During *et al.* 1985) exemplified by the *Bryum erythrocarpum* complex. Such species are usually very short-lived above ground, and population maintenance is predominantly by means of subterraneous tubers on the rhizoids; often, they maintain an enormous diaspore bank in the soil (During and Ter Horst 1983; During *et al.* 1988). Sporophyte production is rare, but if it occurs, large numbers of small spores are produced. This group is called here ephemeral colonists.

With regard to the perennials, at least a separation between stress-tolerant (patient) and competitive (violent) perennials is needed. These two groups differ primarily in growth rate (cf. Furness and Grime 1982), morphological plasticity (Rincon and Grime 1989*a*), and in their degree of stress (e.g. drought) tolerance.

The fugitive strategy type is very rare among bryophytes (*Funaria hygrometrica* is one of the very few examples) and far more common among fungi and bacteria, but it has been maintained because it differs fundamentally from the other types. At the other extreme, a new category of dominants is proposed to accomodate species with large spores that are potentially very long-lived. Again, rather few bryophytes belong to this category; probably, only some *Sphagnum* species should be placed here. This strategy is far more common among flowering plants (notably trees).

Much of the foregoing discussion refers primarily to temperate world bryophytes. Tropical bryophytes, particularly epiphytes and epiphylls, are much more diverse in growth-forms and subtle adaptations in their morphology and life-history. Thus, a variety of devices for absorption and retention of water has developed (Thiers 1988; Gradstein and Pocs 1989); phyllodioecious species with minute, short-lived male gametophytes living on the female plants are found in several groups; and in some ephemeral environments such as fine twigs and living leaves the life cycle may be shortened to the extent that juvenile characters are retained in the adult stage (neoteny; Schuster 1984; Richards 1984; Thiers 1988). Consequently, application of the life strategy concept would be fascinating, but has not yet been attempted (Richards 1988; Gradstein and Pocs 1989).

Based on morphological characteristics and field experience, lists of species with their life-history strategy type have been provided by During (1982), Orbán (1984), Lloret (1987), and During *et al.* (1988). Spectra representing the relative abundance of the strategy types calculated on the basis of such lists clearly correspond to differences in environmental conditions, notably the frequency and length of periods of suitable moisture conditions and the stability of the substrate (During 1982; Lloret 1988).

Anthropogenic influences strongly affect the relative success of species belonging to different strategy types. The fact that the fugitive *Funaria hygrometrica* and a number of colonists s.s. (e.g. *Bryum bicolor, B. argenteum,*

Ceratodon purpureus) and competitive perennials such as *Brachythecium rutabulum* are common in many areas is at least partly due to human activity, and the abundance of many ephemeral colonists and annual short-lived species is related to the wide spread of agriculture, although the recent trend in some countries of over-fertilization of maize fields is clearly detrimental to these bryophytes. The decrease of pioneers and short-lived shuttle species and the concomitant increase of competitive perennials in Dutch chalk grasslands is probably due to the combined effect of a changed management regime and the input of large amounts of atmospheric nitrogen via precipitation and dry deposition (During and Willems 1986).

Although information concerning growth rates and life span of lichens is still rather limited, application of this system to lichens should be possible (cf. Sipman 1983; Rogers 1990). Some ephemeral lichens of seasonally suitable substrates (e.g. *Aphanopsis coenosa*, Poelt and Vezda 1990) might be regarded as annual shuttle species, whereas the ephemeral *Aspicilia excavata*, which only reproduces by means of soredia (Poelt and Vezda 1990), would fit in the ephemeral colonists. Pioneers are rather more abundant in lichens than in bryophytes (cf. Topham 1977). The strategies of colonists s.s. and both stress-tolerant and competitive perennials also seem to be represented well among lichens, and, among epiphytic lichens at least, short-lived and long-lived shuttle species may be quite common (cf. Topham 1977; Sipman 1983; Rogers 1990). Thus, long-lived shuttle species may be represented by species with very large spores and frequent production of soredia or isidia, such as *Pertusaria* spp. and *Ochrolechia* spp.

The value of the system of life-history strategies described above strongly depends on the relevance of the trade-offs upon which it is based. At least for bryophytes, some basic information concerning life span, investment in reproduction, and the relation between spore characteristics and establishment success is becoming available, albeit for only very few species. There is some evidence, that plants without sporophytes grow faster than plants bearing sporophytes in *Scorpidium scorpioides* (A. M. Kooijman and H. J. During unpublished data) and *Plagiothecium undulatum* (Hofman 1991). However, to my knowledge a direct connection between sporophyte production and mortality risk of the gametophytes supporting sporophytes has never been analysed, and the question of whether the high investment in sexual reproduction shown by many species indeed contributes significantly to their long-term fitness is still completely open. The study of Miles and Longton (1990) on germination and establishment clearly shows that meaningful investigations into such aspects require careful studies in the field besides laboratory experiments. Obviously, there is a challenging field of research open here!

1.6 Conclusion

Distribution of growth forms of bryophytes and lichens is strongly correlated with water economy. The life form systems presented also primarily reflect adaptations to frequency and intensity of periods of water shortage.

The triangular C–S–R model of plant strategies provides a meaningful way to ordinate bryophytes and lichens along axes of competitive ability and stress tolerance. Lichens occupy a triangular subspace in the stress-tolerant region of the model, whereas bryophytes are mainly found in a triangular subspace in the region in between the stress-tolerant and the ruderal corners.

The somewhat expanded system of bryophyte life history strategies presented here is still mainly based on untested assumptions, but it may provide a framework for future studies. Applicability of this classification to lichens and tropical bryophytes awaits further investigations.

Acknowledgments

Comments on an earlier version by H. F. van Dobben are gratefully acknowledged.

References

Barkman, J. J. (1958). *Phytosociology and ecology of cryptogamic epiphytes.* Van Gorcum, Assen.

Barkman, J. J. (1988). New systems of plant growth forms and phenological plant types. In *Plant form and vegetation structure*, (eds M. J. A. Werger, P. J. M. van der Aart, H. J. During, and J. T. A. Verhoeven), pp. 9–44. SPB Academic Publishing, The Hague.

Beck, E, Mägdefrau, K., and Senser, M. (1986). Globular mosses. *Flora*, **178**, 73–83.

Birse, E. M. (1957). Ecological studies on growth-form in bryophytes II. Experimental studies on growth-form in mosses. *Journal of Ecology*, **45**, 721–33.

Birse, E. M. (1958a). Ecological studies on growth-form in bryophytes III. The relationship between the growth-form of mosses and ground-water supply. *Journal of Ecology*, **46**, 9–27.

Birse, E. M. ((1958b). Ecological studies on growth-form in bryophytes IV. Growth-form distribution in a deciduous wood. *Journal of Ecology*, **46**, 29–42.

Brown, D. H. (1984). Uptake of mineral elements and their use in pollution monitoring. In *The experimental biology of bryophytes*, (eds A. F. Dyer and J. G. Duckett), pp. 229–55. Academic Press, London.

Buch, H. (1947). Ueber die Wasser- und Mineralstoffversorgung der Moose I, II. *Societas Scientiarum Fennica. Commentationes Biologicae*, **9** (16, 20).

Clayden, S. and Bouchard, A. (1983). Structure and dynamics of conifer–lichen stands on rock outcrops south of Lake Abitibi, Quebec. *Canadian Journal of Botany*, **61**, 850–71.

Collins, N. J. (1976). Growth and population dynamics of the moss *Polytrichum alpestre* in the maritime Antarctic. *Oikos*, **27**, 389–401.

Cook, R. E. (1979). Asexual reproduction: a further consideration. *American Naturalist*, **113**, 769–72.

Cornelissen, J. H. C. and Ter Steege, H. (1989). Distribution and ecology of epiphytic bryophytes and lichens in dry evergreen forest of Guyana. *Journal of Tropical Ecology*, **5**, 131–50.

Creveld, M. C. (1981). *Epilithic lichen communities in the alpine zone of southern Norway*. D.Phil. Thesis, University of Utrecht.

Du Rietz, G. E. (1931). Life-forms of terrestrial flowering plants I. *Acta Phytogeographica Suecica*, **3**, 1–95.

During, H. J. (1979). Life strategies of bryophytes: a preliminary review. *Lindbergia*, **5**, 2–17.

During, H. J. (1980). Life forms and life strategies in Nanocyperion communities from the Netherlands Frisian Islands. *Acta Botanica Neerlandica*, **29**, 483–96.

During, H. J. (1982). Bryophyte flora and vegetation of Lanzarote, Canary Islands. *Lindbergia*, **7**, 113–25.

During, H. J. (1990). Clonal growth patterns among bryophytes. In *Clonal growth in plants: regulation and function*, (eds J. van Groenendael and H. de Kroon), pp.153–76. SPB Academic Publishing, The Hague.

During, H. J. and Ter Horst, B. (1983). The diaspore bank of bryophytes and ferns in chalk grassland. *Lindbergia*, **9**, 57–64.

During, H. J. and Ter Horst, B. (1985). Life span, mortality and establishment of bryophytes in two contrasting habitats. *Abstracta Botanica*, **9**, Supplement 2, 145–58.

During, H. J. and Van Tooren, B. F. (1987). Recent developments in bryophyte population ecology. *Trends in Ecology and Evolution*, **2**, 89–93.

During, H. J. and Van Tooren, B. F. (1988). Pattern and dynamics in the bryophyte layer of a chalk grassland. In *Dynamics of pattern and diversity in plant communities*, (eds H. J. During, M. J. A. Werger, and J. H. Willems), pp. 195–208. SPB Academic Publishing, The Hague.

During, H. J. and Willems, J. H. (1986). The impoverishment of the bryophyte and lichen flora of the Dutch chalk grasslands in the thirty years 1953–1983. *Biological Conservation*, **36**, 43–58.

During, H. J., Schenkeveld, A. J., Verkaar, H. J., and Willems, J. H. (1985). Demography of short-lived forbs in chalk grassland in relation to vegetation structure. In *The population structure of vegetation*, (ed. J. White), pp. 341–70. Junk, Dordrecht.

During, H. J., Bruguès, M., Cros, R. M., and Lloret, F. (1988). The diaspore bank of bryophytes and ferns in the soil in some contrasting habitats around Barcelona, Spain. *Lindbergia*, **13**, 137–49.

During, H. J., Bruguès, M., Cros, R. M., Lloret, F., and Van Tooren, B. F. (1991).

Dynamics of bryophytes in relation to their life strategies. In *Proceedings of the seventh Central and East European Bryological Working Group meeting*, (eds N. Konstantinova and D. G. Horton) (in press).

Ennos, R. A. (1990). Population genetics of bryophytes. *Trends in Ecology and Evolution*, **5**, 38–9.

Frey, E. (1924). Die Berücksichtigung der Lichenen in der soziologischen Pflanzen-Geografie, speziell in den Alpen. *Verhandlungen der naturforschenden Gesellschaft, Basel*, **35**, 303–20.

Furness, S. B. and Grime, J. P. (1982). Growth rate and temperature responses in bryophytes II. A comparative study of species of contrasted ecology. *Journal of Ecology*, **70**, 525–36.

Gadgil, M. and Solbrig, O. T. (1972). The concept of r- and K-selection: evidence from wild flowers and some theoretical considerations. *American Naturalist*, **106**, 14–31.

Giesenhagen, K. (1910). Die Moostypen der Regenwälder. *Annales du Jardin Botanique de Buitenzorg*, Supplement 3, **2**, 711–91.

Gimingham, C. H. and Birse, E. M. (1957). Ecological studies on growth-form in bryophytes I. Correlations between growth-form and habitat. *Journal of Ecology*, **45**, 533–45.

Gimingham, C. H. and Robertson, E. T. (1950). Preliminary investigations on the structure of bryophyte communities. *Transactions of the British Bryological Society*, **1**, 330–44.

Gimingham, C. H. and Smith, R. I. L. (1971). Growth form and water relations of mosses in the maritime Antarctic. *Bulletin of the British Antarctic Survey*, **25**, 1–21.

Gould, S. J. and Lewontin, R. C. (1979). The spandrels of San Marco and the panglossian paradigm: a critique of the adaptationist programme. *Proceedings of the Royal Society, London, Series B*, **205**, 581–98.

Gradstein, S. R. and Pocs, T. (1989). Bryophytes. In *Tropical rain forest ecosystems*, (eds H. Lieth and M. J. A. Werger), pp. 311–25. Elsevier, Amsterdam.

Grime, J. P. (1974). Vegetation classification by reference to strategies. *Nature (London)*, **250**, 26–31.

Grime, J. P. (1977). Evidence for the existence of three primary strategies in plants and its relevance to ecological and evolutionary theory. *American Naturalist*, **111**, 1169–94.

Grime, J. P. (1979). *Plant strategies and vegetation processes*. Wiley, Chichester.

Grime, J. P., Rincon, E. R., and Wickerson, B. E. (1990). Bryophytes and plant strategy theory. *Botanical Journal of the Linnean Society*, **104**, 175–86.

Grubb, P. J. (1986). The ecology of establishment. In *Ecology and landscape design*, (eds A. D. Bradshaw, D. A. Goode, and E. Thorp), pp. 83–98. Blackwell, Oxford.

Grubb, P. J. (1987). Some generalizing ideas about colonization and succession in green plants and fungi. In *Colonization, succession and stability*, (eds A. J. Gray, M. J. Crawley, and P. J. Edwards), pp. 83–102. Blackwell Scientific Publications, Oxford.

Hale, M. E. (1983). *The biology of lichens*, (3rd edn). Edward Arnold, London.

Hawksworth, D. L. and Hill, D. J. (1984). *The lichen-forming fungi*. Blackie, Glasgow.

Hilitzer, A. (1925). La végétation epiphytique de la Bohème. *Publications du Faculté scientifique de l'Université Charles, Prague*, **41**, 1–200.

Hofman, A. (1991). Phylogeny and population genetics of the genus *Plagiothecium* (*Bryopsida*). D. Phil. thesis, University of Groningen.

Jahns, H. M. (1984). Morphology, reproduction and water relations—a system of morphogenetic interactions in *Parmelia saxatilis. Nova Hedwigia, Beihefte*, **79**, 716–37.

Joenje, W. and During, H. J. (1977). Colonisation of a desalinating Wadden-polder by bryophytes. *Vegetatio*, **35**, 177–85.

Kiss, T. (1985). The life-strategy system of lichens—a proposal. *Abstracta Botanica*, **9**, 59–66.

Kiss, T. (1988). Dispersal and growth forms: an approach towards an understanding of the life-strategy concept in lichenology. *Acta Botanica Hungarica*, **34**, 175–91.

Lange, O. L., Kilian, E., and Ziegler, H. (1986). Water vapor uptake and photosynthesis of lichens: performance differences in species with green and blue–green algae as photobionts. *Oecologia*, **71**, 104–10.

Larson, D. W. (1981). Differential wetting in some lichens and mosses: the role of morphology. *The Bryologist*, **84**, 1–15.

Larson, D. W. and Kershaw, K. A. (1976). Studies on lichen-dominated systems XVIII. Morphological control of evaporation in lichens. *Canadian Journal of Botany*, **54**, 2061–73.

Lloret, F. (1987). Estudio briológico del alto valle del Ter. D. Phil. thesis. Autonomous University, Barcelona.

Lloret, F. (1988). Estrategias de vida y formas de vida en briófitos del pirineo oriental (España). *Cryptogamie, Bryologie et Lichénologie*, **9**, 189–217.

Longton, R. E. (1984). The role of bryophytes in terrestrial ecosystems. *Journal of the Hattori Botanical Laboratory*, **55**, 147–63.

Longton, R. E. (1988*a*). *Biology of polar bryophytes and lichens*. Cambridge University Press, Cambridge.

Longton, R. E. (1988*b*). Life-history strategies among bryophytes of arid regions. *Journal of the Hattori Botanical Laboratory*, **64**, 15–28.

Longton, R. E. and Miles, C. J. (1982). Studies on the reproductive biology of mosses. *Journal of the Hattori Botanical Laboratory*, **52**, 219–20.

Longton, R. E. and Schuster, R. M. (1983). Reproductive biology. In *New manual of bryology*, Vol. 1., (ed. R. M. Schuster), pp. 386–462. Hattori Botanical Laboratory, Nichinan.

MacArthur, R. H. and Wilson, E. O. (1967). *Theory of island biogeography*. Princeton University Press, Princeton.

Mägdefrau, K. (1982). Life-forms of bryophytes. In *Bryophyte ecology*, (ed. A. J. E. Smith), pp. 45–58. Chapman and Hall, London.

Mattick, F. (1951). Wuchs- und Lebensformen, Bestand- und Gesellschaftsbildung der Flechten. *Engler's Botanische Jahrbücher*, **75**, 378–423.

Meusel, H. (1935). Wuchsformen und Wuchstypen der europäischen Laubmoose. *Nova Acta Leopoldina*, N.F. **3**, 123–277.

Miles, C. J. and Longton, R. E. (1987). Life history of the moss, *Atrichum undulatum* (Hedw.) P. Beauv. *Symposia Biologica Hungarica*, **35**, 193–207.

Miles, C. J. and Longton, R. E. (1990). The role of spores in reproduction in mosses. *Botanical Journal of the Linnean Society*, **104**, 149–73.

Mischler, B. (1988). Reproductive ecology of bryophytes. In *Plant reproductive ecology. Patterns and strategies*, (eds J. Lovett Doust and L. Lovett Doust), pp. 285–306. Oxford University Press, Oxford.

Noble, I. R. and Slatyer, R. O. (1979). The use of vital attributes to predict successional changes in plant communities subject to recurrent disturbances. *Vegetatio*, **43**, 5–21.

Ochsner, F. (1928). Studien über die Epiphytenvegetation der Schweiz. *Jahrbücher der Sankt Gallischen Naturwissenschaftlichen Gesellschaft*, **63**, 1–106.

Orbán, S. (1984). A Magyarországi mohák stratégiái és T, W, R értékei. *Acta Academiae Paedagogicae Agriensis*, Nova Series, **17**, 755–65.

Ott, S. (1987*a*). Sexual reproduction and developmental adaptations in *Xanthoria parietina*. *Nordic Journal of Botany*, **7**, 219–28.

Ott, S. (1987*b*). Reproductive strategies in lichens. In *Progress and problems in lichenology in the eighties*, (ed. E. Peveling), pp. 81–93. Bibliotheca Lichenologica, **25**. Cramer, Vaduz.

Pianka, E. R. (1970). On r- and K-selection. *American Naturalist*, **104**, 592–7.

Poelt, J. and Vezda, A. (1990). Ueber kurzlebige Flechten. *Bibliotheca Lichenologica*, **38**, 377–94.

Poorter, H. (1990). Interspecific variation in relative growth rate: on ecological causes and physiological consequences. In *Variation in growth rate and productivity*, (eds H. Lambers, M. L. Cambridge, H. Konings, and T. L. Pons), pp. 45–68. SPB Academic Publishing, The Hague.

Proctor, M. C. F. (1980). Diffusion resistances in bryophytes. In *Plants and their atmospheric environment*, (eds J. Grace, E. D. Ford, and P. G. Jarvis), pp. 219–29. Blackwell, Oxford.

Proctor, M. C. F. (1982). Physiological ecology: Water relations, light and temperature responses, carbon balance. In *Bryophyte ecology*, (ed. A. J. E. Smith), pp. 333–81. Chapman and Hall, London.

Proctor, M. C. F. (1984). Structure and ecological adaptation. In *The experimental biology of bryophytes*, (eds A. F. Dyer and J. G. Duckett), pp 9–37. Academic Press, London.

Pyatt, F. B. (1973). Lichen propagules. In *The lichens*, (eds V. Ahmadjian and M. E. Hale), pp. 117–145. Academic Press, New York.

Rabotnov, T. A. (1975). On phytocoenotypes. *Phytocoenologia*, **2**, 66–72.

Ramensky, L. G. (1938). *Introduction to complex soil-geobotanical study of lands*. Selkhozgiz, Moscow. (in Russian).

Raunkiaer, C. (1934). *The life forms of plants and statistical plant geography*. Oxford University Press, Oxford.

Richards, P. W. (1984). The ecology of tropical forest bryophytes. In *New manual of bryology*, Vol. 2., (ed. R. M. Schuster), pp. 1233–70. Hattori Botanical Laboratory, Nichinan.

Richards, P. W. (1988). Tropical forest bryophytes. *Journal of the Hattori Botanical Laboratory*, **64**, 1–4.

Rincon, E. and Grime, J. P. (1989*a*). Plasticity and light interception by six bryophytes of contrasted ecology. *Journal of Ecology*, **77**, 439–46.

Rincon, E. and Grime, J. P. (1989*b*). An analysis of seasonal patterns of bryophyte growth in a natural habitat. *Journal of Ecology*, **77**, 447–55.

Robinson, A. L., Vitt, D. H., and Timoney, K. P. (1989). Community structure and morphology of bryophytes and lichens relative to edaphic gradients in the subarctic forest-tundra of northwestern Canada. *The Bryologist*, **92**, 495–512.

Rogers, R. W. (1988). Succession and survival strategies in lichen populations on a palm trunk. *Journal of Ecology*, **76**, 759–76.

Rogers, R. W. (1989). Colonization, growth, and survival strategies of lichens on leaves in a subtropical rain forest. *Australian Journal of Ecology*, **14**, 327–33.

Rogers, R. W. (1990). Ecological strategies of lichens. *The Lichenologist*, **22**, 149–62.

Schmidt, T. (1918). Die Verbreitung von Samen und Blütenstaub durch die Luftbewegung. *Oesterreichische Botanische Zeitung*, **67**, 313–28.

Schofield, W. B. (1981). Ecological significance of morphological characters in the moss gametophyte. *The Bryologist*, **84**, 149–65.

Schuster, R. M. (1984). Evolution, phylogeny and classification of the Hepaticae. In *New manual of bryology*, Vol. 2., (ed. R. M. Schuster), pp. 892–1070. Hattori Botanical Laboratory, Nichinan.

Segal, S. (1966). *Ecological notes on wall vegetation*. Junk, The Hague.

Sipman, H. J. M. (1983). *A monograph of the lichen family Megalosporaceae*. Bibliotheca Lichenologica, **18**. Cramer, Vaduz.

Sipman, H. J. M. (1989). Lichen zonation in the Parque Los Nevados transect. *Studies on tropical andean ecosystems*, **3**, 461–83.

Sipman, H. J. M. and Harris, R. C. (1989). Lichens. In *Tropical rain forest ecosystems*, (ed. H. Lieth and M. J. A. Werger), pp. 303–9. Elsevier, Amsterdam.

Slack, N. G. (1977). Species diversity and community structure in bryophytes: New York State studies. *Bulletin of the New York State Museum*, **428**, 1–70.

Smith, A. J. E. (1982). Epiphytes and epiliths. In *Bryophyte ecology*, (ed. A. J. E. Smith), pp. 191–227. Chapman and Hall, London.

Southwood, T. R. E. (1988). Tactics, strategies and templets. *Oikos*, **52**, 3–18.

Stearns, S. C. (1976). Life history tactics: a review of the ideas. *Quarterly Review of Biology*, **51**, 3–47.

Stearns, S. C. (1989). Trade-offs in life-history evolution. *Functional Ecology*, **3**, 259–68.

Thiers, B. M. (1988). Morphological adaptations of the Jungermanniales (Hepaticae) to the tropical rainforest habitat. *Journal of the Hattori Botanical Laboratory*, **64**, 5–14.

Topham, P. B. (1977). Colonization, growth, succession and competition. In *Lichen ecology*, (ed. M. R. D. Seaward), pp. 31–68. Academic Press, London.

Van Leerdam, A., Zagt, R. J., and Veneklaas, E. J. (1990). The distribution of epiphyte growth-forms in the canopy of a Colombian cloud-forest. *Vegetatio*, **87**, 59–71.

Van Tooren, B. F. and During, H. J. (1988). Early succession of bryophyte communities on Dutch forest earth banks. *Lindbergia*, **14**, 40–6.

Van Zanten, B. O. (1978). Experimental studies on transoceanic long-range dispersal of moss spores in the southern hemisphere. *Journal of the Hattori Botanical Laboratory*, **44**, 455–82.

Van Zanten, B. O. and Gradstein, S. R. (1988). Experimental dispersal geography of neotropical liverworts. *Nova Hedwigia, Beihefte*, **90**, 41–94.

Van Zanten, B. O. and Pocs, T. (1981). Distribution and dispersal in bryophytes. *Advances in Bryology*, **1**, 479–562.

Vitt, D. H. (1984). Classification of the Bryopsida. In *New manual of bryology*, Vol. 2., (ed. R. M. Schuster), pp. 696–759. Hattori Botanical Laboratory, Nichinan.

Watson, E. V. (1971). *The structure and life of bryophytes* (3rd edn). Hutchinson, London.

Watson, M. A. (1979). Age structure and mortality within a group of closely related mosses. *Ecology*, **60**, 988–97.

Weber, W. A. (1977). Environmental modification and lichen taxonomy. In *Lichen ecology*, (ed. M. R. D. Seaward), pp. 9–29. Academic Press, London.

Whitehouse, H. L. K. (1985). Advances in knowledge of the life strategies of British bryophytes. In *British Bryological Society Diamond Jubilee*, (eds R. E. Longton and A. R. Perry), pp. 43–9. British Bryological Society, Cardiff.

Wyatt, R. and Anderson, L. E. (1984). Breeding systems in bryophytes. In *The Experimental biology of bryophytes*, (eds A. F. Dyer and J. G. Duckett), pp. 39–64. Academic Press, London.

Wyatt, R., Odrzykoski, I. J., and Stoneburner, A. (1989). High levels of genetic variability in the haploid moss *Plagiomnium ciliare*. *Evolution*, **43**, 1085–96.

2

The role of bryophytes and lichens in terrestrial ecosystems

ROYCE E. LONGTON

2.1 Introduction

The importance of bryophytes and lichens in primary succession has long been recognized. Otherwise these plants have generally been regarded as troublesome to identify when compiling releves but of little significance in the functioning of mature communities. This attitude has recently changed with the use of mosses and lichens as pollution monitors, and it has also been realized that they play a significant role in the functioning of undisturbed communities (Longton 1984; Seaward 1988; Slack 1988). Nowhere is the importance of mosses and lichens greater than in polar tundra and in northern forests and mires, which have so far been less modified by human activity than other, more complex ecosystems. Such communities are ideally suited to the study of fundamental ecosystem processes, and their cryptogamic component was intensively investigated during the International Biological Programme (Longton 1988a). Global warming is likely to be most intense at high latitudes, and the role of bryophytes and lichens in boreal and polar communities is therefore emphasized in this account.

2.2 Succession

2.2.1 Primary succession on rock

Vegetation has developed through colonization of bare areas by pioneer organisms, followed by the gradual displacement of the pioneers in a series of successional communities. The component biota are believed to modify edaphic and microclimatic conditions in ways that favour members of the next community, with which they therefore become unable to compete. Succession thus reflects interaction between abiotic factors and the influence of the successional species. It may culminate in the establishment of relatively stable, climax vegetation. Bryophytes and lichens are commonly prominent during succession and undoubtedly influence rates of change, although climax vegetation appears to be determined principally by climate interacting with drainage patterns and the chemistry of surface rocks.

Adaptations enabling lichens to colonize rock surfaces include longevity, tolerance of desiccation and extreme temperatures, and low growth rates commensurate with the slow release of mineral nutrients from the substratum (Topham 1977). Many mosses show similar characteristics. Crustose lichens are often the first visible colonizers, giving way to foliose lichens or small cushion-forming mosses and later to fruticose lichens associated with mats and turfs of larger bryophytes. However, species composition, growth-form representation (Hale 1983; Richards 1984), and the relative abundance of bryophytes and lichens vary in relation to moisture, rock type, and other factors within a climatic region.

Once established, mosses and lichens promote soil formation by accelerating physical and chemical weathering, by trapping wind-blown organic and inorganic material, and by contributing directly to undecomposed organic matter. They appear also to concentrate several essential elements, including K, P, and S (Syers and Iskandar 1973; A. J. E. Smith 1982; Ugolini and Edmonds 1983), and may increase the availability of N, which is typically low in young soils.

Physical weathering is promoted by expansion and contraction of appressed, partially endolithic crustaceous lichens as their water content varies, and by penetration of rhizines of foliose species which leads to the influx of water and thus to frost action. Lichens induce chemical weathering by liberating oxalic acid, carbonic acid and lichen compounds, which may crystallize on the outer surface of hyphae bringing them into direct contact with the rock. Although of high molecular weight, lichen compounds are slightly water-soluble and act as metal chelating agents (Rundel 1978).

SEM studies have confirmed the incidence of weathering beneath lichen thalli (Ascaso 1985; Jones and Wilson 1984; Viles 1987), where etching of the rock creates irregular surfaces susceptible to physical weathering. Garty and Delarea (1987) described how germinating ascospores of *Caloplaca aurantia* establish contact with free-living cells of a green alga, presumably *Trebouxia*, in pits in weathered roof tiles, with dust particles becoming trapped in the young thallus. Penetration of lichens into rock, also with incorporation of rock particles into the thalli, has been demonstrated on Antarctic quartz mica-schist (Walton 1985*a*). Direct evidence of substantial weathering comes from antarctic sandstones, where the surface periodically peels off due to cementing material between the rock crystals being dissolved by substances released from endolithic lichens (Friedmann 1982).

Snails feeding on endolithic lichens in desert limestone ingest up to 9 mm^3 rock per individual per day, resulting in weathering at 0.7–1.1 metric tons ha^{-1} yr^{-1} (Schachak *et al.* 1987).

Whether lichen-associated weathering on such a scale is widespread remains unclear. Free-living fungi show greater ability than either fungi or algae isolated from lichens to chelate ferric iron in culture (Williams and Rudolph 1974). Brodo (1973) considered that weathering, humus forma-tion, and entrapment of wind-blown particles by lichens are very slow. Thalli of *Rhizocarpon geographicum* on Arctic rocks may be several thousand years old (Andrews and Barnett 1979) confirming that lichen colonization does not always initiate rapid succession. Indeed, lichens or bryophytes may retard weathering by protecting surfaces from erosion, by insulating against freeze–thaw cycles, or by absorbing precipitation and further reducing frost shattering (Lindsay 1978). Danin *et al.* (1983) considered that lichen cover accelerates weathering in dry environments but protects the rock surface under wet conditions. The importance of lichens in weathering and pedogenesis is critically assessed by Jones (1988).

While crustose lichens are normally the first visible colonizers of newly exposed rocks, their presence is not always necessary to permit subsequent establishment of mosses, often in cracks or depressions. Moss rhizoids and associated fungi penetrate at least 5 mm into some rocks (Hughes 1982). Moreover, mosses grow faster than lichens, giving a greater capacity to trap wind-blown material and to contribute organic matter to developing soil (A. J. E. Smith 1982). Oosting and Anderson (1939) considered these processes of greater significance than the influence of cryptogams on weathering in their classic study of granite outcrops in North Carolina. The first colonizers, crustose and foliose lichens such as *Verrucaria nigrescens* and *Parmelia conspersa*, were regarded as unimportant in aiding establishment by other species. The effective pioneer was the moss *Grimmia laevigata* which formed spreading cushions. As the margins advanced, *G. laevigata* was displaced in the centre by communities domi-nated successively by the fruticose lichen *Cladonia leparina*, by *Selaginella rupestris* and by *Polytrichum ohioense*. As these communities spread laterally concentric zonation became established. Eventually, soil thickness in the centre of the mats increased, paving the way for angiosperm dominated vegetation. Many variations on this theme have been described (Topham 1977).

2.2.2 Primary succession on inorganic particulates

Several plant forms can act as pioneers on sand, gravel, and glacial till. Mosses and grasses are often the first macrophytes, with little association between the two (Corner and Smith 1973; Fridriksson 1975; Worsley and

Ward 1974). Indeed, on South Georgia some moraines subject to cryoturbatic disturbance are colonized first by grasses, with mosses later becoming established in older grass tufts. In contrast, crustose lichens appear first on more stable soils, subsequently giving way to colonies of mosses in which flowering plants become established (Heilbron and Walton 1984; Smith 1984).

Cryptogams undoubtedly influence pedogenesis on immature mineral soils by contributing organic matter and through their impact on nutrient cycling and their stabilizing effect on soil temperature and moisture regimes (Rouse and Kershaw 1971; Edward and Miller 1977). Usnic acid and other sparingly soluble lichen compounds are mobile, and Dawson *et al.* (1984) suggested that they contribute significantly to podsolization and profile development in coniferous woodland and alpine tundra soils. Mucher *et al.* (1988) showed that the occurrence of cryptogams, including species of *Diploschistes, Xanthoparmelia* (lichens), *Bryum, Desmatodon* (mosses), and *Targionia* (liverwort) was positively associated with the formation of a surface crust which prevents erosion of red earths in Australian rangelands. Soils beneath the mosses and lichens also had water and nutrients concentrated near the surface, and the cryptogams were considered desirable in terms of seed lodgement and germination.

2.2.3 Primary succession in water

Mosses are also important in successional processes that convert water bodies to dry land. In boreal regions, *Sphagnum* spp., associated with sedge rhizomes, form floating mats that extend outwards from the shore, increase in thickness, and eventually support plants characteristic of mires and later of mesic communities. Thus forest may develop on a floating raft of peat (Vitt and Slack 1975; Tallis 1983). Benthic mosses such as *Drepanocladus* and *Scorpidium* spp. are influential in the early stages of other Arctic hydroseres by accelerating accumulation of organic matter and inorganic sediments on lake bottoms (Polunin 1935).

2.2.4 Auto-succession

While climax vegetation often develops through successive replacement of one community by another there is growing evidence that some vegetation develops by auto-succession. Muller (1952) defined this as 'a succession consisting of a single stage, in which pioneer and climax species are the same'. It occurs particularly where climatic severity so restricts the number of species that competition is minimized and displacement fails to occur. Originally described in alpine vegetation in Scandinavia (Muller 1952), auto-succession also predominates among luxuriant cryptogamic communities in the cold Antarctic, where there is little evidence of competition, or replacement of one community by another (Smith 1972; Longton and Holdgate 1979). Nor is this pattern restricted to areas of extreme climatic severity. Oosting and Anderson (1939) noted in North Carolina that successional change was very slow and that 'some places may actually be in a condition of pioneer equilibrium'.

2.2.5 Secondary succession

Bryophytes and lichens are prominent in succession following the destruction of established vegetation. An abundance of *Funaria hygrometrica* on burnt-over ground, and of other small acrocarpous mosses on disturbed roadside verges, is a familiar sight throughout temperate regions. The nutrient relationships of *F. hygrometrica* on soils enriched by fire are discussed by Southorn (1977) and Dietert (1979), but the impact of these pioneer mosses on soil development has not been investigated.

Mosses and lichens are both prominent in secondary succession in the boreal forest, where lightning-induced fire is a recurrent factor. The following (Ahti 1977) is one of several characteristic lichen sequences: 1. Bare soil; 1–3 years after fire; 2. Crustose lichen stage, 3–10 years, characterized by *Lecidea* spp.; 3. Cup lichen stage, 10–30(–50) years, characterized by species of *Cladonia* subgenus *Cladonia*, e.g. *C. crispata*; 4. First reindeer lichen stage, 30(–50)–80(–120) years, characterized by species of *Cladonia* subgenus *Cladina*, e.g. *C. rangiferina*; 5. Second reindeer lichen stage, beginning after 80(–120) years, characterized by *Cladonia* (*Cladina*) *stellaris*. The moss *Ceratodon purpureus* and the thallose hepatic *Marchantia polymorpha* are also among the earliest colonizers of bare soil, while *Polytrichum juniperinum* and *P. piliferum* are associated with *Lecidea* spp. during the crustose lichen stage.

Cladonia spp. or *Stereocaulon paschale* carpet the ground beneath white spruce (*Picea glauca*) in open lichen woodland in the north of the Canadian boreal forest. Lichen woodland may be a climax community on sandy soils under dry, continental climates (Ahti 1977; Johnson 1981), although Klein (1982) considers it dependent on the regular occurrence of fire. In moist habitats (Maikawa and Kershaw 1976), and more generally in oceanic areas, closure of the tree canopy occurs and fruticose lichens become replaced by the large, weft-forming pleurocarpous mosses *Hylocomium splendens*, *Pleurozium schreberi*, and *Ptilium crista-castrensis*, often associated with tall turf-forming mosses (*Dicranum* spp.) and colonized by foliose lichens such as *Peltigera* spp. (Fig. 2.1). These plants may later be overgrown by *Sphagnum* spp. which blanket the forest floor causing waterlogging and reductions in pH, decomposition rates, nutrient availability, and forest productivity (Foster 1985). Paludification, resulting in

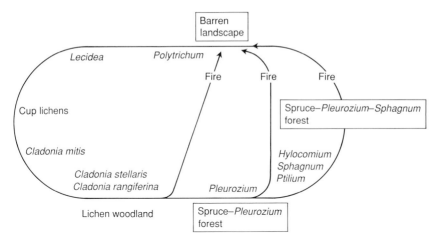

Fig. 2.1 Schematic representation of fire-induced succession in Canadian boreal forests (after Foster 1985).

degeneration of forests following invasion of *Sphagnum* spp., occurs widely in northern regions (Tallis 1983; see also Chapter 7).

Cryptogams modify environmental conditions throughout the post-fire sequence. Before colonization, there is wide diurnal temperature fluctuation at the ground surface, with high maxima inimical to many plants during sunny weather. Colonization by *Polytrichum* and *Lecidea* spp. increases surface albedo, thus reducing net radiation, and accumulating organic matter retains moisture so that absorbed energy is increasingly dissipated as latent heat. Maximum surface temperatures therefore decline, favouring the establishment of fruticose lichens which have a similar, but enhanced, effect. Their elimination following closure of the canopy apparently results from light limitation of photosynthesis (Kershaw 1985). Besides stabilizing soil temperature and moisture regimes, cryptogams are major contributors of organic matter to the soil, leading to acidification of the surface layers, even though accumulation of organic matter may be accompanied by increases in exchangable cations (Filion and Payette 1989).

2.2.6 Cyclic succession

Bryophytes and lichens are commonly involved in cyclic succession. This occurs in subclimax communities where directional succession is reversed, as where thalli of pioneer epilithic lichens, such as *Parmelia saxatilis* and *Placopsis contortuplicata* in the cold Antarctic (Lindsay 1978), slowly expand to form roughly circular colonies which die and become eroded from the centre outwards. This restores a bare surface that may be recolonized by

the same, or ecologically related species. A more dynamic situation was
revealed by John's (1989) analysis of lichens on a rock slide in the Rocky
Mountains. A diversity of crustose lichens and more competitive foliose
species gave total cover as high as 84 per cent, but most species were
represented by a large proportion of small (presumably young) thalli,
suggesting that active recruitment continued. There was little evidence of
succession towards climax forest, and John suggested that cyclic succession,
initiated by death or erosion of older thalli, was maintaining the diverse
community: species distribution was also thought to be influenced by
allelopathic substances released by some of the crustose species. Cryptogams are also involved in cyclic processes that result in temporal change in
species composition at a given point within a climax community, for
example in association with the pattern of growth and regeneration of
Calluna vulgaris bushes in heathland (Fig. 2.2).

2.2.7 Succession and strategies

Both primary and secondary succession are normally marked by a progressive increase in species diversity, although diversity may decline slightly in
the climax community. This applies to the ecosystem as a whole (Odum
1971), and to the cryptogamic component (Magomedova 1980; Filion and
Payette 1989). It implies that competition becomes more intense as
succession proceeds. In a parallel trend, there is a tendency for r-selection
to predominate in pioneer communities, giving way to K-selection in the
climax, a pattern also evident among bryophytes (During 1979) and lichens
(Topham 1977; Rogers 1988). Ahti (1982) considered some species of
Cladonia, such as *C. crispata*, to be r-selected since they are effective
colonizers with a capacity for rapid population growth, early maturation,
and abundant production of small ascospores. Later species, such as *C.
stellaris*, show K-selection as they operate as large, long-lived perennials in
stable communities and show a lower, primarily asexual reproductive effort.
Similarly *Funaria hygrometrica*, a pioneer moss following fire on Spanish
dunes had a higher reproductive effort, in terms of both spore output and
investment in sporophyte tissue, than the perennial *Tortella flavovirens* in
nearby woodland (Longton 1988*b*). In many groups a trend from r-selection
to K-selection is marked by an increase in diaspore size. This pattern is
less clear among bryophytes and lichens, where one species may produce
spores and asexual diaspores in a range of sizes, and many ruderal mosses
show shuttle strategies marked by large spores (During 1979).

Rogers (1990) has recently related an assemblage of over 30 lichen
species to Grime's (1979) triangular ordination of strategies. He confirmed
that species of relatively stable vegetation, such as *Cladonia alpestris*, lie near

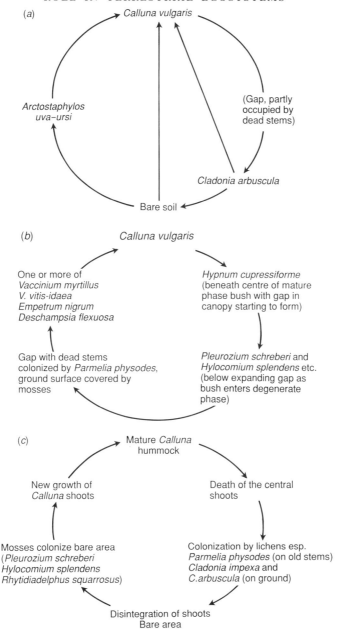

Fig. 2.2 Examples of cyclic succession in *Calluna*-dominated communities in Scotland, indicating the importance of mosses and lichens in (*a*) a *Calluna–Arctostaphylos* community; (*b*) a *Calluna–Vaccinium* community; (*c*) a dune heath. (*Parmelia physodes* = *Hypogymnia physodes*, *Cladonia impexa* = *C. portentosa*, *Hypnum cupressiforme* is probably *H. jutlandicum*). Reproduced from Gimingham (1972) by permission of Chapman and Hall.

the competitor pole of the ordination on the basis of large size and high relative growth rate. His findings also suggest that strategies among pioneer species may vary with substrate, as lithophytes such as *Rhizocarpon obscurata* and epiphylls such as *Porina epiphylla* showed the characteristics of stress tolerators and ruderals respectively. A detailed consideration of bryophyte and lichen strategies in different communities is presented in Chapter 1.

2.3 Production and phytomass

Bryophytes and lichens are dominant in many tundra and mire communities. They also contribute substantially to phytomass in a number of other vegetation types, of which only selected examples can be considered here. Grassland (Van Tooren *et al.* 1988) and coastal desert (Nash *et al.* 1977) are among further types of vegetation with significant cryptogamic components. Methods of assessing cryptogamic production and phytomass, and their reliability, are discussed by Longton (1988*a*), Russell (1988), and Russell and Botha (1988).

Table 2.1   Vegetation zones in polar regions (Longton 1988*a*)

Zone	Highest mean monthly air temperature (°C)	Characteristics of the vegetation
Mild polar	6–10 (to 12)	Extensive grass heath, dwarf shrub heath, mire, and other closed phanerogamic vegetation. *Sphagnum* abundant in many mires, though local in the mild Antarctic. Fellfields on the drier uplands.
Cool polar	3–7	Open fellfields and barrens predominant but mire, dry meadow, and other closed angiosperm-dominated communities locally extensive in favourable habitats. Dwarf shrub heaths of restricted occurrence or absent. *Sphagnum* seldom a major component of mires.
Cold polar	0–2	Closed stands of bryophytes, lichens, or algae extensive where wet or mesic conditions occur, with open cryptogamic vegetation on drier ground. Herbaceous phanerogams subordinate to cryptogams or absent. Liverworts frequent.
Frigid polar	<0	Vegetation largely restricted to scattered colonies of mosses, lichens, or algae, and to endolithic micro-organisms. Phanerogams absent. Liverworts very rare.

2.3.1 Polar tundra

Polar regions are highly variable in climate and vegetation. Four intergrading vegetation zones may be recognized based on growth-form representation. The zonation is correlated with mean summer temperature (Table 2.1).

The frigid Antarctic

The frigid Antarctic has no counterpart in the Arctic. Mean air temperature normally remains below 0°C throughout the year, and aridity is a major limiting factor. The entirely cryptogamic vegetation is largely restricted to sparse, open communities of algae and small turf- and cushion-forming mosses such as *Bryum* and *Grimmia* spp. on the predominantly mineral soil and of lichens, including crustose, foliose, and fruticose species, on rocks (Longton 1979*a*). The few estimates of annual production in open moss communities are all <5 g m^{-2} (Longton 1974; Ino 1983). Total phytomass is only 5–200 g m^{-2} in the typical open communities (Longton 1974; Kappen 1985), but reaches 1000 g m^{-2} in occasional closed stands of mosses and of epilithic fruticose lichens. Green phytomass alone was 1097 g m^{-2} in an exceptional stand of mosses (Seppelt and Ashton 1978). Endolithic lichens are a major component of the biota. Their productivity is estimated at <5 mg m^{-2} yr^{-1} but turnover is slow (Vestal 1988), with organic matter in the surface layers of rock reaching 46–177 g m^{-2} (Friedmann 1982).

Cold polar regions

Mean air temperatures in the more oceanic cold Antarctic reach only 0–2°C in summer, but mean winter temperatures may remain as high as -8 to -10°C. Precipitation is light but frequent. The terrain is rugged with much bare ground in the uplands; there are only two native angiosperms, but luxuriant cryptogamic communities are developed extensively close to sea level. These include deep banks of tall turf-forming mosses, notably *Polytrichum alpestre* and *Chorisodontium aciphyllum*, on mesic slopes with carpets and large hummocks of *Calliergon* spp., *Drepanocladus* spp., and other pleurocarps in seepage areas and permanently wet situations. Associations of *Andreaea* with *Usnea*, *Himantormia* spp. and other fruticose lichens occur on inland rocks, while species of *Caloplaca*, *Xanthoria*, and crustose lichens clothe exposed coastal rocks. Annual production ranges from 200–900 g m^{-2} in closed moss turf and moss carpet communities (Longton 1970; Davis 1981), levels comparable with those in temperate grassland. Phytomass is also large in the moss turfs, reaching 300–1000 g m^{-2} for green shoots, 20 000–30 000 g m^{-2} for phytomass above permafrost, and up to 46 000 g m^{-2} for total phytomass including permanently frozen material to the base of a bank 1 m deep. Estimates of 250 g m^{-2} for annual production and 800–1750 g m^{-2} for phytomass have been reported for fruticose lichens (Smith 1984).

Cool and mild polar regions

Flowering plants are conspicuous in cool and mild polar tundra (Table 2.1), in areas where mean temperatures in summer are typically 3–10°C. Bryophytes and lichens are abundant associates, and are dominant in communities such as *Racomitrium* heath and lichen heath. Production generally increases along a xeric → hydric gradient. It is higher in Antarctic than in Arctic regions (Longton 1988*a*; Russell 1990), probably because of greater precipitation and soil enrichment by marine sources of N and P.

Turf- and cushion-forming mosses including *Andreaea* and *Ditrichum* spp., with mat-forming pleurocarps such as *Hylocomium*, *Hypnum*, and *Racomitrium* spp., grow between cushions of *Dryas* spp. and other flowering plants in xeric fellfields. Associated species include crustose (e.g. *Lecidea*, *Ochrolechia* spp.), fruticose (*Sphaerophorus* spp.) and foliose (*Parmelia*, *Solorina*) lichens. Grass heath is one of the most extensive mesic vegetation types in the Arctic. It is formed by caespitose grasses and sedges, including species of *Kobresia* and *Carex*, associated with abundant cryptogams. Mosses include tall turfs of *Dicranum*, *Ditrichum*, and *Polytrichum*, and mats of *Hypnum* or *Racomitrium* spp., while species of *Cetraria*, *Cladonia*, *Stereocaulon*, and other fruticose lichens are often prominent.

Total annual production in these communities is generally <60 g m^{-2}. The contribution of mosses to both production and phytomass varies between stands but is often substantial, and while lichen production is low, lichens commonly form a significant proportion of above-ground phytomass (Table 2.2). This contribution reaches 44 per cent in mesic graminoid communities in Alaska (Webber 1978). Strikingly high ratios of phytomass:production of up to 70:1 have been recorded for mosses in arctic grass heaths. Annual production in grass heaths, herbfields and other mesic communities on cool Antarctic islands reaches 850–1650 g m^{-2}; bryophyte production has been recorded as only 150–250 g m^{-2}, but mosses again contribute significantly to above ground production and phytomass. These points are illustrated in Table 2.2, which also emphasizes the sparse total production, with insignificant moss and lichen components, in the extensive barrens that characterize the more arid parts of the North American cool Arctic.

Arctic wetlands are dominated by grasses and sedges with rhizomes embedded in an understorey of mosses. These include large, turf-forming acrocarps, e.g. species of *Meesia* and *Cinclidium*, and carpet-forming pleurocarps like *Calliergon* and *Drepanocladus* spp. Hepatics also occur and *Sphagnum* spp. are abundant, particularly in the mild Arctic. Total annual production is normally 100–300 g m², with bryophytes representing 10–45 per cent of the total and exceeding the above-ground vascular plant component at wetter sites. Below ground parts of flowering plants often form the major component of both production and phytomass. High phytomass:production ratios again characterize the bryophytes. Lichen production is generally low and phytomass variable (Table 2.2).

Production in individual colonies of cool Antarctic mosses reaches the remarkably high levels recorded for continuous stands in the cold Antarctic, estimates ranging from 27 g m^{-2} for *Ditrichum strictum* in dry fellfield to 1028 g m^{-2} for *Pohlia wahlenbergii* and *Tortula robusta* in streams (Clarke *et al.* 1971; R. I. L. Smith 1982; Russell 1985). Production in individual colonies is again higher in the cool Antarctic than in comparable arctic situations, with annual production in *Polytrichum alpestre* estimated as 450–500 g m^{-2} in cool Antarctic tussock grassland compared with 100–150 g m^{-2} in mild Arctic spruce woodland (Longton 1979*b*). Similarly, productivity of *Cladonia rangiferina* on South Georgia may reach 100–1130 g m^{-2} yr^{-1} (Lindsay 1975; Smith 1984), whereas maximum values reported by Andreev (1954) for Arctic cladonias were only 17–27 g m^{-2} yr^{-1}.

Adaptations of polar bryophytes and lichens

Polar environments are characterized by a cool, usually short growing season and low solar irradiance. Continental regions experience severe winter cold and low precipitation, while the immature soils are typically deficient in P and available N. These conditions impose severe limitations on the growth of vascular plants, and yet mosses and lichens are abundant and the mosses may be highly productive. Factors underlying the success of bryophytes and lichens in polar regions are discussed in depth by Kershaw (1985) and Longton (1988*a*).

Mosses and lichens are remarkably similar in terms of attributes favourable in polar regimes. Many show a broad response of net assimilation rate (NAR) to temperature, often with maxima at 10–15°C but with substantial rates of both net assimilation and dark respiration maintained at temperatures close to or below 0°C (Fig. 2.3). Light compensation and saturation levels are typically lower than in vascular plants, and both decrease at low temperature. This permits positive net photosynthesis under cool, low-light conditions. Some species become photosynthetically active beneath snow cover in spring, and maintain positive net assimilation for 24 hours per day in mid-summer (Oechel and Sveinbjörnsson 1978), while physiological responses show acclimatization to changing conditions

Table 2.2 Representative data for annual net production and phytomass (g m⁻²) in mild and cool polar vegetation

Vegetation Type	Locality	Annual net production					Phytomass				
		Vascular plants		Bryophytes	Lichens	Total	Vascular plants		Bryophytes	Lichens	Total
		Above-ground	Below-ground				Above-ground*	Below-ground			
ARCTIC											
Wet meadow											
Cotton grass–dwarf shrub tundra	Demster, Alaska	87	—	69	<5	—	66(+102)	2372	4753	69	7362
Wet sedge–moss meadow	Devon I, Canada	46	130	103	0	279	78(+120)	1295	1097	0	2592
Hummocky sedge–moss meadow	Devon I	45	104	33	0	182	86(+187)	2023	908	0	3208
Frost-boil sedge–moss meadow	Devon I	58	119	15	0	193	112(+202)	1332	1100	0	2748
Grass heath											
Graminoid steppe	Elef Ringness I, Canada	13	13	32	<1	58	13(+74)	88	2128	20	2323
Moss-graminoid meadow	King Christian I, Canada	5	5	32	<1	42	41	23	2136	10	2210
Dwarf shrub heath											
Cassiope tetragona heath	Devon I	18	90	20	4	132	159(+228)	1041	423	48	1899
Fellfield											
Cushion plant–lichen fellfield	Devon I	15	3	2	3	23	89(+298)	57	15	49	508
Cushion plant–moss fellfield	Devon I	27	5	20	2	54	126(+192)	50	600	23	991

Table 2.2 (cont.)

Vegetation Type	Locality	Annual net production					Phytomass				
		Vascular plants		Bryophytes	Lichens	Total	Vascular plants		Bryophytes	Lichens	Total
		Above-ground	Below-ground				Above-ground	Below-ground			
Barren											
Papaver radicata barren	Devon I	0.5	1.0	0.1	0	1.5	310(+8.2)	0.9	2.4	0	15
ANTARCTIC											
Grass heath											
Festuca contracta heath	South Georgia	340	350	150	2	842	425(+1598)	1642	500	12	4177
Herbfield											
Acaena magellanica herbfield	South Georgia	885	500	250	0	1635	1300(+517)	7536	221	0	9574
Pleurophyllum hookeri herbfield	Macquarie I	314	550	146	4	1014	139(+266)	1920	393	9	2727

*Living (+standing dead)
Data from various sources as indicated in Longton (1988a)

during the growing season (Fig. 2.3). On the negative side, moderate illumination at low temperature causes photoinhibition in some mosses (Adamson *et al.* 1988) and lichens (Kappen *et al.* in press).

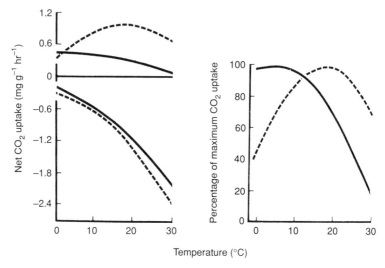

Fig. 2.3 Temperature responses of net photosynthesis and dark respiration in the moss *Polytrichum alpinum* from wet meadow in Alaska in early summer (solid line) and late summer (broken line). Data from Oechel and Sveinbjörnsson (1978).

Many mosses and lichens are poikilohydric (Proctor 1990). They have little access to soil moisture, but lack an effective cuticle, thus enabling them to absorb water through much of their surface. However, this results in rapid water loss under drying conditions, and in compensation the cytoplasm is desiccation-tolerant. The plants become inactive when dry, but resume normal metabolism rapidly on remoistening. This feature may also be adaptive in polar environments as the plants are able to utilize short periods of favourable conditions whenever they occur. Rates of drying, and the length of active periods, are influenced by the spatial organization of individuals within the colony, leading to correlation between growth-form and habitat. Poikilohydry may also enhance frost resistance, by conferring tolerance of cytoplasmic dehydration resulting from extra-cellular ice formation.

These features are also shown by mosses and lichens at lower latitudes, and many ecologically important tundra cryptogams occur in boreal and temperate communities. Lechowicz (1982) found no correlation with latitude of origin in respect of maximum NAR, saturating irradiance, or optimum water content for photosynthesis in lichens, although there was a significant trend of decreasing optimum temperature for net photosynthesis with increasing latitude. Some Antarctic lichens have temperature optima below 0°C, but this results from enhanced respiration at high temperatures (Lange and Kappen 1972) rather than from high NAR under cold conditions.

Some mosses show inherent intraspecific variation between polar and temperate populations. Clinal variation in leaf length and annual shoot elongation in *Polytrichum alpestre* results in short, compact turfs in cold, dry polar environments (Longton 1979*b*). Maximum NAR declines with increasing latitude of origin in *P. commune* and other mosses (Sveinbjörnsson and Oechel 1983; Kallio and Saarnio 1986). Comparable variation occurs within lichen species, for example between populations of *Alectoria ochroleuca* from contrasting arctic environments (Larson and Kershaw 1975), but cultivation difficulties have prevented confirmation of a genetic basis. There is also evidence of adaptive physiological variation between species characteristic of different polar habitats (Kershaw 1985).

2.3.2 Boreal forests

A continuous understorey of bryophytes and lichens is characteristic of boreal forests, where prolonged severe winters alternate with mild summers at mean July temperatures up to 20°C. As a broad generalization, the predominant cryptogams are *Cladonia* spp. and other fruticose lichens on dry soils, particularly in open woodland in the north, and large, weft-forming pleurocarpous mosses, notably *Hylocomium splendens* and *Pleurozium schreberi*, under mesic conditions. *Sphagnum* spp. become abundant at wetter sites.

Phytomass of *Cladonia stellaris* may exceed 300 g m^{-2} in mature lichen woodland, with the lichen carpet over 10 cm deep (Ahti 1977). Maximum production occurs in younger, shallower lichen stands (Andreev 1954), because NAR is depressed in older colonies due to respiration below the level of effective light penetration (Sveinbjörnsson 1987). Ahti (1977) considered that the most productive lichen stands for reindeer grazing were at the first reindeer lichen stage, with a depth of 4–6 cm, phytomass of 50–150 g m^{-2}, and annual production of 6–16 g m^{-2}. Boreal forests may also support abundant epiphytic lichens including foliose species such as *Hypogymnia physodes* and *Platismatia glauca*, and finely branched fruticose species, e.g. *Pseudevernia*, *Usnea* and *Alectoria*, spp. hanging in festoons from trunks, branches, and twigs. Phytomass of arboreal lichens may reach 500 kg ha^{-1} of forest (Scotter 1962).

Abundance of mosses varies between different communities. In Alaska, green phytomass of *Hylocomium splendens* and *Pleurozium schreberi* was 170–290 g m^{-2} under spruce at densities up to 7000 shoots m^{-2}. Mean

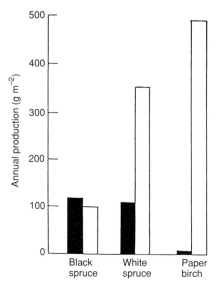

Fig. 2.4 Annual production of mosses (black) and the above-ground component of trees (white) in selected forest types in interior Alaska. Data from Oechel and Van Cleve (1986).

moss density was often less than 10 shoots m^{-2} in successional stands of *Betula* and *Populus* spp., with phytomass only 4–6 g m^{-2}. Annual moss production was negligible in the deciduous forests, but ranged from 70–150 g m^{-2} under spruce, where it commonly exceeded above-ground tree production (Fig. 2.4). Comparable production by the large pleurocarpous mosses has been reported in other coniferous forests (Tamm 1953; Weetman 1968; Pakarinen 1978). As ectohydric mosses, *H. splendens* and *P. schreberi* derive water largely from precipitation, and thus seasonal timing and annual extent of growth vary with rainfall patterns (Busby *et al.* 1978; Longton and Greene 1979; Vitt 1990).

2.3.3 Temperate forests

Many temperate deciduous forests have a sparse growth of cryptogams on the ground, as at sites in New Hampshire where bryophyte phytomass was only 2–3 g m^{-2} (Forman 1969). However, Rieley *et al.* (1979) reported bryophyte phytomass of 1600–2900 g m^{-2} in Welsh oakwoods (*Quercus petraea*), as compared with only 250–300 g m^{-2} for the herb layer. The dominant mosses included ectohydric pleurocarps such as *Pleurozium schreberi* and *Rhytidiadelphus loreus*, and partially endohydric acrocarps including *Dicranum majus* and *Polytrichum formosum*. Production was estimated as 170–210 g m^{-2} for the mosses and 120 g m^{-2} for the herbs.

Shading by shrub and herb layers and dense tree litter, falling in autumn at what is often the beginning of the moss growing season (Pitkin 1975), may be important in limiting cryptogams on deciduous wood floors.

The greatest abundance of bryophytes and lichens in temperate forests is on rotting wood and as epiphytes. European epiphytic communities were described in detail by Barkman (1958). Abundance of epiphytes is greatest in oceanic regions, and Nadkarni (1984) estimated the mean phytomass of bryophytes on *Acer macrophyllum* in temperate rainforest in Washington State as *c*. 800 g m^{-2} of tree surface, 27 kg tree^{-1} or 5000 kg ha^{-2} of forest. Epiphytic lichen phytomass has been estimated at 100–1800 kg ha^{-1} in a range of North American temperate forests, with annual production from 50–500 kg ha^{-1} (Pike 1978). Evergreen *Nothofagus* forests in southern Chile may also have a rich, luxuriant flora of lichens and bryophytes including an abundance of leafy hepatics (Greene *et al.* 1985; Guzman *et al.* 1990).

2.3.4 Mires

Mosses form a continuous understorey in mires from sub-Arctic to temperate and locally tropical regions, and are commonly the predominant peat-formers. Species of *Cladonia*, *Peltigera*, and other lichens are some-times prominent, but probably contribute little to net production. *Sphagnum* spp. are the principal bryophytes in ombrotrophic bogs and nutrient-poor fens, while pleurocarpous mosses in the Amblystegiaceae and Brachythe-ciaceae are characteristic under minerotrophic conditions in richer fens (see Chapter 7). Associated species include graminoids and hepatics, with ericoids particularly in bogs. The gradient from bogs to rich fens is marked by increases in pH, Ca, Mg, and Na, but there is little variation in P and available N (Vitt 1990). Mires generally show a hummock–hollow topography with different mosses arranged along the resulting hydrological gradient.

Clymo and Hayward (1982) note that annual production of *Sphagnum* spp. has been recorded as 100–150 g m^{-2} for hummocks, 500 g m^{-2} in lawns, and 600–800 g m^{-2} in pools over a wide latitudinal range. However, from literature data Moore (1989) detected a weak tendency for production to decrease in line with mean annual temperature towards the north. Although *Sphagnum* production within a mire is, in general, positively correlated with water availability, production on hummocks exceeds that in lawns under some circumstances (Moore 1989; Vitt 1990).

Mean *Sphagnum* production in mires varies from year to year, but is generally 70–400 g m^{-2}, with most values for pleurocarpous mosses in fens in the lower half of this range (Vitt 1990). The ratio of moss:vascular plant production also varies. Annual production of *Sphagnum* spp. at Stordalen, northern Sweden, was 70 g m^{-2}, compared with 59 g m^{-2} for the above-ground plus 24 g m^{-2} for the below-ground angiosperm component (Ros-swall *et al.* 1975). In contrast, annual production of *Sphagnum* and of vascular plants at Moor House, northern England, was estimated as 213 g m^{-2} and 659 g m^{-2} respectively (Forest and Smith 1975; Heal *et al.* 1975). The relative contribution of bryophytes to above-ground standing crop may exceed their contribution to production due to slow decomposition. At Stordalen, mean annual moss production exceeded above-ground vascular plant production by 1.2:1, whereas the green phytomass of the sphagna (300 g m^{-2}) exceeded that of vascular plants by 1.6:1 (Rosswall *et al.* 1975).

2.3.5 Tropical rain forests

General descriptions of tropical forest lichens (Sipman and Harris 1989) and bryophytes (Pócs 1982; Richards 1984; Gradstein and Pócs 1989) indicate that the ecological role of these plants, particularly lichens, remains unclear. The bryophytes and lichens are primarily epiphytic, and they increase in diversity and abundance with altitude. Perhaps in response to the prevailing high humidity, the corticolous bryophytes commonly exhibit growth-forms where the photosynthetic shoots are solitary rather than aggregated into colonies, although they are usually attached to a primary shoot system adherent to the substratum. Solitary growth-forms include the dendroid type (e.g. *Rhodobryum*), the bracket form where the green shoots spread horizontally (e.g. *Leiomela*, *Spiridens*), the hanging type where the long spreading branches grow vertically downwards (e.g. Meteoriaceae), and feather forms where the spreading shoots are more or less pinnately branched (e.g. *Hypopterygium*). The latter should not be confused with weft-forming pleurocarps of the boreal forest, which are sometimes referred to as 'feather mosses'. The bryophyte and lichen communities of tropical forests are described in detail in Chapter 9.

In studies along altitudinal gradients Frahm (1990) recorded bryophyte phytomass as generally below 10–12 g m^{-2} of tree trunk at altitudes up to 1000 m, rising in montane regions to 140 g m^{-2} in Peru and 400 g m^{-2} (exceptionally to 800 g m^{-2}) in Borneo. The differences between localities were thought to reflect local rather than regional factors. For Borneo, these values represent phytomass of under 1 kg ha^{-1} of forest below 1000 m, rising to 200–500 kg ha^{-1} (exceptionally to 1300 kg ha^{-1}) in the upper montane region. Even higher values of 10 300 kg ha^{-1} were recorded for high-altitude epiphytic bryophytes in Tanzania (Pócs 1982). Although representing a small proportion of community phytomass, Pócs estimated that bryophytes absorb up to 30 000 l ha^{-1} of water during a single rain event, a figure exceeding that elsewhere in the canopy. They were therefore considered of great importance in maintaining soil nutrient status, limiting erosion, and in stabilizing the flow in rivers to the benefit of lowland agriculture.

Richards (1984) suggested that the low bryophyte biomass in humid lowland forests is due to a combination of continuous heat and humidity inducing respiration at rates not balanced by day-time photosynthesis in deep shade, and Frahm (1990) has provided supportive experimental evidence. Conversely the cooler, brighter, but equally humid conditions in the elfin woodland canopy are clearly favourable for many species.

2.4 Herbivory

A striking feature emerging from these data is the high phyto-mass:production ratios typical of bryophytes and lichens, particularly in tundra and mires. This is consistent with low rates of decomposition and herbivory (Frankland 1974; Lawrey 1986). Nevertheless, these plants form an important food resource for a wide range, if undoubtedly a minority, of animals, including both generalist and specialist feeders.

2.4.1 Lichens

Lichens are important winter food for many ungulates due to their availability when other food is in short supply (Richardson and Young 1977; Robbins 1987). They represent 60–70 per cent of the winter diet of caribou and reindeer (*Rangifer tarandus*), who reach the lichens by digging craters in soft woodland snow. *Cladonia* and *Cetraria* spp. are preferred to nitrogen-fixing lichens (Kallio 1975). Rundel (1978) cites earlier reports that *Cladonia stellaris* is preferred to *C. arbuscula* and *C. rangiferina*, which contain the bitter-tasting and possibly inhibitory depside fumarprocetraric acid, but other authors consider *C. stellaris* less acceptable than related lichens (Gaare and Skogland 1975). Arboreal species of *Alectoria*, *Evernia*, and *Usnea* are also eaten. Reindeer herders in Fennoscandia fell lichen-covered trees to provide food during adverse conditions (Hustich 1951).

Lichens in some respects represent low quality food, the preferred species being deficient in protein, lipids, and several essential elements. Captive animals lose weight on a lichen diet unless given a nitrogen supplement (Staaland *et al.* 1984). However, the preferred lichens contain abundant carbohydrate (Scotter 1972), are readily digestible by *R. tarandus* which possesses the enzyme lichenase (White and Trudell 1980; Slack 1988), and provide an effective energy source in winter. This and other aspects of its nutritional relationship with lichens (Longton 1988a) led Klein (1982) to postulate that *Rangifer* evolved in response to a lichen-based Arctic food niche unoccupied by other animals.

One reindeer needs at least 2 kg dry weight of lichen daily in winter and browses about 2000 m² in six months. Overgrazing can increase the lichen regrowth period from 2–5 years to 10–15 or even 100 years (Slack 1988). Careful management of lichen pastures is therefore vital, as domesticated reindeer in Eurasia and free-ranging caribou in North America remain an important dietary component of indigenous Arctic people. This task is hindered by human interference in the form of pipelines and highways which alter migration routes. At the same time, increasing sulphur and nitrogen pollution in the Arctic pose a direct threat to the lichens. Accumulation and retention by lichens of radionuclides, from atmospheric weapon tests and the Chernobyl accident, has led to disturbing concentrations in meat and its human consumers (Hanson 1982; MacKenzie 1986), while accumulation of heavy metals by both lichens and mosses has implications for other food webs.

Bryophytes and lichens provide habitats in which a wide array of invertebrates, from protozoans to insects, find concealment and a stable environment, particularly as regards humidity. Some animals, e.g. lepidopteran caterpillars, mimic lichens as defence against predation (Seaward 1988), while both mosses and lichens are found attached to the backs of tropical forest weevils, apparently helping to camouflage the animals (Gressitt *et al.* 1965). The relationships range from accidental to cases where particular animals are seldom found except in association with mosses or lichens. Not all the animals concerned eat cryptogams. Animals regularly feeding on lichens include molluscs and mites, as well as springtails (Collembola), barklice (Psocoptera) and moths. Some feed almost exclusively on lichens (e.g. the oribatid mites *Pirnodus detectidens* and *Dometorina plantivaga*), while others eat lichens only when alternative food is unavailable. Springtails, in particular, may occur on lichens in vast numbers and their grazing can cause major damage. Invertebrates eat lichens by scraping the surface (molluscs), chewing (many insects) or piercing cells and sucking out their contents (prostigmatid mites). Details are given by Gerson and Seaward (1977) and Seaward (1988).

Rundel (1978), while noting that overall levels of herbivory on lichens are low, regarded consumption as irregular, with a variety of vertebrates and invertebrates consuming some species in considerable quantity. Low nutrient content, difficulties in metabolizing the polysaccharide lichenin, the patchiness of lichens as a resource, and morphological features, such as the gelatinous sheaths covering members of the Collemataceae, were considered inhibitory to feeding. Rundel pointed out that lichens with a high nitrogen content, while potentially the most nutritious, were normally among the least grazed, as supported by Lawrey's (1983) field observations and feeding trials. Rundel (1978) reviewed circumstantial evidence that nitrogen-fixing lichens may be protected by terpenes while pulvinic acid derivatives, anthraquinones, depsidones, and depsides such as fumarpro-tocetraric acid were regarded as potentially defensive in some other types of lichen. However, acetone extracts from several lichens, containing depsides, depsidones, and usnic acid, had no significant effect on larvae of the gypsy moth (*Lymantria dispar*), whereas aqueous extracts reduced feeding and growth. The larvae are generalist feeders not normally attacking lichens. Glycosides and polypeptides were suggested as the active compounds in aqueous extracts (Blewitt and Cooper-Driver 1990). Lawrey (1986) concluded that unpalatability of lichens to invertebrates is based in part on chemical defence, that chemical defence is related to the nutritional value of the lichens, but that further research is required to assess the generality of the anti-herbivore role of lichen substances.

2.4.2 Bryophytes

Unlike lichens, mosses are seldom eaten in quantity by *Rangifer tarandus*, and their digestibility by caribou is low (Person *et al.* 1980; White and Trudell 1980; Thomas and Kroeger 1981). However, mosses are eaten in winter by herds on Arctic islands which have no access to abundant lichens, and on Svalbard the caribou appear to have evolved adaptations to a bryophytic diet, notably an enlarged caecum–colon complex (Staaland *et al.* 1979). Mosses are freely consumed by other Arctic and alpine vertebrates, including geese, voles, and particularly lemmings (*Lemmus* spp.).

North American populations of *L. sibiricus* fluctuate through 4–6 year cycles, peak numbers reaching 100–200 individuals ha^{-1} (Batzli *et al.* 1980). Plant consumption may then reach 25 per cent of above-ground production, with mosses such as *Calliergon*, *Dicranum*, and *Polytrichum* spp. forming 5–20 per cent of the diet in summer and 30–40 per cent in winter (Bunnell *et al.* 1975). However, captive lemmings lost weight on a moss diet, due to low ingestion and poor digestibility (Batzli and Cole 1979). Lemmings may eat mosses because of winter availability and relatively high concentrations of calcium, magnesium, and iron. Low digestibility could increase the value of mosses as a mineral source by increasing the intake required to satisfy energy requirements (Batzli and Cole 1979). Prins (1982) has further speculated that Arctic animals benefit from ingesting arachidonic acid, a highly unsaturated fatty acid with a low melting point (−49.5°C). Largely restricted to mosses among plants, this compound is present in unusually high concentrations in the blood of Arctic animals and could protect animal cell membranes, and increase limb mobility, at low temperatures.

Collared lemmings (*Dicrostonyx groenlandicus*) have been observed eating *Polytrichum* gametophytes in summer (Longton 1980), but captive animals ate only capsules when offered fruiting *Funaria arctica* (Pakarinen and Vitt 1974). In temperate regions, grouse chicks and titmice eat moss capsules (Richardson 1981), and capsules are freely eaten in some north American forests (Slack 1988). There are few reports of moss gametophytes being eaten by temperate or tropical vertebrates, and Rieley *et al.* (1979) suggested that preferential grazing of grasses by sheep leads to the abundance of bryophytes in Welsh oakwoods.

Some invertebrates also prefer capsules. Harvester ants (*Messor* sp.) collect capsules of *Crossidium crassinerve* in the Negev Desert, presumably opportunistically as the moss does not fruit every year (Loria and Herrnstadt 1980). In Britain several species of slug (*Arion* spp.) preferred capsules to gametophyte shoots of *Brachythecium rutabulum*, *Mnium hornum*, and *Funaria hygrometrica*, in the field and in laboratory trials. Young green capsules were heavily predated at some sites and were eaten in preference to brown capsules containing spores (Davidson and Longton 1987; Davidson *et al.* 1990). Protonema was eaten freely in the laboratory, but shoot consumption was negligible except for *Funaria hygrometrica*, a finding in agreement with the view that annuals invest less in defense than perennials (Grime 1979).

Studies reviewed by Gerson (1982, 1987) and Slack (1988) have confirmed that moss gametophytes are eaten by a range of invertebrates including species of tardigrades, gastropods, crustaceans, mites, and many groups of insects. Animals whose diet appears to comprise principally bryophytes include a scorpion fly (*Boreus brumalis*: Gerson 1982) and the tardigrade *Echiniscus testudo* (Morgan 1977). Gerson (1982) considered few moss feeders to be host specific, a view challenged by Whitehouse (1985). A common pattern of consumption involves stripping leaf lamina cells while leaving the midrib and any specialized border cells intact (Wyatt and Stoneburner 1989; Davidson *et al.* 1990). Other common moss feeders include aphids and mites that pierce the cells and suck out the contents.

A few quantitative assessments indicate that consumption may be locally high. Rock grasshoppers (*Trimerotropsis saxatilis*) were estimated to eat *Grimmia laevigata* at $391 \text{ mg m}^{-2} \text{ yr}^{-1}$ in south-east USA (Duke and Crossley 1975), while weevils (*Ectomnorrhinus similis*) each consumed 1.67 mg of *Brachythecium rutabulum* per day (37 per cent of body weight) on a cool Antarctic island (Smith 1977). In general, however, consumption of moss gametophytes, including *Sphagnum* spp., appears to be remarkably low (Clymo and Hayward 1982; Longton 1988*a*; Davidson *et al.* 1990). Thus food intake by the microfauna in two cold Antarctic moss communities ranged from $200–300 \text{ g m}^{-2} \text{ yr}^{-1}$, but was principally bacteria, fungi, algae, and dead organic matter: living mosses and lichens contributed $<0.2 \text{ g m}^{-2} \text{ yr}^{-1}$ (Davis 1981).

Bryophyte gametophytes generally have a lower calorific value than leaves of associated angiosperms (Forman 1968; Pakarinen and Vitt 1974), as reflected in lower protein and carbohydrate contents (Skre et al. 1975). However, the differences are generally small, and in the slug-grazing study ash-free calorific values of immature capsules, which were consumed, were no higher than those of the rejected leafy shoots (Davidson et al. 1990). Other factors that could contribute to the low acceptability of bryophytes include the presence of inhibitory compounds and low nitrogen content, particularly in *Sphagnum* (Clymo and Hayward 1982).

In mosses, low digestibility, as demonstrated for caribou and lemmings, is likely to be a major factor restricting consumption. With *Arion* spp., examination of faeces suggested that moss cells passed through the animals intact when starved individuals ate shoots of *Brachythecium rutabulum* and *Mnium hornum* (Davidson et al. 1990). The concentration of phenolic compounds was greater in shoots than in immature capsules of these species, and certain phenolics were released from the shoot only after severe hydrolysis, suggesting an intimate association with the cell wall. One such compound, ferulic acid, was present in the shoots but not in young capsules of *M. hornum* (Davidson et al. 1989). It is also recorded from hydrolysed *Sphagnum* shoots (Bland et al. 1968). Phenolic polymers occur in the walls of other mosses (Erikson and Miksche 1974; Logan and Thomas 1985), and could make the walls indigestible and impenetrable to fungal hyphae. In angiosperms, wall-bound hydroxycinnamic acids, especially ferulic acid, are considered to represent a primitive defence against pathogens and herbivores (Swain 1977; Fry 1983). Such a mechanism in mosses would help to account for the low incidence of herbivory, low rates of decomposition (see below), and the strong representation of arthropods with sucking mouthparts among the invertebrate consumers.

Aqueous and methanol extracts of moss shoots had no inhibitory effect on feeding by *Arion* spp. (Davidson et al. 1990), but antifeedants have been reported from the cytoplasm of numerous hepatics and it is possible that mosses and liverworts show essentially different mechanisms of antiherbivore defence. Asakawa (1990) states that biologically active compounds in hepatics are principally terpenoids and lipophilic aromatic compounds located in oil bodies, e.g. the sesquiterpene lactone plagiochiline A which showed potent antifeedant activity against the African army worm *Spodoptera exempta*. Other examples were summarized by Ando and Matsuo (1984) and Gerson (1982).

2.5 Defence against micro-organisms

Many hepatics harbour endophytic fungi (Pocock and Duckett 1985), and a wide range of fungi has been reported as parasites on mosses and liverworts (Felix 1988; During and Van Tooren 1990). However, both bryophytes and lichens appear to be relatively immune from damaging attack by bacteria and fungi, and more than 50 per cent of species tested from both groups are known to produce inhibitors to the growth of such organisms. Lichen compounds appear to be particularly inhibitory to fungi and gram-positive bacteria, with the most effective substances including usnic acid, pulvinic acid derivatives, and some types of depside and depsidone (Vartia 1973). There is evidence that some compounds are effective against both herbivores and microorganisms (Lawrey 1989). Banerjee and Sen (1979) suggested that bryophytes more commonly contain inhibitors against bacteria than fungi, but antifungal activity has recently been reported in many mosses and liverworts (Ando and Matsuo 1984; Asakawa 1990).

A variety of compounds is again involved, including polyphenolics, fatty acids, and lunularic acid. Little is known about their mode of operation in nature. Lawrey (1986) referred to work by Malicki indicating that usnic acid leached from terricolous lichens inhibits bacterial decomposers in the soil, suggesting a role in regulating decomposition and in nutrient cycling. He also suggested that antibiotic activity of lichens substances may be largely responsible for the remarkable longevity of some lichens, with the defence constitutive rather than induced since the compounds are produced regardless of the presence of micro-organisms or grazers.

2.6 Decomposition

The limited grazing of bryophytes and lichens implies that much of their phytomass passes into the detritus pathway. However, these plants often decompose more slowly than angiosperms. Rates of lichen decomposition in Arctic communities appear generally to be low. Mean loss of lichen phytomass after two years at Fennoscandian tundra sites was only 13 per cent, compared with values up to 37 per cent for angiosperms (Rosswall *et al.* 1975). Slow decomposition has also been reported of tundra lichens in Alaska (Williams *et al.* 1978) and of *Cladonia stellaris* in lichen woodland (Moore 1984). In contrast, rapid decomposition has been reported for *Cladonia mitis* in boreal forest (Fyles and McGill 1987) and for boreal (Wetmore 1982) and temperate epiphytic lichens (Pike 1978; Guzman *et al.* 1990). The latter authors reported decomposition from 40–90 per cent within one year in *Sticta hypochra* and *Pseudocyphellaria* spp. in Chilean *Nothofagus* forests. Current evidence suggests that decomposition rates among lichens may be highest in nitrogen-fixing types with low C:N ratios, as suggested by Crittenden and Kershaw (1978).

Decomposition rates are variable, but often low among bryophytes. Rates in Antarctic species increase along a gradient of increasing soil moisture and tissue mineral content, and of decreasing tissue C:N ratios (Russell 1990). Annual rates in the cold Antarctic range from 0–3 per cent in tall turves of *Polytrichum* and *Chorisodontium* spp. to 14–25 per cent in wet carpets of *Calliergon* and *Drepanocladus* spp. and 40 per cent in *Brachythecium austrosalebrosum* in a minerotrophic flush (Collins 1973). Decomposition accounts for <10 per cent of annual bryophyte production in tundra sites generally (Heal *et al.* 1981). New growth of mosses takes 5–12 years to decompose in Canadian spruce forest, resulting in a thick layer of partially decomposed *Pleurozium schreberi* on the ground (Weetman 1968; Fyles and McGill 1987). Kilbertus (1968) gave a corresponding figure of five years for *Pseudoscleropodium purum* in French pine forests, and considered that this species plays a major role in humus formation despite relatively low annual production (39 g m^{-2}). Pócs (1982) indicated that epiphytic bryophytes contribute significantly to humus formation in Tanzanian elfin forest. A higher proportion of *Sphagnum* production than of the associated angiosperms is preserved as peat (Clymo and Hayward 1982).

The acidic, waterlogged, anaerobic conditions created by *Sphagnum* spp. contribute to slow decomposition in mires. Low nitrogen content could also reduce decomposition rates of *Sphagnum*, and of mosses growing under aerobic non-waterlogged conditions such as the cold Antarctic moss turfs. Walton (1985*b*) considered high levels of holocellulose and crude fibre inhibitory to decomposition of similar vegetation in the cool Antarctic, while Kilbertus *et al.* (1970) found cell walls of *Pseudoscleropodium purum* resistant to attack by micro-organisms, possibly for reasons associated with the low incidence of herbivory in other mosses. Moss tissue may remain alive for many years after it becomes brown (Longton 1972), possibly retaining biologically active compounds.

2.7 Nutrient cycling

Bryophytes and lichens require a similar range of elements to other plants, and through growth and decomposition they inevitably participate in nutrient cycling. They clearly dominate the process in many tundra and mire communities, and their role elsewhere could be distinctive, and possibly disproportional in relation to production. Mineral uptake and release by bryophytes and lichens have been critically discussed by Nieboer *et al.* (1978), Brown (1984), Brown and Beckett (1985), Kershaw (1985), and Brown and Bates (1990).

2.7.1 Nitrogen fixation

Nitrogen fixation is carried out by lichens with cyanobacterial photobionts, and by cyanobacteria and heterotrophic bacteria growing in the moist, often warm, low-oxygen environment within bryophyte colonies. Terricolous lichens such as *Stereocaulon paschale*, and micro-organisms epiphytic on mosses, provide significant inputs of N in dry and wet habitats respectively in tundra and boreal forest. Fixation may reach 50–400 mg N m^{-2} yr^{-1} (Kallio 1975; Alexander *et al.* 1978). It was considered of major importance in nitrogen-deficient northern ecosystems by Dowding *et al.* (1981), although these rates are lower than for biological fixation in temperate regions. In contrast, Gunther (1989) indicated fixation by lichens to be of minor overall significance in Alaskan forests. Bacteria epiphytic on mosses provide a major nitrogen input in mires (Basilier *et al.* 1978; Rosswall and Granhall 1980), and contribute significantly in other communities including grassland (Vlassak *et al.* 1973), epiliths (Snyder and Wullstein 1973; Jones and Wilson 1978), and tropical forests where epiphyllous liverworts are important hosts (Bentley 1987). Epiphytic lichens contribute substantial fixed nitrogen in some temperate and tropical forests and other communities (Forman 1975; Pike 1978).

2.7.2 Other sources of nutrients

Apart from biological nitrogen fixation, mineral nutrients enter terrestrial ecosystems through weathering and aerial deposition of particulates and of ions in solution in precipitation. The influence of cryptogams on weathering has already been considered, and it is suspected that lichens are particularly effective at releasing nutrients from trapped particulates (Nieboer *et al.* 1978). Elements in solution enter plants via the same route as water. They are likely to pass directly into actively growing aerial parts of mosses and lichens, with internal transport in a transpiration stream from below important only in partially endohydric bryophytes.

Lichens, and particularly mosses, can take up nutrients from dilute solutions (Babb and Whitfield 1977; Williams *et al.* 1978), in part because of their high cation exchange capacities. They are therefore likely to enhance retention of nutrients dissolved in precipitation, often enriched as runoff from taller vegetation, by incorporating part of the input directly into shoot tissue. Retention of precipitation in capillary spaces within the colonies, whence it may evaporate or drain slowly into the underlying substratum, could also reduce nutrient losses, both from the new input in precipitation and by leaching from the pool already present in the soil. Humus, to which mosses may contribute substantially, also helps maintain soil fertility by reducing drainage and through chemical association with mineral ions (Longton 1984; Nadkarni 1984).

The relative importance to cryptogams of nutrients dissolved in precipitation or throughfall, as opposed to soil moisture, has been debated ever since Tamm (1953) demonstrated that annual input in precipitation would account for annual accumulation by *Hylocomium splendens* of all major nutrients except N (Nieboer *et al.* 1978; Longton 1980, 1988*a*; Brown 1984; Brown and Bates 1990; Van Tooren *et al.* 1990). The ability of bryophytes to remove elements from rainfall or artificial solutions percolating through the colonies has been confirmed for mosses from the cold Antarctic (Allen *et al.* 1967), from British woodlands (Rieley *et al.* 1979; Bates 1990), and for mats of bryophytes, lichens, and other cryptogamic epiphytes in Amazonian forests (Herrera *et al.* 1978). Both lichens and mosses in boreal woodland have been shown to retain nitrogen compounds applied in solution (Crittenden 1983; Weber and Van Cleve 1984), while epiphytic lichens from temperate forests took up N and the divalent cations Ca^{++} and Mg^{++} during laboratory misting experiments (Pike 1978).

The reverse process may also occur, however, and these results do not imply that bryophytes and lichens obtain no minerals from their substratum. The moisture gradient that commonly develops in lichen colonies will result in upward movement of soil moisture and dissolved nutrients (Nieboer *et al.* 1978). Bates and Farmer (1990) showed that *Pleurozium schreberi* can take up Ca from both a layer of $CaCO_3$ placed on the ground beneath the colony and from dilute solutions sprayed on the leaves. Thus both soil and precipitation are likely to supply nutrients to mosses and lichens, their relative importance depending on age, growth form, and other characteristics of individual species, and on climatic and other external factors (Longton 1980). However, the influence of these plants in retaining part of the atmospheric input is likely to be significant, particularly in tropical forests, boreal and polar communities, ombrogenous mires, and other systems deficient in available N and other essential nutrients.

2.7.3 Nutrient release

Nutrients taken up by bryophytes and lichens may be released and made available to other organisms by leaching, upon death and decomposition, following fire, or to a limited extent by herbivory. However, a proportion is retained indefinitely as undecomposed organic matter, and in some communities bryophytes and lichen compete effectively with vascular plants for nutrients available in limiting quantities.

Some authors consider that lichens conserve much of their N until death and decomposition (Lindsay 1978). However, there is ample evidence that N fixed by lichens, and by bacteria epiphytic on bryophytes, is released and utilized by other organisms. Crittenden (1983) reported leaching of organic nitrogen compounds from *Stereocaulon paschale* by rainfall, and cited evidence for transfer of N to nearby mosses. N is also freely leached from nitrogen-fixing epiphytic lichens (Pike 1978). Alexander *et al.* (1978) considered that N fixed by bacteria on tundra mosses is important in short-term turnover, studies with ^{15}N indicating release of NH_4-N and its transfer to mosses and vascular plants. Biological fixation is likely to be particularly important during primary succession. An association between *Funaria hygrometrica* and cyanobacteria occurred widely during colonization of Surtsey, and growth of the moss was enhanced by nitrogen compounds released by the latter (Rodgers and Henriksson 1976). Jones and Wilson (1978) reported that ^{15}N fixed by *Nostoc* growing on mosses on limestone rock was digested by invertebrates and taken up by mosses and vascular plants in laboratory experiments.

Some elements are also likely to be freely released from mosses and lichens by leaching, losses following thawing or rehydration perhaps being particularly important (Hooker 1977; Brown 1984). Bates (1990) reported that spraying KH_2PO_4 solution on *Pseudoscleropodium purum* resulted in only temporary enhancement of P and K content in the moss. A small proportion of the addition was collected in throughfall beneath the colony, and Bates concluded that most was released from the moss and either utilized by micro-organisms or remained uneluted in the litter.

In other cases nutrients absorbed by cryptogams are made available by decomposition. Weetman (1968) found that tree roots were concentrated at the base of the moss layer in black spruce forest. He considered that decomposing mosses provided a collecting point for elements, particularly N, absorbed by the moss from throughfall. A similar conclusion was reached by Weber and Van Cleve (1984) following studies of ^{15}N uptake by *Hylocomium splendens* and *Pleurozium schreberi*. Involvement of mycorrhizal fungi in transferring P from *H. splendens* to spruce trees was suspected by Chapin *et al.* (1987). Brown and Bates (1990) point out that the P might equally have been taken up by the fungi and passed to the trees in the absence of mosses. The importance of mosses in this process may be determined by the rate at which leaching would occur without the moss layer.

The examples of nutrient release so far considered imply that mosses and lichens enhance nutrient availability to other organisms. This seems certain to apply in some situations, for Sendstad (1981) found that removal of lichen cover, principally *Cetraria delisei*, from lichen heath on Svalbard resulted in reduced concentrations of soil organic matter and of macronutrients. During (1990) considered that nutrient release from bryophytes in chalk grassland, while unimportant over whole sites, might contribute to the vigour and survival of seedlings in dense moss patches. In other cases, however, mosses may decrease nutrient availability. This applies particularly to N, which may be retained in association with phenolic compounds in the cell wall (Berg 1984), and to divalent cations such as Ca^{++} and Mg^{++} that remain strongly bound to anionic exchange sites in the cell wall. These factors apply in dead as well as living tissue. Release of H^+ ions in exchange for divalent cations is one cause of acidity in *Sphagnum*-dominated mires (Clymo and Hayward 1982). Lichens appear to be less effective at retaining Ca^{++} and Mg^{++} than mosses (Nieboer *et al.* 1978).

As examples, Dowding *et al.* (1981) estimated that 50 per cent of the Ca in mesic tundra meadows on Devon Island was in the bryophytes, with a turnover time of 22 years. Absorption of N and other nutrients by *Sphagnum* is considered to reduce their availability to vascular plants in mires (Pakarinen 1978; Woodin *et al.* 1985). Oechel and Van Cleve (1986) suggest that nutrient immobilization by an increasing phytomass of mosses reduces vascular plant productivity as succession proceeds from deciduous to coniferous woodland in Alaska, despite the beneficial effect of bryophytes on nutrient cycling noted above.

There are many intriguing possibilities concerning the influence of bryophytes and lichens on nutrient cycling. That these plants have a significant effect in some communities where they are abundant can hardly be doubted, but further research is required to quantify the effects and to assess their generality.

2.8 Conclusion

This chapter has been mainly concerned with the role of bryophytes and lichens in succession, energy flow, nutrient cycling, and other fundamental ecosystem processes. Although largely unquantified, their influence is likely to be both significant and distinctive, by virtue of moderate levels of production, limited palatability to herbivores and often low rates of decomposition combined with the high water-holding and cation exchange capacities that characterize the resulting biomass. At the same time, bryophytes and lichens also interact directly with other organisms in a variety of ways, for example by providing habitats and camouflage for invertebrates, nesting material for animals, and conditions that may in some cases favour and in others discourage seed germination and establishment (During and Van Tooren 1990). Phenolic compounds from lichens epiphytic on oaks are absorbed by the host leading to reductions in NAR and to defoliation (Giménez and Vicente 1989), and a range of allelopathic compounds are released by both lichens (Lawrey 1986) and bryophytes (Asakawa 1990). *Sphagnum* spp. control the environment of bogs by creating acid, waterlogged, anaerobic conditions and by retaining N and divalent cations. These and other interactions are considered in the reviews of Gerson and Seaward (1977), Richardson and Young (1977), Seaward (1988), and Slack (1988).

These various relationships could alter dramatically in the coming decades even if, as seems likely, SO_2 emissions to which mosses and lichens are particularly sensitive can be controlled. Thus agricultural problems associated with any intensification of the drying trend in sub-Saharan Africa could be exacerbated if reduced humidity or forest clearance in montane regions leads to a reduction in the phytomass, and therefore water-holding capacity, of epiphytic bryophytes. However, the most dramatic effects can be anticipated at high latitudes, particularly if a rise in mean annual temperature of anything approaching the levels of 8–10°C sometimes forecast for northern regions (Garrels 1982) is accompanied by shifts in rainfall patterns and a continuing rise in nitrogen deposition.

The effects of environmental change on the distribution and abundance of mosses and lichens will be determined in part by direct responses, but will also be strongly influenced by associated effects on angiosperms. Thus the abundance of cryptogams in arctic tundra is dependent in large measure on the failure of flowering plants to assume dominance under the cool summer conditions currently prevailing in polar regions. It is unclear how far this results from the effects of low temperature directly on growth or indirectly on rates of nutrient cycling and the resulting deficiencies of available N and P (Chapin 1983). A significant rise in CO_2 concentration, temperature, and nitrogen availability could combine to increase the abundance of angiosperms and thus eliminate the highly distinctive tundra communities composed principally of mosses and lichens. This effect could be enhanced if increasing concentrations of atmospheric CO_2 lead to generally increased rates of biological nitrogen fixation, as demonstrated in the temperate lichen *Lobaria pulmonaria* (Norby and Sigal 1989). Conversely, loss of an insulating layer of cryptogams would accelerate any melting of permafrost initiated by rising temperature, perhaps resulting in extensive disruption of the land surface and its vegetation cover, and providing opportunities for cryptogams in successional communities. Existing cryptogamic vegetation will probably survive longer in the Antarctic where summer temperatures are lower than at comparable Arctic latitudes and barriers to migration more effective.

At the species level some bryophytes may be particularly well adapted to survive changes in climate since, like the majority of tundra species, they range widely in several biomes. Such wide distributions result from both intraspecific genetic differentiation and the wide phenotypic plasticity permitted by individual genotypes, combined with poikilohydry and the consequent ability to switch rapidly between states of metabolic activity and rest as dictated by external conditions at any time of the year (Longton 1988a). Moreover, brophytes can survive as small populations in microhabitats where conditions are far different from those prevailing in an area generally. Thus some Arctic endemics have their closest relations among tropical floras and are viewed as having survived in the Arctic since a period of temperate or subtropical conditions in the early Tertiary (Steere 1978). Other species, however, particularly those that are already rare and have limited dispersal ability through failure to produce spores, could face extinction given a significant shift in climate.

The impact on the global ecosystem of significant changes in the distribution of bryophytes and lichens is difficult to predict. However, it is abundantly clear that release of C from undecomposed mosses in peat could substantially accentuate the process of global warming. Figures in Clymo and Hayward (1982) indicate that peatlands cover about 150×10^6 ha (>1 per cent of the Earth's land surface) and contain some $300\,000 \times 10^6$ metric tons of peat, of which perhaps half is *Sphagnum* with a further component of other mosses. Assuming a 40 per cent carbon content the peat contains $120\,000 \times 10^6$ metric tons of C. This is equivalent to 24 years' emission from fossil fuel at the current rate of 5000×10^6 tons yr^{-1}, and to more than 50 per cent of total emission from this source since 1860, according to estimates in Liss and Crane (1983), Moore *et al.* (1989), and Rotty and Marland (1986). Any disruption of northern mires that results in rapid decomposition of the peat could thus have most serious implications. The message is clear for those who plan to convert peatlands to forest! Again, it is the combination of climatic change, leading to warmer and possibly drier conditions, with increased nitrogen deposition that could be particularly significant. Lee *et al.* (1990) show that nitrogen levels in severely polluted areas are already supraoptimal for *Sphagnum* and could favour growth of angiosperms, while enhanced nitrogen levels will also accelerate decomposition.

Finally, the potential significance of currently accumulating bryophyte phytomass as a carbon sink should not be overlooked. Assuming annual *Sphagnum* production at 200 g m^{-2} over 100 000 ha of mire (Clymo 1983) gives 200×10^6 metric tons dry weight or 80×10^6 metric tons C. Similarly, a conservative estimate of annual moss production in the boreal forest as 100 g m^{-2} over half its area, estimated in Whittaker (1970) as 12×10^6 km^2, gives 240×10^6 tons C. Thus bryophytes appear to fix annually the equivalent of 6.5 per cent of current C emissions from fossil fuel in these two community types alone, and in each case there is probably significant accumulation due to slow decomposition. Success in combating the rise in atmospheric CO_2 concentrations must surely be achieved by implementing a wide range of measures, many with individually small effects. Encouraging production of slowly decomposing bryophyte phytomass, and its subsequent storage in a useful form, e.g. as insulation material, is one such approach.

References

Adamson, H., Wilson, M., Selkirk, P., and Seppelt, R. D. (1988). Photoinhibition in antarctic mosses. *Polarforschung*, **58**, 103–11.

Ahti, T. (1977). Lichens of the boreal coniferous zone. In *Lichen ecology*, (ed. M. R. D. Seaward), pp. 145–81. Academic Press, London.

Ahti, T. (1982). Evolutionary trends in cladoniiform lichens. *Journal of the Hattori Botanical Laboratory*, **52**, 331–41.

Alexander, V, Billington, M., and Schell, D. M. (1978). Nitrogen fixation in arctic and alpine tundra. In *Vegetation and production ecology of an Alaskan arctic tundra*, (ed. L. L. Tieszen), pp. 539–58. Springer, New York.

Allen, S. E., Grimshaw, H. M., and Holdgate, M. W. (1967). Factors affecting the availability of plant nutrients on an Antarctic island. *Journal of Ecology*, **55**, 381–96.

Ando, H. and Matsuo, A. (1984). Applied bryology. *Advances in Bryology*, **2**, 133–224.

Andreev, V. N. (1954). Prirost kormovykh lishainikov i priemy ego regulirovaniia. *Geobotanica* **9**, 11–74.

Andrews, J. T. and Barnett, D. M. (1979). Holocene (Neoglacial) moraine and periglacial lake chronology, Barnes Ice Cap, N.W.T. Canada. *Boreas*, **8**, 341–58.

Asakawa, Y. (1990). Terpenoids and aromatic compounds with pharmacological activity from bryophytes. In *Bryophytes their chemistry and chemical taxonomy*, (eds H. D. Zinsmeister and R. Mues), pp. 369–410. Oxford Science Publications, Oxford.

Ascaso, C. (1985). Structural aspects of lichens invading their substrata. In *Surface physiology of lichens*, (eds C. Vicenti, D. H. Brown, and M. Estrella Legaz), pp. 87–113. Universidad Complutense de Madrid, Madrid.

Babb, T. A. and Whitfield, D. W. S. (1977). Mineral nutrient cycling and limitation of plant growth in the Truelove Lowland ecosystem. In *Truelove Lowland, Devon Island, Canada: a high arctic ecosystem*, (ed. L. C. Bliss), pp. 589–606. University of Alberta Press, Edmonton.

Banerjee, R. D. and Sen, S. P. (1979). Antibiotic activity of bryophytes. *The Bryologist*, **82**, 141–53.

Barkman, J. J. (1958). *Phytosociology and ecology of cryptogamic epiphytes*. Van Gorcum, Assen.

Basilier, K. Granhall, U., and Senström, T. -A. (1978). Nitrogen fixation in wet minerotrophic moss communities of a subarctic mire. *Oikos*, **31**, 236–46.

Bates, J. W. (1990). Interception of nutrients in wet deposition by *Pseudoscleropodium purum*: an experimental study of uptake and release of potassium and phosphorus. *Lindbergia*, **15**, 93–8.

Bates, J. W. and Farmer, A. M. (1990). An experimental study of calcium acquisition and its effects on the calcifuge moss *Pleurozium schreberi*. *Annals of Botany*, **65**, 87–96.

Batzli, G. O. and Cole, F. R. (1979). Nutritional ecology of microtine rodents: digestibility of forage. *Journal of Mammalogy*, **60**, 740–50.

Batzli, G. O., White, R. G., MacLean, S. E., Pitelka, F. A., and Collier, B. D. (1980). The herbivore-based trophic system. In *An arctic ecosystem: the coastal tundra at Barrow, Alaska*, (eds J. Brown, P. C. Miller, L. L. Tieszen, and F. L. Bunnell), pp. 335–410. Hutchinson and Ross, Dowden.

Bentley, B. L. (1987). Nitrogen fixation by epiphylls in a tropical rainforest. *Annals of the Missouri Botanical Garden*, **74**, 234–41.

Berg, B. (1984). Decomposition of moss litter in a mature Scots pine forest. *Pedobiologia* **26**, 301–8.

Bland, D. E., Logan, A., and Menshun, M. (1968). The lignin of *Sphagnum*. *Phytochemistry*, **7**, 1373–7.

Blewitt, M. R. and Cooper-Driver, G. A. (1990). The effects of lichen extracts on feeding by gypsy moths (*Lymantria dispar*). *Bryologist*, **93**, 220–1.

Brodo, I. M. (1973). Substrate ecology. In *The lichens*, (eds V. Ahmadjian and M. E. Hale), pp. 401–41. Academic Press, New York.

Brown, D. H. (1984). Uptake of mineral elements and their use in pollution monitoring. In *The experimental biology of bryophytes*, (eds A. F. Dyer and J. G. Duckett), pp. 229–55. Academic Press, London.

Brown, D. H. and Bates, J. W. (1990). Bryophytes and nutrient cycling. *Botanical Journal of the Linnean Society*, **104**, 129–47.

Brown, D. W. and Beckett, R. P. (1985). Minerals and lichens: acquisition, localization and effect. In *Surface physiology of lichens*, (eds. C. Vicente, D. H. Brown, and M. Estrella Legaz), pp. 127–49. Universidad Complutense de Madrid, Madrid.

Bunnell, F. L., MacLean, S. F., and Brown, J. (1975). Barrow, Alaska, USA. In *Structure and function of tundra ecosystems*, (eds T. Rosswall and O. W. Heal), pp. 73–124. *Ecological Bulletins (Stockholm)*, **20**.

Busby, J. R., Bliss, L. C., and Hamilton, C. D. (1978). Microclimate control of growth rates and habitats of the boreal forest mosses, *Tomenthypnum nitens* and *Hylocomium splendens*. *Ecological Monographs*, **48**, 95–110.

Chapin, F. S. (1983). Direct and indirect effects of temperature on arctic plants. *Polar Biology*, **2**, 47–52.

Chapin, F. S., Oechel, W. C., Van Cleve, K., and Lawrence, W. (1987). The role of mosses in the phosphorus cycling of an Alaskan black spruce forest. *Oecologia (Berlin)*, **74**, 310–15.

Clarke, G. C. S., Greene, S. W., and Greene, D. M. (1971). Productivity of bryophytes in polar regions. *Annals of Botany*, **35**, 99–108.

Clymo, R. S. (1983). Peat. In *Mires: swamp, bog, fen and moor. General studies*, (ed. A. J. P. Gore), pp. 159–224. Elsevier, Amsterdam.

Clymo, R. S. and Hayward, P. M. (1982). The ecology of *Sphagnum*. In *Bryophyte ecology*, (ed. A. J. E. Smith), pp. 229–89. Chapman and Hall, London.

Collins, N. G. (1973). Productivity of selected bryophytes in the maritime Antarctic. In *Proceedings of the conference on primary production and production processes, tundra biome*, (eds L. C. Bliss and F. E. Wielgolaski), pp. 177–83. IBP Tundra Biome Steering Committee, Edmonton.

Corner, R. W. M. and Smith, R. I. L. (1973). Botanical evidence of ice-recession in the Argentine Islands. *British Antarctic Survey Bulletin*, **35**, 83–6.

Crittenden, P. D. (1983). The role of lichens in the nitrogen economy of subarctic woodlands: nitrogen loss from the nitrogen-fixing lichen *Stereocaulon paschale* during rainfall. In *Nitrogen as an ecological factor*, (eds J. A. Lee, S. McNeill, and I. H. Rorison), pp. 43–68. Blackwell, Oxford.

Crittenden, P. D. and Kershaw, K. A. (1978). Discovering the role of lichens in the nitrogen cycle in boreal-arctic ecosystems. *The Bryologist*, **81**, 258–67.

Danin, A., Gerson, R., and Garty, J. (1983). Weathering patterns of hard limestone and dolomite by endolithic lichens and cyanobacteria: supporting edvidence for aeolian action of crustaceous lichens to Terra Rossa soil. *Soil Science*, **1361**, 213–17.

Davidson, A. J. and Longton, R. E. (1987). Acceptability of mosses as food for a herbivore, the slug, *Arion hortensis. Symposia Biologica Hungarica*, **35**, 707–20.

Davidson, A. J., Harborne, J. B., and Longton, R. E. (1989). Identification of hydroxycinnamic and phenolic acids in *Mnium hornum* and *Brachythecium rutabulum* and their possible role in protection against herbivory. *Journal of the Hattori Botanical Laboratory*, **67**, 415–22.

Davidson, A. J., Harborne, J. B., and Longton, R. E. (1990). The acceptability of mosses as food for generalist herbivores, slugs in the Arionidae. *Botanical Journal of the Linnean Society*, **104**, 99–113.

Davis, R. C. (1981). Structure and function of two Antarctic terrestrial moss communities. *Ecological Monographs*, **51**, 125–43.

Dawson, H. J., Hrutfiord, B. F., and Ugolini, F. C. (1984). Mobility of lichen compounds from *Cladonia mitis* in arctic soil. *Soil Science*, **138**, 40–45.

Dietert, M. F. (1979). Studies on the gametophyte nutrition of the cosmopolitan species *Funaria hygrometrica* and *Weissia controversa. The Bryologist*, **82**, 417–31.

Dowding, P., Chapin, F. S., Wielgolaski, F. E., and Kilfeather, P. (1981). Nutrients in tundra ecosystems. In *Tundra ecosystems: a comparative analysis*, (eds L. C. Bliss, O. W. Heal, and J. J. Moore), pp. 647–83. Cambridge University Press, Cambridge.

Duke, K. M. and Crossley, D. A. (1975). Population energetics and ecology of the rock grasshopper *Trimerotropsis saxatilis. Ecology*, **56**, 1106–17.

During, H. J. (1979). Life strategies of bryophytes: a preliminary review. *Lindbergia*, **5**, 2–18.

During, H. J. (1990). The bryophytes of calcareous grasslands. In *Calcareous grasslands—ecology and management* (eds S. H. Hillier, D. W. H. Walton, and D. A. Wells), pp. 35–40. Bluntisham Books, Huntingdon, UK.

During, H. J. and Van Tooren, B. F. (1990). Bryophyte interactions with other plants. *Botanical Journal of the Linnean Society*, **104**, 79–98.

Edward, N, and Miller, P. C. (1977). Validation of a model on the effect of tundra vegetation on soil temperature. *Arctic and Alpine Research*, **9**, 89–104.

Erikson, M. and Miksche, G. E. (1974). On the occurrence of lignin or polyphenols in some mosses and liverworts. *Phytochemistry*, **13**, 2295–9.

Felix, H. (1988). Fungi on bryophytes, a review. *Botanica Helvetica*, **98**, 239–69.

Filion, L. and Payette, S. (1989). Subarctic lichen polygons and soil development along a colonization gradient on eolian sands. *Arctic and Alpine Research*, **21**, 175–84.

Forest, G. I. and Smith, R. A. H. (1975). The productivity of blanket bog types in the northern Pennines. *Journal of Ecology*, **63**, 173–202.

Forman, R. T. T. (1968). Caloric values of bryophytes. *The Bryologist*, **71**, 344–7.

Forman, R. T. T. (1969). Comparison of coverage, biomass, and energy as measures of standing crop of bryophytes in various ecosystems. *Bulletin of the Torrey Botanical Club*, **96**, 582–91.

Forman, R. T. T. (1975). Canopy lichens with blue–green algae: a nitrogen source in a Colombian rainforest. *Ecology*, **56**, 1176–84.

Foster, D. R. (1985). Vegetation development following fire in *Picea mariana* (black spruce)—*Pleurozium* forests in south-eastern Labrador, Canada. *Journal of Ecology*, **73**, 517–34.

Frahm, J.-P. (1990). Bryophyte phytomass in tropical ecosystems. *Botanical Journal of the Linnean Society*, **104**, 22–33.

Frankland, C. J. (1974). Decomposition of lower plants. In *Biology of plant litter decomposition* I, (eds C. H. Dickinson and J. F. Pugh), pp. 3–36. Academic Press, London.

Fridriksson, S. (1975). *Surtsey: evolution of life on a volcanic island*. Butterworth, London.

Friedmann, E. I. (1982). Endolithic microorganisms in the Antarctic cold desert. *Science*, **215**, 1045–53.

Fry, S. C. (1983). Feruloylated pectins from the primary wall: their structure and possible function. *Planta*, **157**, 111–23.

Fyles, J. W. and McGill, W. B. (1987). Decomposition of boreal forest litters from central Alberta under laboratory conditions. *Canadian Journal of Forestry Research*, **17**, 109–14.

Gaare, E. and Skogland, T. (1975). Wild reindeer food habits and range use at Hardangervidda. In *Fennoscandian tundra ecosystems. 2: Animals and systems analysis*, (ed. F. E. Wielgolaski), pp. 195–205. Springer, New York.

Garrels, R. M. (1982). Introduction: chemistry of the troposphere—some problems and their temporal framework. In *Atmospheric chemistry*, (ed. E. D. Goldberg), pp. 3–16. Springer Berlin.

Garty, J. and Delarea, J. (1987). Some initial stages in the formation of epilithic crustose lichens in nature: a SEM study. *Symbiosis*, **3**, 49–56.

Gerson, U. (1982). Bryophytes and invertebrates. In *Bryophyte ecology*, (ed. A. J. E. Smith), pp. 291–332. Chapman and Hall, London.

Gerson, U. (1987). Mites which feed on mosses. *Symposia Biologica Hungarica*, **35**, 721–4.

Gerson, U. and Seaward, M. R. D. (1977). Lichen–invertebrate associations. In *Lichen ecology*, (ed. M. R. D. Seaward), pp. 69–119. Academic Press, London.

Giménez, I, and Vicente, C. (1989). Occurrence of lichen phenolics in tissues of *Quercus rotundifolia* in relation to defoliation produced by epiphytic lichens. *Phyton*, **49**, 111–18.

Gimingham, C. H. (1972). *Ecology of heathlands*. Chapman and Hall, London.

Gradstein, S. R. and Pócs, T. (1989). Bryophytes. In *Tropical rain forest ecosystems*, (eds H. Leith and M. J. A. Werger), pp. 311–24. Elsevier, Amsterdam.

Greene, S. W., Hässel de Menédez, G. G., and Matteri, C. M. (1985). La contribucion de les briofitas en la vagetacion de la transecta. In *Transecta botánica de la Patagonia austral*, (eds O. Boelcke, D. M. Moore, and F. A. Roig), pp. 557–91. Consego Nacional de Investigaciones Cientificas y Tecnicas, Buenos Aires.

Gressit, J. L., Sedlacek, J., and Szent-Ivany, J. J. H. (1965). Flora and fauna on backs of large Papuan moss-forest weevils. *Science*, **150**, 1833.

Grime, J. P. (1979). *Plant strategies and vegetation processes*. Wiley, Chichester.

Gunther, A. J. (1989). Nitrogen fixation by lichens in a subarctic watershed. *The Bryologist*, **92**, 202–8.

Guzman, G., Quilhot, W., and D. J. Galloway. (1990). Decomposition of species of *Pseudocyphellaria* and *Sticta* in a southern Chilean forest. *Lichenologist*, **22**, 325–31.

Hale, M. E. (1983). *The biology of lichens*, (3rd edn). Arnold, London.

Hanson, W. C. (1982). [137]Cs concentrations in northern Alaskan eskimos. *Health Physics*, **42**, 433–47.

Heal, O. W., Flanagan, P. W., French, D. D., and MacLean, S. F. (1981). Decomposition and accumulation of organic matter. In *Tundra ecosystems: a comparative analysis*, (eds L. C. Bliss, O. W. Heal, and J. J. Moore), pp. 587–633. Cambridge University Press, Cambridge.

Heal, O. W., Jones, H. E., and Whittaker, J. B. (1975). Moor House, UK. In *Structure and function of tundra ecosystems*, (eds T. Rosswall and O. W. Heal), pp. 295–320. *Ecological Bulletins (Stockholm)*, **20**.

Heilbron, T. D. and Walton, D. H. W. (1984). Plant colonization of actively sorted stone stripes in the subantarctic. *Arctic and Alpine Research*, **16**, 161–72.

Herrera, R., Jordan, C. F., Klinge, H., and Medina, E. (1978). Amazon ecosystems. Their structure and functioning with particular emphasis on nutrients. *Interciencia*, **3**, 223–32.

Hooker, T. N. (1977). The growth and physiology of antarctic lichens. Unpublished. Ph.D. thesis, University of Bristol.

Hughes, J. G. (1982). Penetration by rhizoids of the moss *Tortula muralis* Hedw. into well cemented oolitic limestone. *International Biodeterioration Bulletin*, **18**. 43–6.

Hustich, I. (1951). The lichen woodlands in Labrador and their importance as winter pasture for domesticated reindeer. *Acta Geographica*, **12**, 1–48.

Ino, Y. (1983). Estimation of primary production in moss community on East Ongul Island, Antarctica. *Antarctic Record*, **80**, 30–8.

John, E. A. (1989). An assessment of the role of biotic interactions and dynamic processes in the organization of species in a saxicolous lichen community. *Canadian Journal of Botany*, **67**, 2025–37.

Johnson, E. A. (1981). Vegetation organization and dynamics of lichen woodland communities in the Northwest Territories, Canada. *Ecology*, **62**, 202–15.

Jones, D. (1988). Lichens and pedogenesis. In *CRC Handbook of lichenology*, III, (ed. M. Galun), pp. 109–24. CRC Press, Boca Raton.

Jones, D. and Wilson, M. J. (1985). Chemical activity of lichens on mineral surfaces—a review. *International Biodeterioration Bulletin*, **21**, 99–104.

Jones, K. and Wilson, R. E. (1978). The fate of nitrogen fixed by a free-living blue–green alga. In *Environmental role of nitrogen-fixing blue–green algae and asymbiotic bacteria*, (ed. U. Granhall), pp. 158–63. *Ecological Bulletins (Stockholm)*, **26**.

Kallio, P. (1975). Kevo, Finland. In *Structure and function of tundra ecosystems*, (eds T. Rosswall and O. W. Heal), pp. 193–223. *Ecological Bulletins (Stockholm)*, **20**.

Kallio, P. and Saarnio, E. (1986). The effect on mosses of transplantation to different latitudes. *Journal of Bryology*, **14**, 159–78.

Kappen, L. (1985). Vegetation and ecology of ice-free areas of northern Victoria Land, Antarctica. I. The lichen vegetation of Birthday Ridge and an inland mountain. *Polar Biology*, **4**, 213–35.

Kappen, L., Breuer, M., and Bölter, M. (1991). Ecological and physiological investigations in continental antarctic cryptogams. III. Photosynthetic production of *Usnea sphacelata*: diurnal courses, models and the effects of photoinhibition. *Polar Biology* (in press).

Kershaw, K. (1985). *Physiological ecology of lichens*. Cambridge University Press, Cambridge.

Kilbertus, G. (1968). Décomposition d'une mousse: *Pseudoscleropodium purum* (Hedw.) Fleisch. dans la nature. *Bulletin de l'École Nationale Supérieure Agronomique de Nancy*, **10**, 20–32.

Kilbertus, G., Mangenot, F., and Reisinger, O. (1970). Décomposition des vegetaux II—Etude aux microscopes electroniques de *Pseudoscleropodium purum* (Hedw.) Fleisch. *Bulletin de l'École Nationale Supérieure Agronomique de Nancy*, **12**, 62–7.

Klein, D. R. (1982). Fire, lichens and caribou. *Journal of Range Management*, **35**, 390–5.

Lange, O. L. and Kappen, L. (1972). Photosynthesis of lichens from Antarctica. In *Antarctic terrestrial biology*, (ed. G. A. Llano), pp. 83–95. American Geophysical Union, Washington.

Larson, D. W. and Kershaw, K. A. (1975). Studies on lichen-dominated systems. XVI. Comparative patterns of net CO_2 exchange in *Cetraria nivalis* and *Alectoria ochroleuca* collected from a raised-beach ridge. *Canadian Journal of Botany*, **53**, 2884–2.

Lawrey, J. D. (1983). Lichen herbivore preference: a test of two hypotheses. *American Journal of Botany*, **70**, 1188–94.

Lawrey, J. D. (1986). Biological role of lichen substances. *The Bryologist*, **89**,

111–22.

Lawrey, J. D. (1989). Lichen secondary compounds: evidence for a correspondence between antiherbivore and antimicrobial function. *The Bryologist*, **92**, 326–8.

Lechowicz, M. J. (1982). Ecological trends in lichen photosynthesis. *Oecologia*, **53**, 330–6.

Lee, J. A., Baxter, R., and Eames, M. J. (1990). Responses of *Sphagnum* species to atmospheric nitrogen and sulphur deposition. *Botanical Journal of the Linnean Society*, **104**, 255–65.

Lindsay, D. C. (1975). Growth rates of *Cladonia rangiferina* (L.) Web. on South Georgia. *British Antarctic Survey Bulletin*, **40**, 49–53.

Lindsay, D. C. (1978). The role of lichens in antarctic ecosystems. *The Bryologist*, **81**, 268–76.

Liss, P. S. and Crane, A. J. (1983). *Man-made carbon dioxide and climatic change: a review of scientific problems.* Geo Books, Norwich, UK.

Logan, K. J. and Thomas, B. A. (1985). Distribution of lignin derivatives in plants. *New Phytologist*, **99**, 571–85.

Longton, R. E. (1970). Growth and productivity of the moss *Polytrichum alpestre* Hoppe in Antarctic regions. In *Antarctic ecology*, Vol. 2, (ed. M. W. Holdgate), pp. 818–37. Academic Press, London.

Longton, R. E. (1972). Growth and production in northern and southern hemisphere populations of the peat forming moss *Polytrichum alpestre* Hoppe with reference to the estimation of productivity. *Proceedings of the Fourth International Peat Congress, Helsinki*, **1**, 259–75.

Longton, R. E. (1974). Microclimate and biomass in communities of the *Bryum* association on Ross Island, continental Antarctica. *The Bryologist*, **77**, 109–27.

Longton, R. E. (1979a). Vegetation ecology and classification in the Antarctic zone. *Canadian Journal of Botany*, **57**, 2264–78.

Longton, R. E. (1979b). Studies on growth, reproduction and population ecology in relation to microclimate in the bipolar moss *Polytrichum alpestre* Hoppe. *The Bryologist*, **82**, 325–67.

Longton, R. E. (1980). Physiological ecology of mosses. In *Mosses of North America*, (eds R. J. Taylor and A. E. Leviton), pp. 77–113. Pacific Division AAAS, San Francisco.

Longton, R. E. (1984). The role of bryophytes in terrestrial ecosystems. *Journal of the Hattori Botanical Laboratory*, **55**, 147–63.

Longton, R. E. (1988a). *Biology of polar bryophytes and lichens.* Cambridge University Press, Cambridge.

Longton, R. E. (1988b). Life-history strategies among bryophytes of arid regions. *Journal of the Hattori Botanical Laboratory*, **64**, 15–28.

Longton, R. E. and Greene, S. W. (1979). Experimental studies on growth and reproduction in the moss *Pleurozium schreberi* (Brid.) Mitt. *Journal of Bryology*, **10**, 321–38.

Longton, R. E. and Holdgate, M. W. (1979). The South Sandwich Islands. IV: Botany. *British Antarctic Survey Scientific Reports*, **94**, 1–53.

Loria, M. and Herrnstadt, I. (1980). Moss capsules as food for the harvester ant,

Messor. The Bryologist, **83**, 524–5.

MacKenzie, D. (1986). The rad-dosed reindeer. *New Scientist*, **1539**, 37–40.

Magomedova, M. A. (1980). Succession of communities of lithophilic lichens in the highlands of northern Ural. *Ekologiya*, **3**, 29–38.

Maikawa, E. and Kershaw, K. A. (1976). Studies on lichen-dominated systems. XIX: The postfire recovery sequence of black spruce-lichen woodland in the Abitau Lake region, N.W.T. *Canadian Journal of Botany*, **54**, 2679–87.

Moore, B., Gildea, M. P., Vorosmarty, C. J., Skole, D. L., Melillo, J. M., Peterson, B. J., *et al.* (1989). Biogeochemical cycles. In *Global ecology towards a science of the biosphere*, (eds M. B. Rambler, L. Margulis, and R. Fester), pp. 113–41. Academic Press, London.

Moore, T. R. (1984). Litter decomposition in a subarctic spruce-lichen woodland, eastern Canada. *Ecology*, **65**, 299–308.

Moore, T. R. (1989). Growth and net production of *Sphagnum* at five fen sites, subarctic eastern Canada. *Canadian Journal of Botany*, **67**, 1203–7.

Morgan, C. I. (1977). Population dynamics of two species of Tardigrada (*Macarobiotus hufelandii* (Schultze) and *Echiniscus (Echiniscus) testudo* (Doyere)) in roof moss from Swansea. *Journal of Animal Ecology*, **46**, 263–79.

Mucher, H. J., Chartres, C. J., Tongway, D. J., and Greene, R. S. B. (1988). Micromorphology and the significance of the surface crusts of soils in rangelands near Cobar, Australia. *Geoderma*, **42**, 227–44.

Muller, C. H. (1952). Plant succession in arctic heath and tundra in northern Scandinavia. *Bulletin of the Torrey Botanical Club*, **79**, 296–309.

Nadkarni, N. M. (1984). Biomass and mineral capital of epiphytes in an *Acer macrophyllum* community of a temperate moist coniferous forest, Olympic Peninsula, Washington State. *Canadian Journal of Botany*, **62**, 2223–8.

Nash, T. H., White, S. E., and Marsh, J. E. (1977). Lichen and moss distribution and biomass in hot desert ecosystems. *The Bryologist*, **80**, 470–9.

Nieboer, E., Richardson, D. H. S., and Tomassini, F. D. (1978). Mineral uptake and release by lichens. An overview. *The Bryologist*, **81**, 226–46.

Norby, R. J. and Sigal, L. L. (1989). Nitrogen fixation in the lichen *Lobaria pulmonaria* in elevated atmospheric carbon dioxide. *Oecologia*, **79**, 566–8.

Odum, E. P. (1971). *Fundamentals of ecology*, (3rd edn). Saunders, Philadelphia.

Oechel, W. C. and Sveinbjörnsson, B. (1978). Primary production processes in arctic bryophytes at Barrow, Alaska. In *Vegetation and production ecology of an Alaskan arctic tundra*, (ed. L. L. Tieszen), pp. 269–98. Springer, New York.

Oechel, W. C. and Van Cleve, K. (1986). The role of bryophytes in nutrient cycling in the taiga. In *Forest ecosystems in the Alaskan taiga*, (eds. K. Van Cleve, F. S. Chapin, P. W. Flanagan, L. A. Viereck, and C. T. Dyrness), pp. 121–37. Springer, New York.

Oosting, H. J. and Anderson, L. E. (1939). Plant succession on granite rock in eastern North Carolina. *Botanical Gazette*, **100**, 750–68.

Pakarinen, P. (1978). Production and nutrient ecology of three *Sphagnum* species in southern Finnish raised bogs. *Annales Botanici Fennici*, **15**, 15–26.

Pakarinen, P. and Vitt, D. H. (1974). The major organic components and caloric values of high arctic bryophytes. *Canadian Journal of Botany*, **52**, 1151–61.

Person, S. J., Pegau, R. E., White, R. G., and Luick, J. R. (1980). In vitro and nylon bag digestibilities of reindeer and caribou forages. *Journal of Wildlife Management*, **44**, 613–22.

Pike, L. H. (1978). The importance of epiphytic lichens in mineral cycling. *The Bryologist*, **81**, 247–57.

Pitkin, P. H. (1975). Variability and seasonality of the growth of some corticolous pleurocarpous mosses. *Journal of Bryology*, **8**, 337–56.

Pocock, K. and Duckett, J. G. (1985). On the occurrence of branched and swollen rhizoids in British hepatics: their relationship with the substratum and associations with fungi. *New Phytologist*, **99**, 281–304.

Pócs, T. (1982). Tropical forest bryophytes. In *Bryophyte ecology*, (ed. A. J. E. Smith), pp. 59–104. Chapman and Hall, London.

Polunin, N. (1935). The vegetation of Akpotok Island, Part II. *Journal of Ecology*, **23**, 161–209.

Prins, H. H. Th. (1982). Why are mosses eaten in cold environments only? *Oikos*, **38**, 374–80.

Proctor, M. C. F. (1990). The physiological basis of bryophyte production. *Botanical Journal of the Linnean Society*, **104**, 61–77.

Richards, P. W. (1984). The ecology of tropical forest bryophytes. In *New manual of bryology*, (ed. R. M. Schuster), Vol. 2, pp. 1233–70. Hattori Botanical Laboratory, Nichinan.

Richardson, D. H. S. (1981). *The biology of mosses*. Blackwell, Oxford.

Richardson, D. H. S. and Young, C. M. (1977). Lichens and vertebrates. In *Lichen ecology*, (ed. M. R. D. Seaward), pp. 121–44. Academic Press, London.

Rieley, J. O., Richards, P. W., and Bebbington, A. D. L. (1979). The ecological role of bryophytes in a north Wales woodland. *Journal of Ecology*, **67**, 497–527.

Robbins, C. T. (1987). Digestibility of an arboreal lichen by mule deer. *Journal of Range Management*, **40**, 491–2.

Rodgers, G. A. and Henriksson, E. (1976). Associations between the blue–green algae *Anabaena variabilis* and *Nostoc muscorum* and the moss *Funaria hygrometrica* with reference to the colonization of Surtsey. *Acta Botanica Islandica*, **4**, 10–15.

Rogers, R. W. (1988). Succession and survival strategies in lichen populations on a palm trunk. *Journal of Ecology*, **76**, 759–76.

Rogers, R. W. (1990). Ecological strategies of lichens. *Lichenologist*, **22**, 149–62.

Rosswall, T., Flower-Ellis, J. G. K., Johansson, L. G., Ryden, B. E., and Sonnesson, M. (1975). Stordalen (Abisco), Sweden. In *Structure and function of tundra ecosystems*, (eds T. Rosswall and O. W. Heal), pp. 265–94. *Ecological Bulletin (Stockholm)*, **20**.

Rosswall, T. and Granhall, U. (1980). Nitrogen cycling in a subarctic ombrotrophic mire. In *Ecology of a subarctic mire*, (ed. M. Sonesson), pp. 209–34. *Ecological Bulletins (Stockholm)*, **30**.

Rotty, R. M. and Marland, G. M. (1986). Fossil fuel comsumption: recent amounts, patterns and trends of CO_2. In *The changing carbon cycle: a global analysis*, (eds J. R. Trabalka and D. E. Reichle), pp. 474–90. Springer, New York.

Rouse, W. R. and Kershaw, K. A. (1971). The effects of burning on the heat and water regimes of lichen-dominated subarctic surfaces. *Arctic and Alpine Research*, **3**, 291–304.

Rundel, P. W. (1978). The ecological role of secondary lichen substances. *Biochemical Systematics and Ecology*, **6**, 157–70.

Russell, S. (1985). Bryophyte productivity at Marion Island. In *Antarctic nutrient cycles and food webs*, (eds W. R. Siegried, P. R. Condy, and R. M. Laws), pp. 200–3. Springer, Berlin.

Russell, S. (1988). Measurement of bryophyte growth. 1: Biomass (harvest) techniques. In *Methods in bryology*, (ed. J. M. Glime), pp. 249–57. Hattori Botanical Laboratory, Nichinan.

Russell, S. (1990). Bryophyte production and decomposition in tundra ecosystems. *Botanical Journal of the Linnean Society*, **104**, 3–22.

Russell, S. and Botha, C. E. J. (1988). Measurement of bryophyte growth. 2: Gas exchange techniques. In *Methods in bryology*, (ed. J. M. Glime), pp. 259–74. Hattori Botanical Laboratory, Nichinan.

Scotter, G. W. (1962). Productivity of arboreal lichens and their possible import-ance to barren ground caribou (*Rangifer arcticus*). *Archivum Societatis Zoologicae Botanicae Fennicae Vanamo*, **16**, 155–61.

Scotter, G. W. (1972). Chemical composition of forage plants from the Reindeer Preserve, North West Territories. *Arctic*, **25**, 21–7.

Seaward, M. R. D. (1988). Contribution of lichens to ecosystems. In *CRC Handbook of lichenology*, II, (ed. M. Galun), pp. 107–29. CRC Press, Boca Raton, Florida.

Sendstad, E. (1981). Soil ecology of a lichen heath at Spitzbergen, Svalbard: effects of artificial removal of the lichen plant cover. *Journal of Range Management*, **34**, 442–5.

Seppelt, R. D. and Ashton, D. H. (1978). Studies on the ecology of the vegetation of Mawson Station, Antarctica. *Australian Journal of Ecology*, **3**, 373–88.

Shachak, M., Jones, C. G., and Granot, Y. (1987). Herbivory on rocks and weathering of a desert. *Science*, **236**, 1098–9.

Sipman, H. J. M. and Harris, R. C. (1989). Lichens. In *Tropical forest ecosystems*, (ed. H. Leith and M. J. A. Werger), pp. 303–8. Elsevier, Amsterdam.

Skre, O., Berg, A., and Wielgolaski, F. E. (1975). Organic compounds in alpine plants. In *Fennoscandian tundra ecosystems. 1: Plants and micro-organisms*, (ed. F. E. Wielgolaski), pp. 339–50. Springer, New York.

Slack, N. G. (1988). The ecological importance of lichens and bryophytes. In *Lichens, bryophytes and air quality*, (eds T. H. Nash and V. Wirth), pp. 1–53. Cramer, Berlin–Stuttgart.

Smith, A. J. E. (1982). Epiphytes and epiliths. In *Bryophyte ecology*, (ed. A. J. E. Smith), pp. 191–227. Chapman and Hall, London.

Smith, R. I. L. (1972). The vegetation of the South Orkney Islands with particular reference to Signy Island. *British Antarctic Survey Scientific Reports*, **68**, 1–124.

Smith, R. I. L. (1982). Growth and production in South Georgian Bryophytes. *Comité National Français des Rescherches Antarctiques*, **51**, 229–39.

Smith, R. I. L. (1984). Terrestrial plant biology of the sub-Antarctic and Antarctic.

In *Antarctic ecology*, (ed. R. M. Laws), pp. 61–162. Academic Press, London.

Smith, V. R. (1977). Notes on the feeding of *Ectomnorrhinus similis* Waterhouse (Curculionidae) adults on Marion Island. *Oecologia*, **29**, 269–73.

Snyder, J. M. and Wullstein, L. H. (1973). Nitrogen fixation in granite outcrop pioneer systems. *The Bryologist*, **76**, 197–9.

Southorn, A. L. D. (1977). Bryophyte recolonization of burnt ground with particular reference to *Funaria hygrometrica*. II: The nutritional requirements of *Funaria hygrometrica*. *Journal of Bryology*, **9**, 361–74.

Staaland, H., Jacobsen, E., and White, R. G. (1979). Comparison of the digestive tract in Svalbard and Norwegian reindeer. *Arctic and Alpine Research*, **11**, 457–66.

Staaland, H., Jacobsen, E., and White, R. G. (1984). The effect of mineral supplements on nutrient concentrations and pool sizes in the alimentary tract of reindeer fed on lichens and concentrates during winter. *Canadian Journal of Zoology*, **62**, 1232–41.

Steere, W. C. (1978). *The mosses of arctic Alaska*. Cramer, Vaduz.

Sveinbjörnsson, B. (1987). Reindeer lichen productivity as a function of mat thickness. *Arctic and Alpine Research*, **19**, 437–41.

Sveinbjörnsson, B. and Oechel, W. C. (1983). The effect of temperature preconditioning on the temperature sensitivity of CO_2 flux in geographically diverse populations of the moss *Polytrichum commune* Hedw. *Ecology*, **64**, 1100–8.

Swain, T. (1977). Secondary compounds as protective agents. *Annual Reviews of Plant Physiology*, **28**, 479–501.

Syers, J. K. and Iskandar, I. K. (1973). Pedogenetic significance of lichens. In *The lichens*, (ed. V. Ahmadjian and M. E. Hale), pp. 225–48. Academic Press, New York.

Tallis, J. H. (1983). Changes in wetland communities. In *Mires: swamp, bog, fen and moor: general studies*, (ed. A. J. P. Gore), pp. 311–47. Elsevier, Amsterdam.

Tamm, C. O. (1953). Growth, yield and nutrition in carpets of a forest moss (*Hylocomium splendens*). *Meddelanden Från Statens Skogsforskningsinstitut*, **43**, (1), 1–140.

Thomas, D. C. and Kroeger, P. (1981). Digestibility of plants in ruminal fluids of barren-ground caribou. *Arctic*, **34**, 321–4.

Topham, P. (1977). Colonization, growth, succession and competition. In *Lichen ecology*, (ed. M. R. D. Seaward), pp. 31–68. Academic Press, London.

Ugolini, F. C. and Edmonds, R. L. (1983). Soil biology. In *Pedogenesis and soil taxonomy*. I: *Concepts and interactions*, (eds L. P. Wilding, N. E. Smeck, and G. F. Hall), pp. 193–231. Elsevier, Amsterdam.

Van Tooren, B. F., den Hertog, J., and Verhaar, J. (1988). Cover, biomass and nutrient content of bryophytes in Dutch chalk grasslands. *Lindbergia*, **14**, 47–58.

Van Tooren, B. F., Van Dam, D., and During H. J. (1990). The relative importance of precipitation and soil as sources of nutrients for *Calliergonella cuspidata* (Hedw.) Loeske in chalk grassland. *Functional Ecology*, **4**, 101–7.

Vestal, J. R. (1988). Primary production of the cryptoendolithic microbiota from the Antarctic desert. *Polarforschung*, **58**, 193–8.

Viles, H. (1987). A quantitative scanning electron microscope study of evidence for lichen weathering of limestone, Mendip Hills, Somerset. *Earth Surface Processes and Landforms*, **12**, 467–73.

Vartia, K. O. (1973). Antibiotics in lichens. In *The lichens*, (eds V. Ahmadjian and M. E. Hale), pp. 547–64. Academic Press, New York.

Vitt, D. H. (1990). Growth and production dynamics of boreal mosses over climatic, chemical and topographic gradients. *Botanical Journal of the Linnean Society*, **104**, 35–59.

Vitt, D. H. and Slack, N. G. (1975). An analysis of *Sphagnum*-dominated kettle-hole bogs in relation to environmental gradients. *Canadian Journal of Botany*, **53**, 332–59.

Vlassak, K., Paul, E. A., and Harris, R. E. (1973). Assessment of biological nitrogen fixation in grassland and associated sites. *Plant and Soil*, **38**, 637–49.

Walton, D. W. H. (1985*a*). A preliminary study of the action of crustose lichens on rock surfaces in Antarctica. In *Antarctic nutrient cycles and food webs*, (eds W. R. Siegfried, P. R. Condy, and R. M. Laws), pp. 180–5. Springer, Berlin.

Walton, D. W. H. (1985*b*). Cellulose decomposition and its relationships to nutrient cycling at South Georgia. In *Antarctic nutrient cycles and food webs*, (eds W. R. Siegfried, P. R. Condy, and R. M. Laws), pp. 191–9. Springer, Berlin.

Webber, P. J. (1978). Spatial and temporal variation of the vegetation and its production, Barrow, Alaska. In *Vegetation and production ecology of an Alaskan arctic tundra*, (ed. L. L. Tieszen), pp. 37–112. Springer, New York.

Weber, M. G. and Van Cleve, K. (1984). Nitrogen transformations in feather moss and forest floor layers of interior Alaska black spruce ecosystems. *Canadian Journal of Forestry Research*, **14**, 278–90.

Weetman, G. (1968). The relationship between feather moss growth and the nutrition of black spruce. *Proceedings of the Third International Peat Congress*, (eds C. Lafleur and J. Butler), pp. 366–70. International Peat Society, Quebec.

Wetmore, C. M. (1982). Lichen decomposition in a black spruce bog. *Lichenologist*, **14**, 267–71.

White, R. G. and Trudell, J. (1980). Habitat preference and forage consumption by reindeer and caribou near Atkasook, Alaska. *Arctic and Alpine Research*, **12**, 511–29.

Whitehouse, H. L. K. (1985). Advances in knowledge of the life strategies of British bryophytes. In *British Bryological Society Diamond Jubilee*, (eds R. E. Longton and A. R. Perry), pp. 43–9. British Bryological Society, Cardiff.

Whittaker, R. H. (1970). *Communities and ecosystems*. Macmillan, Toronto.

Williams, M. E. and Rudolph, E. D. (1974). The role of lichens and associated fungi in the chemical weathering of rock. *Mycologia*, **64**, 648–60.

Williams, M. E., Rudolph, E. D., Schofield, E. A., and Prasher, D. C. (1978). The role of lichens in the structure, productivity and mineral cycling of the wet coastal Alaskan tundra. In *Vegetation and production ecology of an Alaskan arctic tundra*, (ed. L. L. Tieszen), pp. 185–206. Springer, New York.

Woodin, S., Press, M. C., and Lee, J. A. (1985). Nitrate reductase activity in *Sphagnum fuscum* in relation of wet deposition of nitrate from the atmosphere. *New Phytologist*, **99**, 381–8.

Worsley, P. and Ward, M. R. (1974). Plant colonization of recent 'annual' moraine ridges at Austre Ostinbreen, North Norway. *Arctic and Alpine Research*, **6**, 217–30.

Wyatt, R. and Stoneburner, A. (1989). Bryophytophagy of *Rhizomnium punctatum* by larvae of the crane fly *Tipula oropezoides*. *Bryologist*, **92**, 308–9.

3

Controls on growth and productivity of bryophytes: environmental limitations under current and anticipated conditions

BJARTMAR SVEINBJÖRNSSON
AND WALTER C. OECHEL

3.1 Introduction

Where they occur in abundance, bryophytes play a key role in the structure and function of ecosystems (Oechel and Lawrence 1985; Chapin *et al.* 1987). Their relative effect on the ecosystems in which they occur, including effects on nutrient cycling, soil temperature, soil moisture, soil decomposition, vascular plant biomass, and vascular plant productivity, may be far greater than the fraction of the total plant biomass they represent (Oechel and Lawrence 1985; Oechel and Van Cleve 1986; Chapin *et al.* 1987).

Bryophytes are particularly important in the functioning of Arctic, taiga, and northern bog ecosystems. The low temperatures and abundant moisture common in these ecosystems favour bryophyte productivity which may easily exceed the productivity of the vascular plants, including that of the black spruce trees in the taiga (Oechel and Van Cleve 1986). Also, mosses have been shown to be major controllers of the successional development of upland forest in the Alaskan taiga (Bonan and Korzuhin 1989) and of primary succession on river floodplain following silt deposition and the formation of new habitat for colonization (Viereck 1970; Van Cleve and Viereck 1981; Van Cleve *et al.* 1983).

As bryophytes proliferate, they may affect environmental conditions which can favour their own growth and development. Bryophytes are particularly important in the development and functioning of northern ecosystems which are systems likely to be affected by global change. These systems are potentially sensitive to global change for several reasons including the fact that they are often permafrost dominated, that permafrost development interacts with moss development and abundance, that the presence of permafrost affects many environmental and ecosystem variables, that with increasing CO_2 levels northern ecosystems are expected to undergo the largest increase in temperature of all terrestrial regions, and that the anticipated temperature rise is sufficient to cause the deepening or eventual loss of permafrost over large areas.

Because of the importance of bryophytes in northern ecosystems, and because of the sensitivity of northern ecosystems to global change, we will concentrate here on the physiology and ecosystem dynamics of bryophytes from these regions. However, many of the physiological attributes and responses to global change discussed below will also apply to bryophytes occurring in other regions particularly where they tend to be dominant such as in Antarctic and wet temperate regions.

As global environmental conditions change, there will be changes in the nature of plant competition and species composition. This will result in part from differential species responses to environmental change (Bazzaz 1990). Bryophytes from many habitats may be particularly well suited to benefit from expected increases in atmospheric CO_2. The abundant supply of nutrients and moisture, remove many of the potential resource limitations to responses to elevated CO_2 (Oechel and Strain 1985), and allow a significant potential response of growth and photosynthesis. Moss physiology, morphology, and anatomy are such that they may be responsive to elevated CO_2. For example, photosynthetic response to elevated CO_2 is less likely to be constrained by limitations of allocation and sink activity, and not at all (at least for gametophytes) by stomatal response to elevated CO_2 as is the case for vascular plants. However, future conditions of drying, warming, and/or increased light intensity could adversely affect moss growth and development.

In this chapter, we evaluate the current environmental controls on the growth and productivity of bryophytes as well as impacts of global change on these factors. In addition, the role of mosses on ecosystem functioning under present conditions is considered together with the likely impacts of global change on various growth forms and physiological types in mosses.

To analyse the effect of global change on mosses, we consider the current understanding of environmental controls on water and carbon dioxide exchange. Much has been written and recently reviewed on the responses of mosses to variations in light, temperature, water, and nutrients

(Smith 1982). Less has been written on the likely effects of elevated atmospheric CO_2 on bryophytes and on the ecosystem attributes and effects of mosses. This chapter will therefore emphasize the likely response of bryophytes to elevated CO_2 and those aspects of moss growth and development that affect ecosystem structure and function. In particular, we evaluate the effect of elevated CO_2 on bryophyte photosynthesis and growth. In doing this we will analyse CO_2 gradients which presently exist in the field, since understanding the response of bryophytes to these may presage their responses to elevated CO_2 levels. We will evaluate the likely effect of changes in other environmental attributes on bryophyte photosynthesis, growth and survival. We will also evaluate ecosystem level interactions which may affect moss development, and the effect of changes in bryophyte growth and productivity on ecosystems. This analysis includes a consideration of the nature and amount of vascular plant litter production, its effects on bryophyte growth, productivity, and distribution, and how it might be expected to change with global change. We also analyse current environmental conditions in the field and their effect on moss growth and development.

3.2 Effects of litter on establishment and growth

Bryophytes are often affected by vascular plant litter and they themselves produce litter affecting the vascular plants. This relationship may become more important in the future as increased temperature may stimulate soil nutrient mineralization, increase nutrient availability to vascular plants, and hence their growth and canopy foliage density.

Leaf litter from broad-leaved trees may exclude the development of dense moss carpets. In the taiga, the early successional stages are characterized by the predominantly deciduous shrub communities and forests (birch or aspen on upland sites, willows, alders, and poplars on floodplain sites). In these deciduous communities, the absence of feather mosses which dominate the forest floors of later successional stages otherwise characterized by black and white spruce, has been attributed to the heavy production of broad-leaf litter (Oechel and Van Cleve 1986). The first occurrence of moss carpets, in the later hardwood stages of taiga succession, is on topographically raised points such as tree stumps and fallen logs blown free of litter.

The litter may affect the moss in various ways. The most obvious effect is shading, the broad-leaf litter allowing insufficient light to the moss canopy. Certainly, the broad-leaf litter shed in annual pulses stays at least for a while on top of a moss, while spruce needles because of their geometry and more continuous shedding, work their way in between the moss shoots.

More elaborate explanations would include inhibitory chemical release, either directly from the broad-leaved plants or from the decomposer micro-organisms active in the litter. Conditions which change the nature and distribution of broad-leaf vegetation may obviously have a major impact on moss vegetation.

There are undoubtedly positive aspects of some litter falling on the bryophyte canopy. Nutrients released from tree litter, in addition to those in dripwater from the tree canopy, may be taken up by the bryophytes on their way to the soil (Tamm 1964; Bunnell 1981; Oechel and Van Cleve 1986) benefiting the bryophytes while severely limiting the tree growth. There is little information in the literature, other than from the taiga, on litter negatively affecting bryophyte growth (see, though, Grime 1979). Rincon (1988) has found that of five bryophyte species of British grasslands, only one was at all inhibited by litter, while the other species were either stimulated or not affected by the litter types. The stimulation of grassland moss growth was considered to be due to mineral nutrient acquisition from the decomposing litter.

Living bryophyte mats and their litter significantly affect tree perform-ance. First, in some instances, moss carpets facilitate seed germination by providing a damp surface (Black and Bliss 1980; Zasada 1986) while in other more exposed conditions they hinder seed germination (Zasada 1986). Second, thick moss carpets may make necessary contact of seedling roots with mineral soil impossible, leading to seedling death (Kataeva and Korzuhin 1987, as cited in Bonan and Korzuhin 1989) while thin ones may stabilize water availability by reducing soil water loss (Black and Bliss 1980). Third, moss carpets may intercept atmospheric nutrients and thus reduce their availability to the vascular plants (Oechel and Van Cleve 1986). But the moss mat may also aid in vascular plant nutrition. Thus, it intercepts easily leached nitrate in precipitation and gradually releases the nitrogen upon its own decomposition as ammonium, which is more easily absorbed by trees such as black spruce in Quebec (Weetman and Timmer 1967; see also Press and Lee 1982). Persistent moss removal has also been considered detrimental to the nutrient relations of sessile oak forests in Wales (Rieley et al. 1979). Bryophytes and their litter thus play a key role in forest tree functioning.

3.3 Microclimate

Except for oceanic and other perhumid climatic regions, the microclimate in which bryophytes function can be quite different from that in the general community (Proctor 1982; Sveinbjörnsson and Oechel 1981a). The struc-ture of the vascular plant community greatly affects conditions for the ground layers of bryophytes, and in forests the trees themselves may become moss habitat.

In a black spruce dominated forest in west-central Alberta, wind speeds at 25 cm above a feathermoss canopy were found to be 12 to 35 per cent of those above *Tomenthypnum nitens* in nearby fens with scattered trees (Busby *et al.* 1978). On the tundra, wind speed at the moss surface varies between 10 and 40 per cent of that at 25 cm (Fig. 3.1) and is mostly between 0.1 and 1 m s^{-1}.

Light penetration to the bryophyte surface may be close to 100 per cent in open habitats, as in *Tomenthypnum nitens* on fens, while it is much lower (about 20 per cent) at the feathermoss surface in the black spruce forest (Busby *et al.* 1978). On another black spruce forest floor, on the North Shore in Quebec, dominated by the feather mosses *Pleurozium schreberi* and *Ptilium crista-castrensis*, light penetration was about 12 per cent (Weetman and Timmer 1967). Near Barrow, Alaska, *Polytrichum commune* and *Dicranum elongatum* growing on polygon rims on the tundra received 100 per cent of the incoming solar radiation while *Polytrichum alpinum* in the graminoid understory in mesic meadows received only about 50 per cent of the incoming solar radiation (Sveinbjörnsson and Oechel 1981*a*). In addition, the tree or graminoid canopies of taiga and tundra remove a greater proportion of low light intensities than high light intensities. Thus, proportional light penetration of the graminoid tundra canopy to the moss layer during mornings and evenings is only about two thirds of that during the midday hours (Sveinbjörnsson and Oechel 1981*a*, Fig. 3.2). Increased vascular plant canopy density may accentuate this pattern.

Precipitation reaching the bryophyte surface is presumably quite variable, especially in heterogeneous habitats such as in open canopy forests. In coniferous forests, drip-line precipitation is considerable while stemflow is negligible; the reverse condition is true of deciduous forests. Floors in black spruce forests on the North Shore in Quebec, dominated by feather mosses, received 53 to 64 per cent of the rainfall on nearby clearcuts (Weetman and Timmer 1967). Bryophytes may obtain water by fog and dew deposition which, while not recorded, may be considerable. On the tundra near Barrow, Alaska, between 0.1 and 0.25 mm of fog and dew were deposited per day in the 1973 growing season (Sveinbjörnsson and Oechel 1981*a*).

The temperature of the bryophyte tissues, like that of vascular plants is a function of the balance between energy or radiation interception on one hand and its dissipation (or receipt) through convection and conduction, latent heat of evaporation and condensation, and long wave radiation on the other (see e.g. Hoffman and Gates 1970; Proctor 1982). Bryophyte tissues, being closer to the ground, are exposed to less wind than the taller

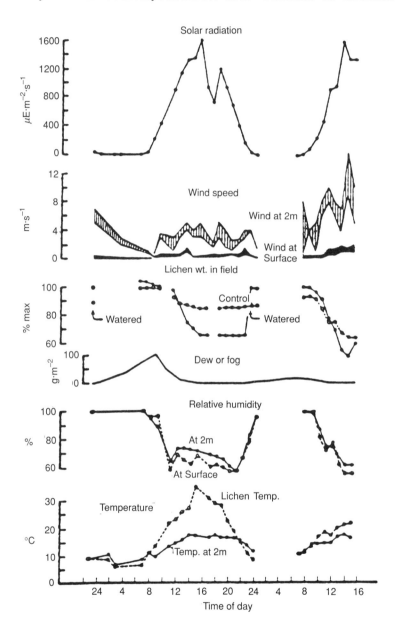

Fig. 3.1 Microclimatic conditions and lichen water content on a peat mound (palsa) edge near the airport on Nunivak Island, Alaska. Note the high humidity and dew deposition at night and early morning in spite of considerable wind speeds. The ground cover was mixed prostrate shrubs, lichens, and bryophytes (from Sveinbjörnsson 1990, reproduced by permission from *Rangifer*).

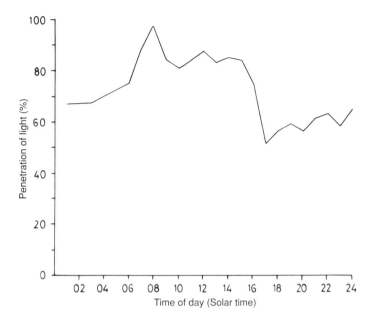

Fig. 3.2 Seasonal mean (*c.* 20 000 observations) diurnal variation of photosynthetically active light penetration through the vascular plant canopy at one location in a non-polygonized mesic meadow near Barrow, Alaska (from Sveinbjörnsson and Oechel 1981*a*, reproduced by permission from *Oikos*).

vascular plants, and therefore lose less heat through convection and evaporation than the vascular plants. Consequently, bryophyte temperatures generally exceed air temperatures during daytime hours. During the wet summer of 1973, seasonal mean maximum bryophyte gametophyte surface temperatures on the tundra near Barrow, Alaska, ranged from 4.5°C to over 6°C above weather station air temperature, while minimum temperatures were less than one degree lower than the air temperatures (Sveinbjörnsson and Oechel 1981*a*). However, bryophyte temperatures generally exceed air temperatures most when the bryophyte is dry and physiologically inactive (Proctor 1982).

3.4 Carbon dioxide

Bryophytes are exposed to a wide range of CO_2 concentrations in the field. Soil carbon dioxide production of tundras and taigas is substantial and can result in high CO_2 concentrations around the bryophyte canopy, especially during the night and early morning hours, before reduction by rapid photosynthesis of the green tissues. Carbon dioxide concentrations exceeding 1000 p.p.m. have been measured in and around the bryophyte canopy under snow in early spring (Oechel, unpublished data). This high CO_2 concentration is the result of plant and soil respiration with more than half coming from soil respiration (Peterson and Billings 1975). This CO_2 build-up can be rapidly depleted by photosynthesis during periods of adequate light.

Despite the rapid draw-down of elevated CO_2 in the canopy of vascular plants, and possibly even in the bryophyte canopy, it seems certain that arctic bryophytes may occur in areas of locally elevated CO_2. Even when the ecosystem is a net CO_2 sink with respect to the atmosphere, bryophytes may be experiencing elevated CO_2 due to being bathed in CO_2 emanating from the ground. The measured mean daily dark efflux from the tundra subsurface and consequently the flux through the bryophyte mat (Peterson and Billings 1975; Oberbauer et al 1986) ranged from about 50 mg CO_2 m^{-2} h^{-1} to 357 mg CO_2 m^{-2} h^{-1}, dramatically increasing the supply of CO_2 for photosynthesis over what it would otherwise have been if all CO_2 was supplied from the atmosphere and after being intercepted by the vascular plant canopy.

Temporal and spatial patterns of CO_2 efflux characterize the tundra (Peterson and Billings 1975). A diurnal pattern follows closely the average soil temperature at 0, 10, and 20 cm depths. In addition, higher rates of efflux are found later in the season than earlier. Finally, there were significantly higher soil respiration rates from polygon troughs than from meadow sites. Clearly these CO_2 flux patterns may significantly alter the hitherto established patterns of CO_2 flux in bryophytes (Oechel and Sveinbjörnsson 1978; Sveinbjörnsson and Oechel 1981a, b).

A similar situation is found in the taiga. In the white spruce forest near Fairbanks, Alaska, the floor has a carpet of feather mosses. Here, daily total soil respiration averaged 450 mg CO_2 m^{-2} h^{-1} (Gordon et al. 1987).

Unfortunately, while many of these studies show the magnitude of the soil efflux and indicate that the CO_2 concentration around photosynthetically active bryophyte tissues may differ substantially from that in the air above the vascular plant canopy, they do not quantify the actual CO_2 concentration around the bryophyte shoots through the day and season.

However, a few studies give some indication of the magnitude and pattern of CO_2 concentration near the bryophyte tissue in the field. Silvola (1986) studied CO_2 dynamics at 0.3 m and 1.5 m above drained peat mires reclaimed for forestry in Eastern Finland. He found that daytime values ranged from 290–330 p.p.m. and that they started rising in the evening, reaching 450–600 p.p.m. at sunrise. Wind reduced the CO_2 concentration in a linear fashion up to 3 m s^{-1} at 1.5 m height. These high CO_2 values appear to be in part a consequence of rapid microbial soil respiration due to the draining of the peatland. Although no information is presented for the immediate bryophyte environment, the study indicates significantly higher soil respiration rates following lowering of the water table. Similar soil respiration increases by water table lowering have also been demonstrated for soil cores from Barrow, Alaska (Peterson *et al.* 1984). Here it was concluded that microbial respiration increase was the cause of the increased CO_2 efflux, as root respiration was unaffected by the position of the water table.

Bazzaz and Williams (1991) studied CO_2 concentrations above a deciduous forest floor in New England. They monitored a height profile starting at 5 cm and reaching 12 m above the forest floor. Their data show markedly higher CO_2 concentrations at 5 and 20 cm than at greater heights, and demonstrate a midday minimum. The seasonal march of midday values shows this to be a summer phenomenon. They point out the importance of this CO_2 elevation for tree seedlings but the implication for forest floor bryophytes which are closer to the ground and presumably at more elevated CO_2 concentration is obvious.

While these studies are interesting and indicate that the bryophyte mat may experience CO_2 concentrations significantly different from the general atmosphere, they do not necessarily represent the bryophyte mat environment. The only published value that we have identified is that of Bazzaz *et al.* (1970) who placed a 'divot' of *Polytrichum commune* from an Indiana oak forest floor under a 1000 foot candle lamp at 20°C. They collected air from halfway between the soil and the gametophyte tips using a needle and syringe. Analysis of the air showed its CO_2 concentration to be 470 p.p.m. A study of a black spruce forest in south-central Alaska with forest floor mats of *Pleurozium schreberi* and *Ptilium crista-castrensis* (Butler and Sveinbjörnsson, unpublished data; Fig. 3.3) found that indeed there are strong CO_2 concentration gradients through the bryophyte mat.

These studies show that bryophytes in many biomes are presently exposed to widely varying CO_2 conditions. Global warming will probably increase soil, decomposing log, and tree bark CO_2 efflux through increased respiration of soils and the tree stems (Oberbauer *et al.* 1986). It is likely that bryophytes have adapted to the fluctuating and elevated present conditions and will be benefited by future increases in CO_2 around the photosynthesising tissues.

3.5 Environmental controls on water relations and photosynthesis

3.5.1 Water relations

Bryophyte water relations differ from those of vascular plants in several respects. Bryophyte water conduction is both internal and external, with

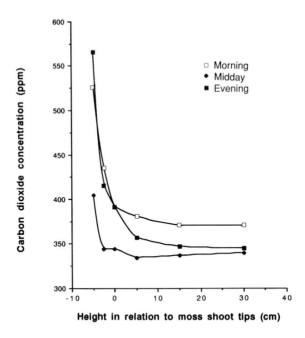

Fig. 3.3 A profile of carbon dioxide concentrations in air, in and above a feather moss mat growing on a black spruce forest floor in Anchorage, Alaska. The measurements were made three times during the day (Butler and Sveinbjörnsson, unpublished data).

the latter being generally more important. Only certain bryophytes, such as *Polytrichales*, have specialized water conducting tissues (hydroids) analogous to xylem in vascular plants; these have below-ground parts that also function as storage and vegetative regeneration organs, as in vascular plants (Collins and Oechel 1974; Sveinbjörnsson and Oechel 1981*b*). These bryophytes are generally metabolically active during the drier time of the growing season when other bryophytes are inactive (Callaghan *et al.* 1983).

Other bryophytes such as those in the carpets of feather mosses in the boreal understorey are active only when wetted by precipitation and apparently cannot wick up water from the water table as do fen and bog mosses like *Sphagnum* spp. and *Tomenthypnum nitens* (Busby *et al.* 1978; Titus *et al.* 1983). While the latter have a great water holding capacity (up to 93 g H_2O g dry wt^{-1}; Oechel and Sveinbjörnsson 1978), they also lose water more easily than the first mentioned group including the *Polytrichales* which fold leaf margins over the leaf and appress their leaves to the stem when dry (Oechel and Sveinbjörnsson 1978).

Various other morphological traits affect water relations. Bryophyte

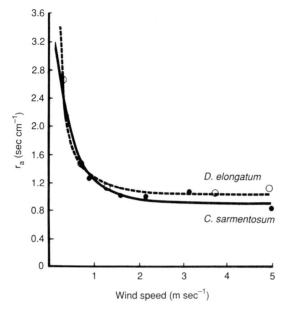

Fig. 3.4 The relationship between wind speed and air resistance (r_a) in *Dicranum elongatum* and *Calliergon sarmentosum* from Barrow, Alaska. Air resistance is based on ground area covered rather than on leaf areas. Lines are fitted by regression analyses (from Oechel and Sveinbjörnsson 1978, reproduced by permission from Springer-Verlag).

cushions, such as those of *Dicranum angustum* on polygon rims in the Arctic, can be very tightly packed, and consequently their boundary layer resistance can be greatly increased. Boundary layer resistance may be even further increased by dead leaf tips like those of *Grimmia pulvinata* (Proctor 1982). These cushions are often hemispherical in shape, thus reducing the exposed surface.

Yet another growth-form is found in extremely wet areas with good aeration, such as on lake margins, in small ponds, and on mist shrouded objects like trees and rocks. This growth-form, exemplified by *Calliergon sarmentosum*, is characterized by thin stems and small leaves; i.e. large surface to volume ratio and a low boundary layer resistance which can be increased during drying by coalescing of several shoots (Oechel and Sveinbjörnsson 1978).

The boundary layer resistance values reach a plateau of minimum resistance at wind speeds between 0.5 and 1 m s⁻¹ (Oechel and Sveinbjörnsson 1978; Fig 3.4). This represents the upper range of wind speeds in the bryophyte canopy, which means that for much of the time, wind directly and linearly regulates water loss from the bryophyte mat. Any change in wind in

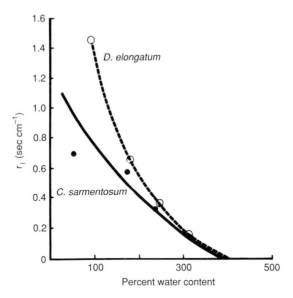

Fig. 3.5 The relationship between water content and leaf resistance (r_l) in *Dicranum elongatum* and *Calliergon sarmentosum* from Barrow, Alaska. Leaf resistances are based on ground area rather than on leaf areas. Lines are fitted by regression analyses (from Oechel and Sveinbjörnsson 1978, reproduced by permission from Springer-Verlag).

future climates, whether caused by changes in atmospheric circulation or vascular plant obstruction, will thus directly affect bryophyte water loss.

Bryophytes only possess stomata on the sporophytes, and these do not appear to be very active in regulation of water loss, judging from the very high resistance values measured (Proctor 1982). Consequently, in the absence of water loss reduction mechanisms such as those of *Polytrichum* and *Marchantia*, and given their generally thin walls and resultant low internal resistance (Fig. 3.5) (Oechel and Sveinbjörnsson 1978; Proctor 1982), water loss from bryophyte tissues is rapid.

3.5.2 CO_2 exchange and growth: responses to water, light, and temperature

Environmental constraints on photosynthesis of bryophytes are qualitatively similar to those of vascular plants, except for their water relations. Some bryophytes can photosynthesize to very low water contents equalling water potentials of -3 to -8.5 MPa (Proctor 1982), which is as low as has been found for vascular plants of the warm desert (Larcher 1983). Many drought resistant bryophytes can take up atmospheric water at high humidities (Lange 1969) while others such as *Conocephalum conicum* (Slavik 1965) and

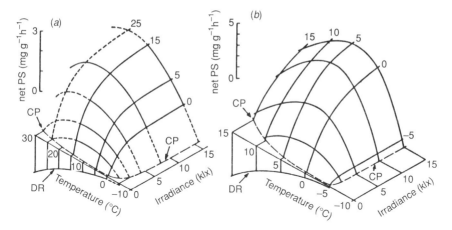

Fig. 3.6 Response surfaces of net photosynthesis to irradiance and temperature at normal atmosphere CO_2 content (*c*.320 ppm) for (*a*) *Hylocomium splendens*, based on data of Stålfelt (1937), and (*b*) *Racomitrium lanuginosum* (S. Finland) based on data of Kallio and Heinonen (1975). Smooth curves have been drawn through the experimental points or interpolated where appropriate; broken lines are extrapolated beyond the range of the data. The dark respiration curve (DR) is shown at zero irradiance. Irradiance curves intersect the line marked CP at the light compensation point for that temperature; temperature curves intersect the same at upper and lower compensation points characteristic for a particular irradiance (from Proctor 1982, reproduced by permission from Chapman and Hall).

Calliergon sarmentosum (Oechel and Sveinbjörnsson 1978), are drought-sensitive and die when mildly droughted. Where conditions continue to be or become perhumid, drought-sensitive bryophytes are likely to be stimulated by the elevated atmospheric CO_2 concentration, particularly given their fast and very opportunistic growth and low resistance to CO_2 diffusion. Bryophytes currently in high CO_2 environments such as those relying on ground or bark respiration as a CO_2 source will benefit less, except to the extent that they receive more CO_2 through increased substrate respiration.

Light responses of net photosynthesis in bryophytes reflect habitat exposure, and thus shade plant responses with low light compensation and saturation requirements are common (e.g. Miyata and Hosokawa 1961; Hosokawa *et al.* 1964; Oechel and Collins 1976; Oechel and Sveinbjörnsson 1978). These relationships interact both with temperature (Fig. 3.6) and CO_2 concentration, with decrease in temperature and increase in CO_2 lowering the light requirements (Kallio and Heinonen 1975; Oechel and Sveinbjörnsson 1978; Silvola 1985). Increased shading by vascular plants is partly but not completely compensated by light acclimatization (Oechel

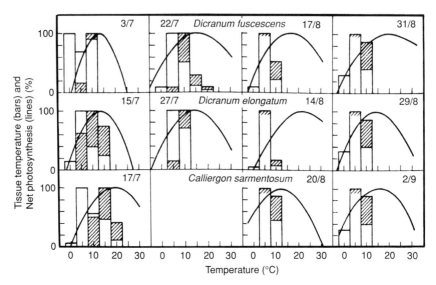

Fig. 3.7 Temperature and photosynthesis of mosses growing on the tundra near Barrow, Alaska. Open bars represent the relative frequency of temperatures in 5°C blocks through the day for 5 days previous to the day of measurement. Hatched bars represent the relative frequency of temperatures at peak photosynthetic periods (1000–1400 h). Bars are shaded at the top only when daily and midday intervals showed the most frequent temperatures in the same temperature range. Curved lines represent the relative net photosynthetic response to temperature on dates indicated. Curves are based on regressions (from Oechel 1976, reproduced by permission from *Photosynthetica*).

and Sveinbjörnsson 1978). Therefore replacement by more light tolerant species may take place where a denser overstorey and greater shading occurs.

Many bryophytes can photosynthesize to lower temperatures than can vascular plants (Kallio and Kärenlampi 1975; Oechel and Sveinbjörnsson 1978) and photosynthesis under snow can significantly contribute to the annual carbon balance (Tieszen 1974; Collins and Callaghan 1980). Temperature optima of bryophytes are generally broad and adjustable through rapid acclimatization, although thermal ecotypes are known to exist (Hicklenton and Oechel 1976). Nevertheless, temperature optima often exceed the most frequent temperatures experienced by bryophytes *in situ* (Fig. 3.7).

Upper temperature compensation points of bryophyte photosynthesis, which are affected more by prevailing temperature than the lower temperature compensation points, seldom appear to be exceeded, as warm periods are generally dry and the bryophyte is quickly rendered inactive (except for

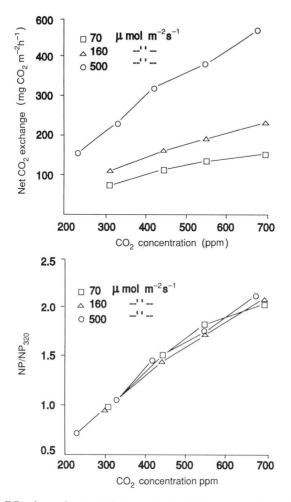

Fig. 3.8 CO_2-dependent net photosynthesis of *Dicranum majus* at three irradiance levels (12°C) and the relation of the CO_2-exchange at various CO_2 concentrations compared with that at 320 ppm (from Silvola 1985, reproduced by permission from *Lindbergia*).

endohydric types like *Polytrichum*). Sometimes, however, tissue temperatures do exceed the upper temperature compensation point of photosynthesis and this appears to result in respiratory carbon loss. This may be the explanation for weight loss during the warm dry season of *Hylocomium splendens* in mountain birch forests in Swedish Lapland (Callaghan *et al.* 1978). In areas where warming will take place without concomitant water stress, it is likely that supraoptimal temperatures will become more frequent and in some cases lead to weight loss and eventual death.

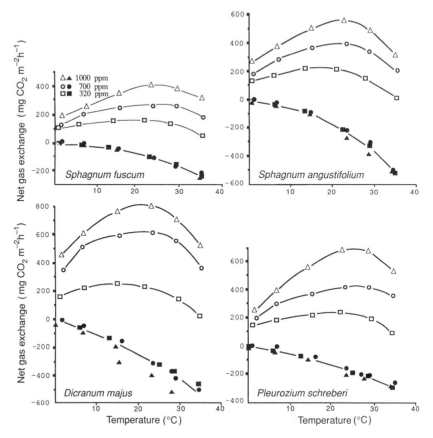

Fig. 3.9   Temperature dependence of dark respiration and net photosynthesis at various CO_2 concentrations in certain forest and peat mosses in Finland. Black symbols—darkness, open symbols—500 μmol m^{-2} s^{-1}. Measurements were performed at constant CO_2 concentrations with changing temperature, and for each species the changes of CO_2-exchange of one sample (177 cm^2) are shown (from Silvola 1985, reproduced by permission from *Lindbergia*).

3.5.3. Responses to elevated CO_2

Short-term effects of CO_2 concentration on bryophyte photosynthesis are the best known parts of this puzzle. Bazzaz *et al.* (1970) showed an increase in photosynthesis of *Polytrichum juniperinum* populations when exposed to 450 p.p.m. CO_2 in air in comparison to 300 p.p.m. CO_2 in air.

The most comprehensive data sets on photosynthetic response surfaces in relation to CO_2 concentration are those of Aro *et al.* (1984) and Silvola (1985). The former studied *Hypnum cupressiforme* and *Dicranum scoparium* collected on an oakwood floor on a north-facing hillslope in France. The latter studied *Pleurozium schreberi*, *Dicranum majus*, *Sphagnum fuscum*, *S.* *angustifolium*, *S. nemoreum*, and *S. magellanicum* from drained afforested peatlands in eastern Finland. Both found significant increase in photosynthesis up to the highest CO_2 concentration used (2500 p.p.m. and 1000 p.p.m. respectively), as Proctor (1982) had also shown using other species.

Elevated atmospheric CO_2 may not only increase the photosynthetic rate, but it may help adjust the photosynthetic response to a changing climate. Light requirements to maintain similar photosynthetic rates decrease with elevated CO_2. Thus, in *Dicranum majus* higher photosynthetic rate was obtained at a photon flux density of 70 μmol quanta m^{-2} s^{-1} and 640 p.p.m. CO_2 than at 160 μmol quanta m^{-2} s^{-1} and 320 p.p.m. CO_2 (Silvola 1985). The proportional effect of increasing CO_2 appears to be the same at all light levels between 70 and 500 μmol m^{-2} s^{-1} (Fig. 3.8). This effect would help offset the effect of decreasing light levels possibly resulting from increasing cloudiness and denser vascular plant canopies following global warming. Silvola (1985) showed that elevated CO_2 tends to increase the temperature optimum for photosynthesis (Fig. 3.9), a beneficial effect if global temperatures increase with elevated CO_2. He studied the effect of temperature on photosynthesis at different CO_2 concentrations and found that the optimum temperature rose with rising CO_2 level from 15°C at 320 p.p.m. to 20–25°C at 1000 p.p.m. CO_2. This increase in temperature optima for photosynthesis, also demonstrated in vascular plants (Oechel and Strain 1985), represents a beneficial effect if global temperatures increase as well as CO_2 concentrations.

No reports of bryophyte photosynthesis under naturally varying CO_2 concentrations have been identified. A number of studies have been carried out on the response to natural variations in light and temperature (e.g. Oechel and Sveinbjörnsson 1978; Callaghan *et al.* 1978), but these have not measured CO_2 concentration around the shoots or controlled concentrations at natural field levels.

There will undoubtedly be ecotypic and environmental constraints on the response to elevated atmospheric CO_2. Bazzaz *et al.* (1970) for example, found that the increase in photosynthetic rate by raised CO_2 levels was greater in a forest than in an alpine population of the moss *Polytrichum* *juniperinum*. It is likely that ecotypes exist in relation to environments of origin, with respect to most environmental parameters including CO_2 levels (Sveinbjörnsson and Oechel 1983).

These enhancement effects of high CO_2 noted in short term experiments, may be changed or may disappear during long-term exposures to elevated CO_2 (Oechel and Strain 1985; Tissue and Oechel 1987; Grulke *et al.* 1990). Unfortunately, only one long-term growth experiment has been reported on the effects of CO_2 concentration on bryophytes. This is a growth experiment on the liverwort *Marchantia polymorpha* (Björkman *et al.* 1968). Although not stated, it appears that this population from California was watered with a nutrient solution. Under these conditions, there was a 10 day dry weight gain of 7, 55, and 78 mg/plant at 110, 320, and 640 p.p.m. CO_2 respectively.

Responses in the field to elevated atmospheric CO_2 may well depend on resource availabilities. Tissue and Oechel (1987) found that photosynthetic rates of *Eriophorum vaginatum* exposed in the field to 340, 510, and 680 p.p.m. CO_2 initially responded to the elevated CO_2. However, photosynthetic rates dropped to pre-treatment levels within three weeks, and remained at that level throughout the course of the manipulation. There was no significant effect on stomatal resistance or total non-structural carbohydrate levels. Photosynthetic rates appeared to be controlled by sink strength, which, in turn, is controlled by nutrient availability.

Arctic and subarctic bryophytes may be less nutrient limited than co-occurring vascular plants. In tested tundra moss species (Oechel and Sveinbjörnsson 1978; Bigger and Oechel 1982), maximum photosynthesis has other overriding restraints. Conversely, *Hylocomium splendens* and *Sphagnum nemoreum* in the Alaskan taiga show increased photosynthesis when fertilized (Oechel and Van Cleve 1986). Growth of *Sphagnum riparium* in northermost Sweden is nutrient limited (Sonesson *et al.* 1980), as is that of *Sphagnum nemoreum* in the Alaskan taiga (Skre and Oechel 1979). *Hylocomium splendens* in Sweden is positively affected by drip water from the spruce canopy (Tamm 1964) and thus apparently nutrient-limited. The nutrient stimulation of growth in forest mosses is by no means universal, as *Pleurozium schreberi* in the Alaskan taiga (Oechel as cited by Brown 1982) and *Pseudoscleropodium purum* in a forest in England (Bates 1987) did not show significant response to nutrient addition. Apparently nutrient limitations for growth and photosynthesis vary by species and they will therefore be differentially affected by changes in nutrient availability, if any are brought about by global change.

Mosses at Barrow, Alaska, showed a significant positive correlation between total non-structural carbohydrate levels and maximum photosynthetic rates (Sveinbjörnsson and Oechel, in preparation) and therefore are not likely to be end-product-inhibited or controlled by sink demand. We hypothesize, therefore, that the long-term response of bryophytes will be more similar to the short-term response than is the case for arctic vascular plants. However, those bryophytes such as liverworts with fairly water-repellent upper surfaces and more porous lower thalli like the *Marchantiales* (Proctor 1982) probably receive most of their CO_2 from the soil and will continue to do so unaffected by atmospheric CO_2 levels.

Net photosynthesis of bryophytes may be significantly increased by elevated CO_2 concentration as there is no evidence of end-product inhibition, but rather the reverse. The pattern of the diurnal and seasonal progression of photosynthesis and respiration is almost certainly different from that hitherto reported, because *in situ* CO_2 concentrations at bryophyte level have not been taken into account. Bryophytes living in environments with fairly stable and/or low atmospheric CO_2 concentrations will no doubt benefit much from future increases in the atmospheric CO_2 concentrations. However, even those mosses and liverworts that mostly receive their CO_2 from the substrate, be it soil or tree bark, will probably also experience increased CO_2 supply as increased temperatures will increase plant and soil respiration rates (Peterson and Billings 1975; Sveinbjörnsson, unpublished data).

Bryophytes will probably benefit more than vascular plants from longer growing seasons in future warmer climates. This is because their growth and photosynthesis is largely opportunistic and no senescence, as found in deciduous vascular plants, is observed (Oechel and Sveinbjörnsson 1978; Callaghan *et al.* 1978; Lindholm 1990). The increased CO_2 levels will compensate for reduced light in the late fall (Silvola 1986) and increase temperature optima for photosynthesis. Temperature optima for net photosynthesis have been found to increase during the growing season in most bryophytes (Oechel 1976; Oechel and Sveinbjörnsson 1978; Sveinbjörnsson and Oechel, in preparation). The late season high photosynthetic temperature optima, which now appear maladaptive, may be better suited in future warmer climates.

Many bryophytes reproduce through asexual reproduction, and in fact this is the only reproductive mode they have in many harsh environments (Sveinbjörnsson and Oechel 1981*b*). *Eriophorum vaginatum* tiller production was significantly increased by exposure to high CO_2 concentration (Tissue and Oechel 1987) and similar increases have been found in other Arctic species (Oechel, unpublished data). The asexual reproduction of many bryophytes is by processes analogous to tillering (Sveinbjörnsson and Oechel 1981*b*; Callaghan *et al.* 1978). We hypothesize, therefore, that bryophyte shoot density will increase, making the bryophyte mat more effective in intercepting and utilizing CO_2 emanating from the soil.

3.6 Effects of bryophytes on functioning of northern ecosystems

Soil temperatures are strongly affected by bryophytes, as stated above (Oechel and Van Cleve 1986). The bryophyte mat characteristic of many tundra and taiga communities acts as insulation, especially when dry, as has long been recognized by the use of moss for home insulation. This property is largely caused by the preponderance of large empty cells (e.g. hyalocyst cells in *Sphagnum* mosses) a fact that also relates to its great water-holding capacity (Proctor 1982). When the top of the bryophyte mat is dry, as during warm weather, the mat retards heat penetration into the ground, whereas conduction is greatest during wet and cool weather when the gradients are small or even reversed as in early and late winter (Oechel and Van Cleve 1986). A thick bryophyte mat therefore generally corresponds to colder soils and in northern taigas and tundras this leads to maintenance and even growth of permafrost.

Table 3.1 Comparison of fluxes with nutrient-sequestering potential of moss at the Washington Creek study site

		N	P	K	Ca	Mg
Nutrient addition in throughfall	mEq/m^2	13.5	0.3	3.3	11.5	3.3
Nutrient addition in litterfall	mEq/m^2	10.7	0.3	0.8	17.5	1.6
Total	mEq/m^2	24.2	0.6	4.1	29	4.9
Seasonal uptake in green moss tissue	mEq/m^2	92.1	4.8	15.6	13.5	12.3
		Sphagnum spp.		*H. splendens*	*P. schreberi*	*P. commune*
Cation exchange capacity of selected moss species (total of green and brown tissue)	mEq/m^2	2010.6		235.3	3791.2	537.1

(From Oechel and Van Cleve 1986, reproduced by permission from Springer-Verlag)

The reduced soil temperature caused by the bryophyte mat, coupled with high diffusive resistance to water vapour from the soil, generally means that the soils underneath are continuously wet. However, the viscosity of water is highly temperature-dependent, increasing with decreasing temperature, and water uptake may, therefore, be hindered, resulting in lowered vascular plant water potentials (Stoner and Miller 1975). The change in bryophyte mat thickness will therefore obviously affect the vascular plants with a feedback effect on the bryophytes.

The bryophyte mat is an effective nutrient filter, and taiga mosses in Alaska show a potential for retention of elements greater than the total flux to the forest floor (Oechel and Van Cleve 1986; Table 3.1). The phosphate absorption capacity of *Hylocomium splendens* exceeds that of black spruce roots (Chapin *et al.* 1987; Fig. 3.10). Part of the reason for the efficiency of nutrient filtering by bryophytes appears to be their charged cell wall surfaces, such as negatively charged carboxyl groups of the pectates in *Sphagnum* (Clymo 1964; Craigie and Maass 1966), combined with a very large surface to volume ratio increased by special surface enlarging

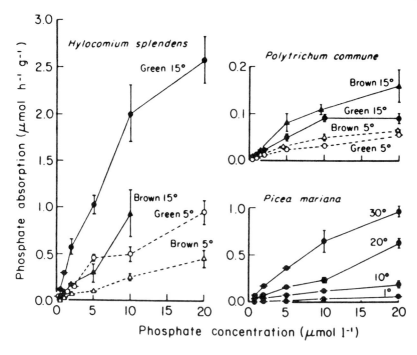

Fig. 3.10   Rate of phosphate absorption from solutions by green and brown parts of two mosses, *Hylocomium splendens* and *Polytrichum commune*, at 5 and 15°C, and by mycorrhizal fine roots of black spruce, *Picea mariana*, at four temperatures. Note difference in scale for *Polytrichum*. Data are means □ and S.E., $n = 4$ (from Chapin *et al.* 1987, reproduced by permission from Springer-Verlag).

structures such as paraphyllia in *Hylocomium splendens* and aerial rhizoids in *Dicranum* species (Tamm 1953), The future nutritional and filtering effects of the bryophytes are presently unknown, but increased rainfall coupled with increased temperature in the taiga may lead to conditions akin to those in the wet coastal forests of north-west North America, which are known for their abundant bryophyte understorey which significantly affects nutrient cycling through its slow decay rates (Binkley and Graham 1981).

3.7 Summary

Bryophytes thrive in oceanic or perhumid climates and cool shaded microclimates of the vascular plant understorey. Increased global temperatures may increase vascular plant growth and canopy foliage density through stimulated soil nutrient mineralization. The nature of the vascular plant canopy under which bryophytes grow affects their light regime, temperature, moisture, and nutrient supply. Leaf litter from broad-leaved trees may exclude the development of dense moss carpets.

Bryophytes have very limited means for soil water uptake, little regulation of water loss, and rely mostly on nutrient supply from atmospheric deposition, tree drip, or litter decay. The boundary layer resistance of a bryophyte mat is affected by canopy density and morphological features. It decreases linearly with increasing wind speeds up to 0.5 to 1.0 m s^{-1}, while wind speeds at the moss level seldom exceed 1 m s^{-1}. Internal resistance to water loss (and CO_2 diffusion) is low but variable; fast growing thin-stemmed and thin-walled species have the lowest resistances. Photosynthesis is strongly dependent on water content.

Photosynthesis of bryophytes has been shown to acclimate rapidly to changes in their light and temperature environments and this acclimation will partly offset negative impacts of a changing environment. In addition, many bryophytes are exposed to varying and elevated CO_2 levels due to the respiration of their substrate, soil or tree bark. Short-term experiments of elevated CO_2 exposure show an increase in net photosynthesis, and there is evidence that ecotypes from CO_2 rich environments respond more than those from CO_2 poor environments. Increased CO_2 levels reduce light requirements and increase temperature optima for photosynthesis both of which allow adjustment to a warmer and darker future under a denser overstorey.

An increase in photosynthesis of vascular plants stimulated by elevated CO_2 is short-lived and its reduction is thought to be end product-controlled due to lack of nutrients. Conversely, bryophytes have little assimilate translocation, little sink differentiation, often show no fertilizer response, and show a positive relationship between non-structural carbohydrate concentration and photosynthetic capacity. Their long-term responses are, therefore, likely to resemble the short-term responses. The only long-term CO_2 enrichment experiment indicates that weight gain is strongly affected by ambient CO_2 level. Doubling of CO_2 concentrations is expected to approximately double the bryophyte net CO_2 balance assuming no change in temperature and light.

A warmer, darker, moister, and CO_2-richer future environment may stimulate bryophyte growth, for example through increased branching, resulting in thicker and denser bryophyte carpets in the understorey. This thicker carpet will be a more effective energy insulator and filter, both of water and nutrients from above and CO_2 from below. Thus, it will counteract or at least dampen the stimulating effect of the global warming on the vascular plants of the overstorey. The balance of these interacting components of the ecosystem will partly depend on the environmental rate of change and partly on their relative rates of response.

References

Aro, E.-M., Gerbaud, A., and Andre, M. (1984) CO_2 and O_2 exchange in two bryophytes, *Hypnum cupressiforme* and *Dicranum scoparium*. *Plant Physiology* **75**, 431–5.

Bates, J. W. (1987) Nutrient retention by *Pseudoscleropodium purum* and its relation to growth. *Journal of Bryology*, **14**, 565–80.

Bazzaz, F. A. (1990). The response of natural ecosystems to the rising global CO_2 levels. *Annual Review of Ecology and Systematics*, **21**, 164–96.

Bazzaz, F. A., Paolillo, D. J., and Jagels, R. H. (1970). Photosynthesis and respiration of forest and alpine populations of *Polytrichum juniperinum*. *The Bryologist*, **73**, 579–85.

Bazzaz, F. A. and Williams, W. E. (1991). Atmospheric CO_2 concentrations within a mixed forest: implications for seedling growth. *Ecology*, **72**, 12–16.

Bigger, C. M. and Oechel, W. C. (1982). Nutrient effect on maximum photosynthesis in arctic plants. *Holarctic Ecology*, **5**, 158–63.

Binkley, D. and Graham, R. L. (1981). Biomass, production, and nutrient cycling of mosses in an old-growth Douglas-fir forest. *Ecology*, **62**, 1387–9.

Björkman, O., Gauhl, E., Hiesey, W. M., Nicholson, F., and Nobs, M. A. (1968) Growth of *Mimulus*, *Marchantia*, and *Zea* under different oxygen and carbon dioxide levels. *Yearbook of the Carnegie Institute, Washington*, **67**, 477–8,

Black, R. A. and Bliss, L. C. (1980). Reproductive ecology of *Picea mariana* (Mill.) B. S. P., at tree line near Inuvik, Northwest Territories, Canada. *Ecological Monographs*, **50**, 331–54.

Bonan, G. B. and Korzuhin, M. D. (1989). Simulation of moss and tree dynamics in the boreal forest of interior Alaska. *Vegetatio*, **84**, 31–44.

Brown, D. H. (1982). Mineral nutrition. In *Bryophyte ecology*, (ed A. J. Smith), pp. 383–444. Chapman and Hall, London.

Bunnell, F. L. (1981). Ecosystem synthesis—a 'fairytale'. In *Tundra ecosystems: a comparative analysis*, (eds L. C. Bliss, O. W. Heal, and J. J. Moore), pp. 637–46 Cambridge University Press, Cambridge.

Busby, J. R., Bliss, L. C., and Hamilton, C. D. (1978). Microclimate control of growth rates and habitats of the boreal forest mosses, *Tomenthypnum nitens* and *Hylocomium splendens*. *Ecological Monographs* **48**, 95–110.

Callaghan, T. V., Collins, N. J., and Callaghan, C. H. (1978). Comparative strategies of photosynthesis, growth and reproduction in two mosses, *Hylocomium splendens* (Hedw.) B. & S. and *Polytrichum commune* Hedw. in Swedish Lapland. IV: Strategies of growth and population dynamics of tundra plants. *Oikos* **31**, 73–88.

Chapin, F. S. III, Oechel, W. C., Van Cleve K., and Lawrence, W. (1987). The role of mosses in phosphorus cycling of an Alaskan black spruce forest. *Oecologia* **74**, 310–15.

Clymo, R. S. (1964). The origin of acidity in *Sphagnum* bogs. *The Bryologist*, **67**, 427–31.

Collins, N. J. and Callaghan, T. V. (1980). Predicted patterns of photosynthetic production in maritime antarctic mosses. *Annals of Botany* **45** 601–20.

Collins, N. J. and Oechel, W. C. (1974). The pattern of growth and translocation of photosynthate in a tundra moss, *Polytrichum alpinum*. *Canadian Journal of Botany*, **52**, 355–63.

Craigie, J. S. and Maass, W. S. G. (1966). The cation exchanger in *Sphagnum* spp. *Annals of Botany*, N. S. **30**, 153–4.

Gordon, A. M., Schlentner, R. E., and Van Cleve, K. (1987). Seasonal patterns of soil respiration and CO_2 evolution following harvesting in the white spruce forests of interior Alaska. *Canadian Journal of Forest Research*, **17**, 304–10.

Grime, J. P. (1979). *Plant strategies and vegetation processes*. Wiley, New York.

Grulke, N. E., Reichers, G. H., Oechel, W. C., Hjelm, U., and Jaeger, C. (1990). Carbon balance in tussock tundra under ambient and elevated atmospheric CO_2. *Oecologia*, **83**, 485–94.

Hicklenton, P. R. and Oechel, W. C. (1976). Physiological aspects of the ecology of *Dicranum fuscescens* in the subarctic. I: Acclimation and acclimation potential of CO_2 exchange in relation to habitat, light and temperature. *Canadian Journal of Botany*, **54**, 1104–19.

Hoffman, G. R. and Gates, D. M. (1970). An energy budget approach to the study of water loss in cryptogams. *Bulletin of the Torrey Botanical Club*, **97**, 361–6.

Hosokawa, T., Odani, N., and Tagawa, H. (1964). Causality of the distribution of corticolous species in forests with special reference to the physio-ecological approach. *The Bryologist*, **67**, 396–441.

Kallio, P. and Heinonen, S. (1973). Ecology of *Rhacomitrium lanuginosum* (Hedw.) Brid. *Reports of the Kevo subarctic Research Station*, **10**, 43–54.

Kallio, P. and Kärenlampi, L. (1975). Photosynthesis in mosses and lichens. In *Photosynthesis and productivity in different environments*, IBP 3, (ed. J. P. Cooper), pp. 393–423. Cambridge University Press, Cambridge.

Lange, O. L. (1969). CO_2-Gaswechsel von Moosen nach Wasserdampfaufrahme aus der Lauftraum. *Planta (Berlin)*, **89**, 90–4.

Larcher, W. (1983). *Physiological plant ecology*. Springer, New York.

Lindholm, T. (1990). Growth dynamics of the peat moss *Sphagnum fuscum* on hummocks on a raised bog in southern Finland. *Annales Botanici Fennici*, **27**, 67–78.

Miyata, I. and Hosokawa, T. (1961). Seasonal variations of the photosynthetic efficiency and chlorophyll content of epiphytic mosses. *Ecology*, **42**, 766–75.

Oberbauer, S. F., Oechel, W. C., and Riechers, G. H. (1986). Soil respiration of Alaskan tundra at elevated atmospheric carbon dioxide concentrations. *Plant and Soil*, **96**, 145–8.

Oechel, W. C. (1976). Seasonal patterns of temperature response of CO_2 flux and acclimation in arctic mosses growing *in situ*. *Photosynthetica*, **10**, 447–56.

Oechel, W. C. and Collins, N. J. (1976). Comparative CO_2 exchange patterns in mosses from two tundra habitats at Barrow, Alaska. *Canadian Journal of Botany*, **54**, 1355–69.

Oechel, W. C. and Lawrence, W. T. (1985). Taiga. In *Physiological ecology of North American plant communities*, (eds B. F. Chabot and H. A. Mooney), pp. 66–94. Chapman and Hall, New York.

Oechel, W. C. and Strain, B. R. (1985). Native species responses to increased carbon dioxide concentration. In *Direct effects of carbon dioxide on vegetation* (eds B. R. Strain and J. D. Cure), pp. 118–154. US Department of Energy, Office of Basic Energy Sciences, Carbon Dioxide Research Division, State-of-the-Art Report (DOE/ER-0238).

Oechel, W. C. and Sveinbjörnsson, B. (1978). Primary production processes in arctic bryophytes at Barrow, Alaska. In *Vegetation and production ecology of an Alaskan arctic tundra* (ed. L. L. Tieszen), pp. 269–98. Springer, New York.

Oechel, W. C. and Van Cleve, K. (1986). The role of bryophytes in nutrient cycling in the taiga. In *Forest Ecosystems in the Alaskan taiga* (eds K. Van Cleve, F. S. Chapin III, P. W. Flanagan, L. A. Viereck and C. T. Dyrness) pp. 121–37. Springer, New York.

Peterson, K. M. and Billings, W. D. (1975). Carbon dioxide flux from tundra soils and vegetation as related to temperature at Barrow, Alaska. *American Midland Naturalist*, **94**, 88–98.

Peterson, K. M., Billings, W. D., and Reynolds, D. N. (1984). Influence of water table and atmospheric CO_2 concentration on the carbon balance of arctic tundra. *Arctic Alpine Research*, **16**, 331–5.

Press, M. C. and Lee, J. A. (1982). Nitrate reductase activity of *Sphagnum* species in the south Pennines. *New Phytologist*, **92**, 487–94.

Proctor, M. C. F. (1982). Physiological ecology: Water relations, light and temperature responses, carbon balance. In *Bryophyte ecology*, (ed. A. J. E. Smith), pp. 333–81. Chapman and Hall, London.

Rieley, J. O., Richards, P. W., and Bebbington, A. D. L. (1979). The ecological role of bryophytes in a North Wales woodland. *Journal of Ecology*, **67**, 497–527.

Rincón, E. (1988). The effect of herbaceous litter on bryophyte growth. *Journal of Bryology*, **15**, 209–17.

Silvola, J. (1985). CO_2 dependence of photosynthesis in certain forest and peat mosses and simulated photosynthesis at various actual and hypothetical CO_2 concentrations. *Lindbergia*, **11**, 86–93.

Silvola, J. (1986). Carbon dioxide dynamics in mires reclaimed for forestry in eastern Finland. *Annales Botanici Fennici*, **23**, 59–67.

Skre, O. and Oechel, W. C. (1979). Moss production in a black spruce *Picea mariana* dominated forest near Fairbanks, Alaska. In *Proceedings of the First Fennoscandian Tree-Line Conference, Kevo, Finland, 6–14 September 1977* (eds. M. Sonesson and P. Kallio).

Slavik, B. (1965). The influence of decreasing hydration level of photosynthetic rate in the thalli of the hepatic *Conocephalum conicum*. In *Water stress in plants*, (ed B. Slavik), pp. 195–201. Junk, The Hague.

Smith, A. J. E. (ed.) (1982). *Bryophyte ecology*. Chapman and Hall, London.

Sonesson, M., Person, S., Basilier, K., and Stenström, T.-A. (1980). Growth of *Sphagnum riparium* Ångstr. in relation to some environmental factors in the Stordalen mire. In *Ecology of a subarctic mire*, (ed M. Sonesson), *Ecological Bulletin (Stockholm)* **30**, 191–207.

Stoner, W. A. and Miller, P.C. (1975). Water relations of plant species in the wet coastal tundra at Barrow, Alaska. *Arctic Alpine Research*, **7**, 109–24.

Sveinbjörnsson, B. and Oechel, W. C. (1981*a*). Controls on CO_2 exchange in two *Polytrichum* moss species. 1. Field studies on the tundra near Barrow, Alaska. *Oikos*, **36**, 114–28.

Sveinbjörnsson, B. and Oechel, W. C. (1981*b*). Controls on CO_2 exchange in two *Polytrichum* moss species. 2. The implication of below-ground plants parts on the whole-plant carbon balance. *Oikos*, **36**, 348–54.

Sveinbjörnsson, B. and Oechel, W. C. (1983). The effect of temperature preconditioning on the temperature sensitivity of net CO_2 flux in geographically diverse populations of the moss *Polytrichum commune, Ecology*, **64**, 1100–8.

Tamm, C. O. (1953). Growth, yield and nutrition in carpets of a forest moss (*Hylocomium splendens*). *Meddelanden från Statens Skogsforskningsinstitut*, **42**, 1–140.

Tamm, C. O. (1964). Growth of *Hylocomium splendens* in relation to tree canopy. *Bryologist*, **67**, 423–6.

Tieszen, L. L. (1974). Photosynthetic competence of the subnivean vegetation of an arctic tundra. *Arctic Alpine Research*, **6**, 253–6.

Tissue, D. T. and Oechel, W. C. (1987). Response of *Eriophorum vaginatum* to elevated CO_2 and temperature in the Alaskan tussock tundra. *Ecology*, **68**, 401–10.

Titus, J. E., Wagner, D. J., and Stephens, M. D. (1983). Contrasting water relations of photosynthesis for two *Sphagnum* mosses. *Ecology*, **64**, 1109–15.

Van Cleve, K., Dyrness, C. T., Vierack, L. A., Fox, J., Chapin, F. S. III, and Oechel, W. C. (1983). Taiga ecosystems in interior Alaska. *Bioscience*, **33**, 39–44.

Van Cleve, K. and Viereck, L. A. (1981). Forest succession in relation to nutrient cycling in the boreal forest of Alaska. In *Forest succession, concepts and application*, (eds D. C. West, H. H. Shugart, and D. B. Botkin), pp. 185–210. Springer, New York.

Viereck, L. A. (1970). Forest succession and soil development adjacent to the Chena river in Interior Alaska. *Arctic Alpine Research*, **2**, 1–26.

Weetman, G. F. and Timmer, V. (1967). Feather moss growth and nutrient content under upland black spruce. *Technical Report no. 503*. Pulp and Paper Institute of Canada, Pointe Claire.

Zasada, J. (1986). Natural regeneration of trees and tall shrubs on forest sites in interior Alaska. In *Forest ecosystems in the Alaskan taiga*, (eds K. Van Cleve, F. S. Chapin III, P. W. Flanagan, L. A. Viereck, and C. T. Dyrness), pp. 44–73. Springer, New York.

4

Bryophyte distribution patterns

WILFRED B. SCHOFIELD

4.1 Introduction

The floristic regions of the world have been assessed recently by Takhtajan (1986). This assessment is based entirely on vascular plants, and essentially on flowering plant distributions, but is comprehensive and encyclopaedic in its treatment. Schuster (1983*a, b*) presents recent expositions for the bryophytes and, although he provides examples of general distribution patterns, he makes no attempt at a detailed mapping of floristic kingdoms, subkingdoms, regions, and their subdivisions. It is apparent, however, from comparison of the two analyses, that a considerable parallel of patterns exists, strongly suggesting that the historical and biological phenomena that determined vascular plant distribution patterns have also been a major factor in moulding bryophyte distribution patterns. An elementary summary is presented by Schofield (1985) while Zanten and Pócs (1981) concentrate on the factors that have determined distribution patterns.

Miller (1982) provided a map showing the bryophyte floristic kingdoms as delineated by Herzog (1926). Miller also presented brief descriptions of these kindgoms, incorporating data that have accumulated during the past 55 years. Further refinement indicates that some of the boundaries recognized by Herzog need alteration. A map is provided here to reflect more recent understanding based on data now available. The boundaries of the floristic kingdoms and regions presented here should be treated as tentative. Considerable improvement will result when detailed data become available, especially when those data are synthesized by researchers with the breadth of experience to do so. Raw lists of taxa exist for many regions, as is shown by Greene and Harrington (1988, 1989). These lists, however, are often founded on determinations made at a time when concepts of some taxa were vague, when the expertise of the student was limited, or when inaccuracies occurred because the student provided a provisional identification which has not been verified. The indices to distribution maps of bryophytes provided by Sjödin (1980*a, b*) are extremely useful.

The database for bryogeography has several serious limitations. Collections in herbaria, however, provide the final vouchers for all published information. Even when errors in determination have been made, it is possible to correct these errors by studying voucher material. Exposing errors in identification is, however, exceedingly time-consuming, and often has to await monographic studies involving careful assessment of the available collections. Such revisions are considered old-fashioned by an influential portion of the research community and are often left to scholars who can pursue productive studies in spite of scorn from researchers, who are naïve concerning the fundamental impetus of scholarship.

The database suffers further from several serious barriers. A considerable body of distributional information has already been destroyed through the activity of humans in altering the environment, a process that has increased with technological advances in machinery. Furthermore, this destruction of the environment and the basic information that it contains continues so efficiently that all future students of natural history will be hampered in studies of distribution, evolution, and applied plant science.

It is somewhat dangerous to make assumptions concerning the evolution of modern distributions based on information extracted from modern patterns. It would be more satisfactory to have a reasonably comprehensive fossil record originating from diverse time periods founded on a wide scattering of localities. The fossil record, however, is disappointingly inadequate, even for relatively recent time. The most promising information has been from amber deposits in which the material, though fragmentary, is often beautifully preserved, and the similarity to modern taxa can be placed reasonably securely (Grolle 1981). The subfossil material available for the past million years is useful for interpretation of recent events, but even it is disappointingly incomplete. This is not surprising in view of the relatively rapid decomposition of most bryophytes.

What restricts ranges of bryophytes? The main factors are availability of water and suitable conditions for growth and reproduction. As is becoming apparent through the limited studies of bryophyte physiology that are available, each bryophyte gametophore is confined within a limited spectrum of conditions that permit it to survive. Bryophytes have evolved several devices to circumvent these constraints, especially in survival of the gametophore during unfavourable periods and through production of vegetative diaspores. While these devices may reduce evolutionary opportunities to some extent, they also provide for the survival of the genetic diversity for more efficient evolutionary selection when sexual reproduction becomes possible. Bryophytes are unique among land plants in the selective pressures acting predominantly on the usually perennial gametophore rather than on the short-lived sporophyte whose survival is controlled by the vitality of the gametophore.

Scott (1988) has presented an intriguing scenario of the effectiveness of expansion of bryophyte ranges, even through vegetative means, given a long period for dispersal. Even expansion of a population by only 1 cm yr^{-1} would permit outward spread from an original point to at least 10 km within a million years. Assuming that some bryophyte taxa have been in existence for many millions of years, as is apparent in well-preserved material of hepatics in amber, it would seem surprising that more taxa are not cosmopolitan.

Bryophytes tend to show wider geographic ranges than other terrestrial plants. This is especially noticeable at the family and generic level, but many species even show extremely wide distributions within each major climatic region. When suitable environments show marked disjunctions, some bryophytes follow these same disjunctions (Schofield and Crum 1972). The wider ranges of bryophytes appear to be the result of two interrelated features: (1) the purported ancient age of the bryophytes as a group of plants and the consequent availability of an extremely long period to achieve a wide range, and (2) the extremely small diaspores that are produced in vast numbers and are readily air borne, thus favouring their wide dispersal.

Unequivocal evidence of an extremely ancient origin of the bryophytes remains very tenuous, although certainly bryophyte-like fossils are present among the earliest land plants of Devonian time. A serious difficulty is the absence of convincing gametangia and attached sporophytes. It is therefore insecure to state that these presumptive gametophores are, in fact, gametophytic. Furthermore, if these fossils prove to be identified convincingly as gametophores, there remains the possibility that they belong to non-bryophytic land plants in which gametophores structurally resembled bryophytes.

As to ready dispersibility of small diaspores, there is little controversy. The uncertainty remains, however, that this dispersibility is an important factor in the wide range of a taxon. There are a number of bryophytes that show a very wide geographic range (e.g. the moss *Rhytidium rugosum* throughout the Holarctic), yet produce small diaspores infrequently. In others innumerable diaspores are produced (e.g. the moss *Dawsonia lativaginata*), yet the distribution is narrowly endemic.

4.2 Cosmopolitan distributions

Many moss families show an essentially cosmopolitan distribution. Absences are related to absence of appropriate habitats for bryophytes.

Most cosmopolitan families show no latitudinal limitation, but many are climatically restricted in some parts of their range. The Sphagnaceae and Andreaeaceae, for example, are widespread in cool temperate regions and are found, at tropical latitudes, to be best represented at sites providing more temperate climatic and habitat conditions. Some families, on the other hand, including Bryaceae, Dicranaceae, Fissidentaceae, Funariaceae, and Hypnaceae, appear to have little climatic or latitudinal limitation. In each of these representative families the greatest taxon richness is at tropical latitudes. Other families, including Aulacomniaceae, Encalyptaceae, Grimmiaceae, and Polytrichaceae are taxon richest in temperate climates.

For the hornworts, if a single family, Anthocerotaceae, is recognized, wide latitudinal range is shown, with greatest taxon richness shown in the tropics, and exclusion from high polar climates.

In the hepatics, similar patterns exist, with Aneuraceae, Dilaenaceae, Jungermanniaceae, Lepidoziaceae, Lophocoleaceae, and Marchantiaceae showing extremely wide ranges, while Marsupellaceae and Scapaniaceae tend to be most richly represented in temperate to frigid environments. Families that show greatest diversity at tropical latitudes include Frullaniaceae, Lophocoleaceae, Plagiochilaceae, and Radulaceae; these have few representatives that extend into frigid climates.

A number of families of bryophytes are essentially pantropical in their distribution. In the Northern Hemisphere these families extend occasionally into temperate climates, but do not reach boreal regions. In the Southern Hemisphere, they extend well into austral climates south of 40° latitude. Among examples of the mosses are Calymperaceae, Pterobryaceae, Racopilaceae, and Rhizogoniaceae, and among the hepatics Adelanthaceae, Isotachidaceae, and Trichocoleaceae. Although the Lejeuneaceae are predominantly tropical, a few representatives extend into cold temperate to polar climates in both the northern and southern hemispheres.

Similar broad patterns are exhibited by many bryophyte genera. Some show no latitudinal restriction, even extending to Antarctica, including the mosses *Amblystegium*, *Andreaea*, *Bartramia*, *Brachythecium*, *Bryum*, *Ceratodon*, *Schistidium*, and *Tortula*, and the hepatics *Cephalozia* and *Lophozia* (Greene 1986). Many bryophyte genera are widespread in the world, but absent from Antarctica. Among the mosses there are more than 60 genera, and among the hepatics at least 30 genera.

Pantropical bryophyte genera are also numerous, some of which have temperate species. Among the mosses are more than 50 genera of which *Breutelia*, *Calymperes*, *Daltonia*, *Ectropothecium*, *Macromitrium*, *Racopilum*, and *Vesicularia* are examples. Representing the hepatics are at least 20 genera of which *Acrobolbus*, *Colura*, *Herbertus*, *Pallavicinia*, and *Symphyogyna* are examples. The hornwort genera *Dendroceros* and *Megaceros* are pantropical, extending in the southern hemisphere to temperate climates at southern latitudes above 40°. Bryophyte genera that are rich in species tend to show very wide distribution within a climatic zone and even through different climates. The hornwort genus *Anthoceros*, for example, is found throughout the world, except in extremely frigid or arid zones. Among the hepatics, the genera *Aneura*, *Bazzania*, *Calypogeia*, *Fossombronia*, *Frullania*, *Lepidozia*, *Marchantia*, *Plagiochila*, *Scapania*, and many others are found in most parts of the world. The same is true for the moss genera *Barbula*, *Bartramia*, *Ditrichum*, *Fissidens*, *Hypnum*, *Sphagnum*, *Thuidium*, *Tortula*, and many others.

Even at the species level a number of taxa are cosmopolitan in so far as latitude is concerned. *Bryum argenteum*, *Ceratodon purpureus*, *Grimmia donniana*, *Polytrichastrum alpinum*, *Polytrichum strictum*, *Racomitrium lanuginosum*, *Schistidium rivulare*, and *Schistidium apocarpum*, among the mosses, *Anthoceros punctatus* among the hornworts, and among the hepatics *Anastrophyllum minutum*, *Anthelia juratzkana*, and *Blepharostoma trichophyllum* are widespread in mountains at most latitudes. *Blasia pusilla*, *Frullania tamarisci* (*sens. lat*), *Marchantia polymorpha*, *Plagiochila porelloides*, and *Radula complanata* are widely distributed in north temperate climates and *Caudalejeunea lehmanniana*, *Cololejeunea minutissima*, *Frullania squarrosa*, and *Leptolejeunea elliptica* are essentially pantropical.

4.3 Bryophyte floristic kingdoms

In the map presented here (Fig. 4.1), the floristic kingdoms are based on floristic similarity, with emphasis on families and genera unique to the kingdoms, and on the degree of endemism, at the species level. Boundaries are difficult to draw precisely, and some kingdoms are far more distinctive than others. Indeed, based on the data available for bryophytes, the Himalayan mountainous area shows greater floristic distinctiveness than the South African kingdom. Since no modern synthesis is available for the Himalayas, however, it seems prudent to maintain this flora within the holarctic, with which it shows greatest affinity.

In the written treatment that follows, general comments are provided for each of the kingdoms. Only the holarctic kingdom is treated in somewhat greater detail.

4.3.2 Holantarctic kingdom

This kingdom, derived from widely separated fragments of Gondwana, shows an unusual intensity of endemism, particularly in genera of hepatics.

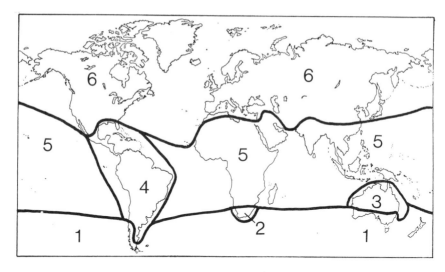

Fig. 4.1 Bryofloristic kingdoms of the world. 1: Holantarctic; 2: South African; 3: Australian; 4: Neotropical; 5: Palaeotropical; 6: Holarctic.

This aspect of its bryoflora has been treated in detail, with numerous supporting distribution maps, by Schuster (1969, 1976, 1979, 1982, 1983a). The information for mosses is more scattered and less comprehensive and the number of endemic genera is relatively low.

In New Zealand, among the hepatics there are at least 10 endemic genera, most of which are monotypic and many of which appear to have narrow ranges. Representative are *Allisonia, Austrolophocolea, Herzogianthus, Jubulopsis, Neogrollea, Steereomitrium, Verdoornia,* and *Xenothallus.* Genera shared only with Tasmania include *Austrometzgeria, Chaetophyllopsis, Eoisotachis, Eotrichocolea, Isolembidium,* and *Trichotemnoma.* Those shared only with New Caledonia include *Chloranthelia* and *Goebeliella.* Further hepatic genera are found in all three areas and extend occasionally as far as Fiji (e.g. *Schusterella*). The genus *Neohattoria* extends northward to Japan.

Endemic genera of mosses are somewhat fewer in New Zealand; like the hepatics, they are mainly monotypic. *Bryodixonia, Crosbya, Cryptopodium, Dichelodontium, Pterobryidium* and *Tetraphidopsis* are representative. When genera shared with only Tasmania and southern Australia are added (*Braithwaitea, Cladomnion, Cyathophorum, Mesotus, Pleurophascum,* and *Sauloma*) this portion of the kingdom is expanded.

Disjuncts with southern South America from New Zealand include the hepatic genera *Archeophylla, Lepidogyna, Phyllothallia,* and *Monoclea,* the latter also recurring in Central America. Moss genera disjunctive with South America involve many genera found otherwise only in southern

Australia, Tasmania, and New Zealand, and include *Catagonium*, *Cyrtopus*, *Goniobryum*, *Lepyrodon*, *Ptychomnion*, and *Weymouthia*.

The cooler humid climatic portions of southern South America and adjacent islands include a rich diversity of endemic hepatic genera, probably exceeding 20, among which the following are representative: *Evansianthus*, *Fulfordia*, *Greeneothallus*, *Grollea*, *Herzogiaria*, *Roivainenia*, and *Vetaforma*. These are mainly monotypic. As noted earlier, several bryophyte genera are disjunct with Australasia. At least 15 mainly monotypic moss genera are confined to southern South America, including *Atrichopsis*, *Catagoniopsis*, *Costesia*, *Cryphidium*, *Duseniella*, *Platyneurum*, *Skottsbergia*, and *Valdiviella*.

Among the oceanic islands and Antarctica a number of holantarctic genera show a relatively continuous range, but most are also widely distributed elsewhere.

As indicated earlier, this kingdom is composed of widely separated fragments of Gondwana. The floristic affinities appear to be greatest between those fragments that have had most recent direct connections or between which modern dispersal events are possible. Floristic richness is greatest in those areas that show greatest habitat and climatic diversity combined with greatest time of isolation from other floras.

4.3.2 South African kingdom

Although relatively small in area and neither topographically nor climatically remarkably diverse, the southern portion of Africa contains a highly distinctive bryoflora, especially in the mosses. The hepatic flora appears to show no generic endemism, but the endemic species, according to Arnell's (1963) flora, number approximately 130, or more than 40 per cent of the flora. The checklist of Magill and Schelpe (1979) and available volumes of the moss flora (Magill 1981, 1987) show the endemic monotypic family Wardiaceae and at least 10 endemic genera, most of which are monotypic. A high proportion of those discovered recently are annuals or short-lived perennials and are exceedingly tiny, a feature that also characterizes many genera in the Australian flora. Representative endemic genera include *Cladophascum*, *Cygnicollum*, *Leucoperichaetium*, *Physcomitrellopsis*, and *Quathlamba*.

Affinities of the South African flora are predominantly with continental Africa, but a number of taxa are disjunct in South America (e.g. *Eustichia*, *Tristichium*), Australia (e.g. *Bryobartramia*, *Carrpos*, *Goniomitrium*), and the Indian subcontinent.

The Gondwanan character of the bryoflora is considerably less than that exhibited by other Southern Hemisphere areas. It is possible that many of these elements (especially those of humid temperate environments) were eliminated from the bryoflora through climatic changes of the past million years or earlier. These changes appear to have favoured the explosion of endemism in the vascular plants. A high proportion of the endemic bryophytes, too, appear to have been selected mainly to survive in a climate in which available moisture is restricted to a very limited period.

4.3.3 Australian kingdom

Recently, two important checklists have been published presenting a modern database for hepatics and hornworts (Scott and Bradshaw 1986) and mosses (Streimann and Curnow 1989). Scott (1988) indicates that the status of understanding of the bryoflora remains so unsatisfactory that any attempt to draw reliable conclusions concerning the structure and affinities must be extremely tentative. In spite of these constraints, it is possible to suggest the basic structure of the bryoflora and note certain tendencies in the affinities.

The Australian bryoflora shows lower endemism than does the vascular flora. Furthermore, a large portion of the southern part of Australia has a bryoflora that is included in the holantarctic kingdom. Sainsbury and Allison (1939) had noted this, and research since that time has expanded the area in Australia that should be included in this floristic kingdom (Stoneburner *et al.* 1990). The Gondwanan nature of the bryoflora of Australia is emphasized with each new contribution to the floristic bryology of the country. I. G. Stone, in particular, continues to document taxa of particular significance. As Scott (1988) has indicated, any figures concerning endemism in the Australian flora are suspect, because many taxa described from Australia have not been assessed critically since they were described. Based on available information, however, a number of endemic mosses are well documented: *Calymperastrum* (1 species), *Mesochaete* (2), *Touwia* (1), *Wildia* (1), and *Viridivellus* (1), the only representative of the Viridivelloraceae. In a number of moss genera that have been thoroughly studied recently, endemism is high, for example *Acaulon* (6 of 9 species), *Archidium* (6 of 9 species), and *Pleuridium* (5 of 8 species).

Among the hepatics, Schuster (1982) has noted no endemic genera or families in that part of Australia outside the holantarctic kingdom. It is important to record, however, that tropical Australia lacks even a modest analysis of the hepatic flora.

Affinities of the Australian bryoflora, excluding the holantarctic kingdom, are mainly with the palaeotropics; e.g. the moss genera *Bescherellia*, *Bryobrothera*, *Bryobartramia*, *Mniomalia*, *Nanobryum*, and *Stoneobryum*.

4.3.4 Neotropical kingdom

This is an area of great size and great topographical, geological, and climatic complexity. It is also an area in great need of detailed bryological exploration, although, as Miller (1982) states, checklists are available for the moss floras of most of the countries. Much of the terrain is difficult and costly to reach and, regrettably, immense areas are being rapidly denuded of all vegetation before any analyses are made.

Endemism appears to be extremely high. In the hepatics, at least 50 genera are endemic, most of which are monotypic (Schuster 1982; Grolle 1969); of these, at least 20 belong to the Lejeuneaceae. Representative genera include *Alobiella*, *Chaetocolea*, *Chonocolea*, *Cronisia*, *Funicularia*, *Gymnocoleopsis*, *Haesselea*, *Lephonardia*, *Phycolepidozia*, and *Vanaea*. Endemic mosses include the families Hydropogonaceae with *Hydropogon* and *Hydropogonella* and Phyllodrepaniaceae with *Phyllodrepanium*, as well as at least 50 genera. Representative endemic genera include *Allioniella*, *Cladostomum*, *Diploneuron*, *Gertrudiella*, *Kingiobryum*, *Mandoniella*, *Spiridentopsis*, *Thamniopsis*, and *Williamsiella*. Most of the endemic moss genera are monotypic and appear to show very restricted ranges.

Affinities of the neotropical flora have been discussed in detail for the hepatics by Grolle (1969) and Gradstein *et al.* (1983), who concentrated on the disjunctions. The latter authors noted that 35 macrodisjunct hepatic species from South America are known also from Africa. These include both lowland (e.g. *Aneura pseudopinguis*, *Cololejuneunea cardicarpa*, *Lophocolea martiana*) and montane taxa (e.g. *Andrewsianthus jamesonii*, *Herbertus subdentatus*, *Marsupella africana*).

The neotropical bryoflora is the richest surviving remnant of the Gondwanan flora, showing its maximum development in humid high elevation cloud forests. With additional research, especially by experienced field bryologists, even the lowland flora is likely to reveal highly intriguing distribution patterns. As Gradstein *et al.* (1983) have shown so well, a number of taxa are predisposed to long-distance transport, i.e. possessing small diaspores and occupying open sites (e.g. the hepatics *Acrolejeunea emergens*, *Isotachis obertii*, etc.), while those confined to one continent (i.e. South America), often lack the ecological and diaspore features that would enhance their wider distribution (e.g. *Dicranolejeunea axillaris*, *Frullania brasiliensis*, etc.). Undoubtedly similar examples could be cited for the mosses.

4.3.5 Palaeotropical kingdom

This represents an immense area. The largest continental portion, Africa, is in need of bryogeographic analysis, although portions have received floristic treatments.

The African bryoflora has been relatively well explored. The long human habitation of the continent has resulted in the destruction of a considerable portion of the natural flora. Added to this, there is a vast area of terrain unsuitable for bryophytes. There appear to be few endemic hepatic genera (e.g. *Paraschistochila*) and no hornwort genera restricted to Africa. As in the neotropics, endemism tends to be higher in montane than lowland areas. Approximately 20 moss genera are confined to Africa. These are mainly monotypic and include *Bryotestua*, *Henicodium*, *Hylocomiopsis*, and *Neorutenbergia* and *Rutenbergia* of the endemic African family Rutenbergiaceae, and *Tisserantiella*. The apparent species endemism, although not completely assessed, is likely to be considerably diminished by synonymization into pantropical species. Affinities of the African flora appear to be approximately equal between the neotropics and the Indomalayan area.

The Indomalayan area and the oceanic islands of the Pacific show considerably greater floristic richness, with at least 60 genera of mosses confined to the area. Pleurocarpous genera, mainly epiphytes and epixylic mosses, predominate. Representative moss genera, mainly monotypic, include *Archboldiella*, *Crepidophyllum*, *Dimorphocladium*, *Ectropotheciopsis*, *Franciella*, *Hageniella*, and *Piloecium*. Predictably, the Malaysian flora shows strongest affinities with the Indian subcontinent and north-eastern tropical Australia, and the oceanic islands show a diminished representation of that flora determined, in large part, on the dispersibility of the taxa.

Although the bryoflora of New Caledonia does not show the extraordinary endemism exhibited by the vascular plants, the hepatic family Perssoniellaceae with *Perssoniella* is endemic, as are the four moss genera *Crytopodendron*, *Franciella*, *Leratia*, and *Parisia*. Species endemism in the moss flora is approximately 50 per cent (Pursell and Reese 1982). Affinities of the flora are modestly with Australia, but greater with the oceanic islands and the Malaysian area. New Caledonia harbours a number of Gondwanan relicts common to New Zealand. It is of interest that the lowland bryoflora of the Hawaiian Islands is essentially of palaeotropical elements or endemics derived from the Indomalayan flora, while most of the montane flora shows affinities with the holarctic with few holantarctic representatives.

The palaeotropic bryoflora, like that of the neotropics, represents mainly fragments and derivatives of the Gondwanan flora, enriched, especially at higher elevations, by fragments of the holarctic and holantarctic flora, most of which probably arrived through long-distance dispersal relatively recently, i.e. within the past two to three million years.

4.3.6 Holarctic kingdom

The holarctic kingdom occupies the largest land areas in the world. These continental land masses are relatively continuous in contrast to the areas occupied by other large floristic kingdoms, except the neotropics. The holarctic presents a vast array of environments, some occupying extensive areas.

Hepatic families confined to the kingdom include the Blasiaceae, Conocephalaceae, Gyrothyraceae, Mesoptychiaceae, and Monoseleniaceae, most of which contain a single genus with one or two species. The hepatic families Antheliaceae, Lophoziaceae, and Scapaniaceae are almost entirely holarctic and are rich in species, and in the Lophoziaceae, in genera. The bryophyte family Takakiaceae is mainly holarctic. Moss families confined to the holarctic include the Andreaeobryaceae, Bryoxiphiaceae, Catoscopi-aceae, Disceliaceae, Pleuroziopsidaceae, and Voitiaceae. The families Climaciaceae, Fontinalaceae, Tetraphidaceae, Theliaceae, and Timmi-aceae are predominantly holarctic.

At least 40 hepatic genera are confined to the area, including many that are monotypic, e.g. *Arnellia*, *Bucegia*, *Conocephalum*, *Douinia*, *Hattoria*, *Ricciocarpos*, and *Schofieldia*. Among the endemic moss genera (at least 80), most of which are monotypic, are the widespread *Abietinella*, *Catoscopium*, *Myurella*, *Paludella*, *Pleurozium*, *Ptilium*, *Rhytidiadelphus*, and *Tetraphis* and several local genera confined to eastern or western North America, to south-east Asia, or to the Himalayas.

In contrast to most other floristic kingdoms, the holarctic has a high proportion of temperate elements with occasional disjunctions from the tropics and some disjunctions to the holantarctic. The latter are at the species level, and probably originated during the Pleistocene glaciations. Widespread holarctic species are generally involved. The relatively uniform floristic character of the northern portions of this kingdom reflects the recent colonization of the area following the retreat of the glaciers and the vastness of the open terrain in the northern regions which enhances both dispersibility and colonization of diaspores. The coniferous forest, though composed of different tree species in the separate continents, offers similar habitats to the many bryophyte species that the areas have in common. Southwards, floristic, spatial, and climatic isolation are greater and ende-mism increases considerably.

The following separation of this kingdom into floristic regions attempts to reflect the distinctiveness of some geographic areas (Fig. 4.2). The Arabian and Saharan regions are excluded because they appear to show greater affinities with the palaeotropics.

Arctic region

Steere (1953, 1977, 1979) considered the Arctic bryoflora briefly. Although an illustrated moss flora exists for the Arctic USSR, no equivalent publication is available for the North American Arctic. This is being

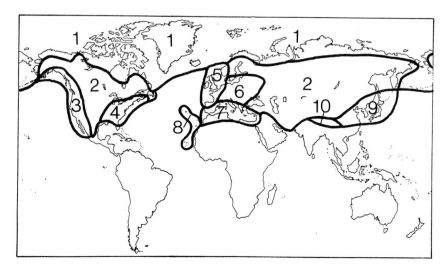

Fig. 4.2 Bryofloristic regions of the Holarctic bryofloristic kingdom. 1: Arctic; 2: Boreal; 3: Pacific North American; 4: Eastern North American; 5: Atlantic European; 6: Central European; 7: Mediterranean; 8: Macaronesian; 9: Southeast Asian; 10: Himalayan.

remedied, however, by the serial publication of a flora edited by Mogensen (1985–87). It is possible to make a number of generalizations based chiefly on the moss flora; these generalizations appear to apply also to the hepatics.

Genera confined mainly to the Arctic region include the hepatics *Arnellia, Cryptocolea,* and *Mesoptychia.* A number of hepatic genera are rich in Arctic species, including *Lophozia* and *Scapania.* Among essentially Arctic moss genera are *Aplodon, Oreas,* and *Philocrya,* while other genera also important in Arctic regions are important in the boreal region as well. These include the mosses *Bryobrittonia, Catoscopium, Encalypta, Paludella, Ptilium, Scorpidium, Tomenthypnum,* and many others. The bryoflora of the Arctic shows endemism of less than 10 per cent, with the residue of the flora predominantly of circumpolar species. In the North American Arctic, a number of species of boreal forests extend into the tundra and the same is true for the Eurasian Arctic. The most dramatic enrichment by temperate species is in southern Iceland, where Johansson (1983) notes *Dicranum tauricum, Plagiomnium undulatum, Pleuridium subulatum, Porella cordaeana,* and *Riccia sorocarpa.* Excluding circumpolar species, bryophytic affinities of the Icelandic flora are closer to Europe.

Affinities of the Arctic region are primarily with the adjacent boreal region and with mountain chains that extend southward from the Arctic. In North America the extension of an Arctic bryoflora to western Newfoundland coincides with the southerly flow of cold ocean currents (Labrador Current) that produce cold summers.

The Arctic bryoflora is relatively recent, composed mainly of fragmentary elements derived from boreal and montane areas and, to a lesser degree, from relictual populations that survived the Pleistocene glaciations in the Arctic.

Boreal region

No detailed bryofloristic analysis appears to have been made for the boreal region, although numerous publications are available, enumerating the taxa that make up the flora of portions of the region. Steere (1965) and Schofield (1972, 1980) presented a discussion of the boreal bryoflora of North America. In the coniferous forest and peatland, independent of the species of seed plants, the bryophyte species found throughout the boreal zone exceed 60 per cent of the flora.

Minor endemics in the boreal kingdom include the hepatic genera *Mylia* and *Ricciocarpos*, and a few moss genera, including *Catoscopium*, *Cephalocladium*, *Pseudoditrichum*, and *Scorpidium*, all of which are monotypic. Only *Cephalocladium* and *Pseudoditrichum* do not extend into arctic regions.

Although not confined to the boreal region, a number of genera are important components of the flora, including the mosses *Drepanocladus*, *Hylocomium*, *Pleurozium*, *Ptilium*, *Rhytidium* and *Warnstorfia*. The hepatic genera *Scapania* and *Lophozia*, as in the Arctic, are diverse in the boreal region. Disjuncts in the boreal region include those from steppe climates of western North America and Eurasia. These include the mosses *Aloina bifrons*, *Desmatodon guepinii*, *Phascum vlassovii*, *Pterygoneurum kozlovii*, and *Tortula intermedia* as well as others. The disjunct range of the genus *Aschisma* with vicariad species in Kansas and the steppes of Eurasia provides another example.

As in the Arctic, temperate climate taxa have often extended their range into the southern portion of the boreal region. In North America, southern Newfoundland harbours taxa of more southern distribution, including the hepatics *Frullania selwyniana*, *Herbertus aduncus*, and *Lejeunea lamacerina*, and the mosses *Atrichum crispum*, *Brotherella recurvans*, *Ctenidium malacodes*, and *Hypnum imponens*. A few extend to southern Greenland. There is a similar incursion of a limited number of species of more general southern distribution into the southern portion of the boreal region of Eurasia. An example is *Buxbaumia aphylla* that follows the Yenisei River to the Arctic. The incursion of central European taxa into the western portion of the boreal region makes the demarcation between the two regions somewhat difficult. From steppe and grassland areas to the semi-desert of both North

America and Eurasia, the bryofloras are dominated by widespread holarctic species, with few taxa confined to the area.

Pacific North American region

This is a very sharply defined bryofloristic region marked in its northern portion by a wet climate, and in its southern part by a summer dry climate. These climatic regions, plus high mountains isolate the area from adjacent ones and present a wide diversity of bryophyte habitats. This region is discussed most recently by Schofield (1984). It shows an endemism of approximately 10 per cent, including the monotypic hepatic family Gyrothyraceae and a number of monotypic genera of mosses: *Alsia*, *Bestia*, *Bryolawtonia*, *Dendroalsia*, *Leucolepis*, *Pseudobraunia*, *Rhytidiopsis*, *Roellia*, *Trachybryum*, and *Tripterocladium* and the monotypic hepatic genera *Geothallus*, *Gyrothyra*, and *Schofieldia*. These genera are sharply isolated taxonomically and imply a long residence in the general region. The endemic species are inter-mixed among predominantly holarctic taxa, but a number are important components of the bryophyte vegetation. Extremely widespread and abundant are *Frullania californica*, *Gyrothyra underwoodiana*, *Porella navicularis*, *Radula bolanderi*, and *Scapania americana* among the hepatics, and *Bryolawtonia vancouveriensis*, *Claopodium bolanderi*, *Coscinodon calyptratus*, *Heterocladium procurrens*, *Hypnum circinale*, *Lescuraea baileyi*, *Leucolepis menziesii*, *Plagiomnium insigne*, *Porotrichum bigelovii*, *Rhytidiopsis robusta*, *Roellia roellii*, and *Sphagnum mendocinum* among the mosses.

The northern portion of the region shows marked disjunctions with species of Europe (Schofield 1988) and south-east Asia (Schofield 1984, 1988; Iwatsuki 1972). The European disjunction is with the Atlantic European region, and is confined mainly to near the coast, especially in hyperoceanic climates. The taxa represented include the hepatics *Anastrepta orcadensis*, *Anastrophyllum donianum*, *Bazzania pearsonii*, *Marsupella adusta*, *Mastigophora woodsii*, and *Scapania ornithopodioides*; and the mosses *Andreaea megistospora*, *Campylopus schwarzii*, *Daltonia splachnoides*, *Dicranum tauricum*, *Hookeria lucens*, *Paraleptodontium recurvifolium*, *Plagiothecium undulatum*, and *Zygodon gracilis*.

The southern portion, of drier climates, shows a less diverse bryoflora. It is also less thoroughly documented, and because of extreme disturbance through human activities, it is likely that many taxa have been extinguished without documentation. Only two monotypic endemic genera have been noted from this area, the hepatic *Geothallus* and the moss *Bestia*. Other monotypic genera are also frequent in this area, extending northward to edaphically suitable coastal sites. The mosses *Alsia*, *Bryolawtonia*, *Dendroalsia*, *Pseudobraunia*, and *Trachybryum* are examples. Most of the genera that show species endemism tend to be tolerant of long periods of desiccation, including the hepatics *Asterella*, *Frullania*, *Mannia*, and *Riccia*, and the mosses *Entosthodon*, *Pottia*, *Triquetrella*, and *Weissia*. This southern climatic region shows a number of taxa disjunctive to the Mediterranean region and has been interpreted as a relictual fragment of a flora that expanded its range along the shores of the Tethys Sea. The paucity of strictly Madrean (Sonoran) elements, so richly represented in the vascular plants, indicates that this area should not be separated into its own bryofloristic region. It is possible, however, that concentrated exploration, especially in the winter and spring, would reveal a greater number of endemics from the area.

The tropical and subtropical affinities of this region are with the Old World, especially south-east Asia. Lack of strong affinity with the neotropics probably results from an arid or semi-arid barrier in western North America that prevented the entrance of the taxa of moister tropical climates. The taxa of subtropical affinities appear to be relicts surviving at or near sites that have been continuously available for many millenia, e.g. the hepatics *Apotreubia nana*, *Chandonanthus hirtellus*, *Cololejeunea macounii*, *Herbertus sakuraii*, and *Plagiochila semidecurrens*; and the mosses *Didymodon nigrescens*, *Gollania turgens*, *Hypopterygium fauriei*, *Limbella tricostata*, and *Sphagnum junghuhnianum*.

Since much of the region was glaciated, it is apparent that refugia south of the glacial boundary, as well as small unglaciated coastal pockets near the north of this boundary were important in preserving this bryoflora during the Pleistocene. The mild humid climate, especially favourable to bryophytes, has enhanced their survival.

Eastern North American region

This region has the longest history of bryological exploration in North America. The most recent analyses of the bryoflora are by Anderson and Zander (1973), Crum (1972), Pursell and Reese (1970), and Schofield (1969, 1980) on the mosses. The most useful assessment of the hepatics is that of R. M. Schuster, some of the information of which is presented in Schuster (1983a), but most of which is scattered in his invaluable *The Hepaticae and Anthocerotae of North America*.

Generic endemism in the bryophytes of the region is restricted to the monotypic moss genera *Brachelyma*, *Bryoandersonia*, *Bryocrumia*, *Donrichardsia*, *Neomacounia*, and *Platylomella*. The moss flora has approximately 70 per cent of the taxa that are either cosmopolitan or predominantly circumpolar holarctic, while about 15 per cent of the species are endemic. The remaining species are either disjunctive in the Northern Hemisphere, or are neotropical in main range. It is probable that the hepatic flora shows a similar structure.

Within the region, taxa of subtropical range or affinity extend northward in gorges of the southern Appalachians as initially noted by Sharp (1938, 1939, 1941) and discussed in detail more recently by Anderson and Zander (1973), Crum (1972), Billings and Anderson (1966). Disjunctions with east Asia are detailed by Iwatsuki (1958, 1972). These northern outliers are interpreted as relicts that have persisted over many millenia.

Among the endemic moss genera *Bryocrumia*, *Donrichardsia*, and *Neomacounia* are known only from the type localities, and only *Donrichardsia* has been collected recently. *Brachelyma*, *Bryoandersonia*, and *Platylomella* are relatively widely distributed. Several genera of mosses show high endemism: *Archidium (3 of 6 species)*, *Bruchia* (6 of 8 species), and *Fontinalis* (8 of 14 species). Among the hepatics only *Plagiochila* has high endemism in the region (7 of 15 species). If tropical Florida were added to the region, endemism would be greatly increased, but much of Florida, as well as a portion of the Gulf Coastal Plain is essentially neotropical in its bryoflora. The hepatic flora especially reflects this affinity.

It is apparent that the bryoflora of this region is greatly influenced by the influx of taxa of circumpolar holarctic distribution, some of which probably entered the region during the Pleistocene glaciations. These taxa are especially abundant and richly represented in the coniferous forest and on outcrops at high elevations. Remnants of the preglacial flora have persisted in the Southern Appalachians or have invaded Florida and the gulf Coastal Plain from populations adjacent to these areas.

A component of the flora of the Southern Appalachians is related to the flora of the neotropics and represents persistent remnants of that flora in protected sites, especially in escarpment gorges. Representative neotropical mosses include *Fissidens polypodioides*, *Macrocoma sullivantii*, *Philonotis longiseta*, *Schlotheimia regifolia*, *Tortula propagulosa*, and *T. caroliniana*, and the hepatics *Calypogeia peruviana*, *Lejeunea laetevirens*, and *Radula mollis*.

Two dramatic disjunctions are apparent. That with western Europe is represented by the mosses *Andreaea crassinervia*, *Bartramidula cernua*, *Bryhnia novae-angliae*, and *Hygrohypnum montanum* and the hepatics *Cephaloziella massalongoi*, *Plagiochila exigua*, and *Radula voluta*. The second disjunction is that of surviving remnants of a bryoflora, probably dating to Tertiary time, that are found in especially rich concentration in the Southern Appalachian Mountains. Among these remnants are taxa identical or closely related to species confined otherwise to south-eastern Asia. These are interpreted as inhabitants originally of a temporate or tropical climate that reach their greatest diversity, in North America, in humid gorges. A few species extend northward to south-eastern Canada, but few extend southward to subtropical Florida and the Gulf Coastal Plain. They appear to represent an older floristic element in North America than the neotropical species. Anderson (1971), Iwatsuki (1972), and Anderson and Zander (1972) provide the most comprehensive recent treatments of this flora. The mosses show the greatest habitat range including taxa on decaying wood (e.g. *Brothera leana, Buxbaumia minakatae*), epiphytes (e.g. *Drummondia prorepens, Orthotrichum sordidum, Schwetschkeopsis fabronia, Solmsiella biseriata*), and species on mineral soil (e.g. *Fissidens garberi, F. hyalinus*). At least 30 species of mosses show this disjunction, while vacariad species or differing subspecies or varieties in the two disjunctive portions of the range add another 11 examples. The hepatics are fewer, are confined mainly to shaded cliff niches, and generally represent vicariad taxa, especially different subspecies, in the disjunctive areas. *Plagiochila*, for example, is represented by four species, each with a vicariad subspecies. *Porella japonica* is also represented by different subspecies, while *Cephaloziella spinicaulis* appears to be identical in the two areas. Billings and Anderson (1966) have provided a well-documented study demonstrating the importance of humid microclimatic conditions in permitting many of these species to survive so many millenia. Unfortunately flooding and hydroelectric power development have destroyed some of these significant populations.

Atlantic European region

The Atlantic region of Europe has had a long history of bryological research. Detailed treatments of the structure, affinities, and genesis of the flora (Greig-Smith 1950; Ratcliffe 1968; Størmer (1969) have noted that the region is circumscribed mainly by the strongly oceanic equable climate and that a surprising number of the taxa in the flora appear to be subtropical in affinity. Where this flora survived during glaciation remains conjectural, but the existence of portions of this flora in the Himalayas, Japan, and north-western North America strongly implies that they are fragments of a very ancient flora. Regrettably, solid evidence for this in the form of subfossil material is entirely lacking. As Ratcliffe (1968) has shown, based on extensive field experience and careful documentation, this floristic element is most strongly represented in the British Isles, reaching its maximum development in Ireland and Scotland. Størmer (1969) has provided an analysis for Norway, another country that habours many of these species.

Some 'Atlantic' bryophytes are found, usually rarely, disjunctively in eastern North America. Representative species are the mosses *Leptodontium flexifolium, Zygodon conoideus* and *Sphagnum strictum*, and the hepatics *Porella pinnata* and *Radula voluta*. Outside Europe, however, they usually show greatest representation in coastal northwestern North America. Representative mosses include *Campylopus schwarzii, Cynodontium jenneri, Daltonia splachnoides*, *Hookeria lucens*, and *Paraleptodontium recurvifolium*; and numerous hepatics, including *Anastrophyllum donianum*, *Bazzania pearsonii*, *Mastigophora woodsii*, *Pleurozia purpurea*, and *Scapania ornithopodioides*. Nicholson (1930) noted a number of these hepatics as rare components in the Himalayas; some are exceedingly rare in Japan.

Among the species confined to Atlantic Europe are the mosses *Bryum riparium*, *Campylopus shawii*, *Fissidens celticus*, *Grimmia retracta*, and *Racomitrium ellipticum*; and the hepatics *Gymnomitrion crenulatum* and *Lepidozia pearsonii*.

Disjunctive to the Macaronesian Region, and especially well represented in the Azores, are the mosses *Cyclodictyon laete-virens* and *Ulota vittata*, and the hepatics *Acrobolbus wilsonii*, *Lejeunea diversiloba*, and *Marchesina mackaii*.

Central European region

This region includes most of Europe, excluding the Atlantic and Mediterranean regions. Endemism is relatively low with the moss genera *Leptobarbula* and *Trochobryum*. The moss genus *Cinclidotus* occurs mainly in central Europe. *Cryptothallus* is the only hepatic (also disjunct in Greenland). This region is extracted from the boreal region mainly on the basis of its greater floristic richness, a reflection of the more favourable climate (Størmer 1983). The floristically most distinctive portion is in the mountains, especially the Alps and Carpathians. Mediterranean and Atlantic species occur occasionally in these mountain regions. The hepatic genus *Bucegia* of Romania has disjunct populations in western North America, and the steppe moss genus *Aschisma* was noted earlier.

Mediterranean region

This region is circumscribed mainly by summer dry–winter wet climates and proximity to the Mediterranean Sea. Much of the terrain has been altered by many centuries of human activity. Fortunately many of the bryophytes survive in sites that are avoided by human exploitation, so pockets of the flora persist. Some Mediterranean taxa extend also to Atlantic Europe. These are treated as 'Mediterranean–Atlantic' by Ratcliffe (1968). Some are found also in other regions with summer dry–winter wet climates, including western North America and south-eastern Australia. These are considered as relicts of the circum-Tethyan subkingdom by Frey and Kürschner (1983). Størmer (1983) suggests that 'some of these species are presumably Tertiary relicts, which have survived the last glaciations here'. Characteristic mosses are *Anacolia webbii*, *Brachymenium commutatum*, *Cheilothela chloropus*, *Gigaspermum muretii*, *Homalia lusitanica*, *Homolothecium aureum*, *Pyramidula alqeriensis*, and *Tortula handelii*; hepatics include *Cephal-*

ozia calycula, Exormotheca bullosa, Fossombronia echinata, and numerous species of *Riccia*.

Macaronesian region

Sergio (1984) provided a recent geographical assessment of the bryoflora of the Macaronesian islands. She concluded that the circumpolar boreal species were most richly represented at high elevations in the Azores but gave no estimates of the proportions of major elements. She states that endemics constitute 9 per cent of the flora, including the monotypic moss genera *Alophosia, Andoa,* and *Tetrastichium*. These are presumed to be palaeoendemics and fragments of a Gondwanan flora for which there is also greatest affinity in the endemic species. Based on the lists provided by Eggers (1982), it is possible to estimate that approximately 45 per cent of the mosses and 50 per cent of the hepatics exhibit a circumpolar holarctic distribution. Utilizing the assessment of Duell (1984) and that of Sergio (1984), it appears that greatest affinity of the residue of the Macaronesian bryoflora is with Europe, of which Mediterranean and Atlantic species are most richly represented in spite of the affinity of the endemic species mainly with taxa outside Europe.

Among the tropical–subtropical taxa, Sergio (1984) indicates that most are part of the African flora, but several are common to the American tropics and most are found also in either Mediterranean or Atlantic Europe. Much of the native vegetation of these islands has been either seriously disturbed or extinguished, and thus the bryoflora persists in pockets; taxa that were restricted in range may have vanished before any documentation.

South-east Asian region

This region shows the greatest generic endemism in mosses in the holarctic kingdom, with at least 24 genera, most of which are monotypic. The endemic hepatic genera are fewer, with, for example, *Cryptocoleopsis, Hattoria, Makinoa, Neotrichocolea, Nipponolejeunea, Trichocoleopsis,* and *Tuzibeanthus*. Among the endemic moss genera, Deguchi and Iwatsuki (1984) cite for Japan *Bissetia, Cratoneurella, Dolichomitra, Dolichomitriopsis, Eumyurium, Orthoamblystegium,* and *Palisadula*, all of which are pleurocarpous and most of which are monotypic and found mainly in broad-leaf deciduous forests.

Disjunctions from this flora to eastern or western North America have been noted earlier. Affinities with the palaeotropical kingdom are especially pronounced in the subtropical and warm temperate portions of the area. The mountain bryoflora contains elements common to the Himalayas and also multicentric patterns for high precipitation areas with temperate climates, e.g. western Europe, western North America, the Alps, and the Himalayas.

This region shows, in the diversity of its affinities and in the richness of its distinctive endemics, that it is a relict Laurasian flora, possibly the best example in existence. The long history of diverse habitats continuously available over many millenia probably explains why this flora has persisted.

Himalayan region

This region shows such a rich representation of endemic bryophyte genera that a case could be made to separate it into an independent floristic kingdom. Further research must precede such a decision. It is possible that this flora is a remnant of the Gondwanan flora that moved northward with the Indian tectonic plate and has persisted since that time in montane and adjacent regions. The lowland flora, on the other hand, is predominantly of palaeotropical taxa. Northward, the holarctic elements have combined with the Gondwanan elements.

Among the eight endemic hepatic genera are *Aitchisoniella*, *Delavayella*, *Sauchia*, and *Stephensoniella*, all of which are monotypic. At least seventeen endemic moss genera are also mainly monotypic and include *Bryowijkia*, *Lyellia*, *Orontobryum*, *Osterwaldiella*, and *Pleuridiella*. Additional genera, mainly of montane fog forests, extend beyond the Himalayas to the Malaysian archipelago and occasionally to oceanic islands or as far north as Japan. The moss genera include *Desmotheca*, *Endotrichella*, *Exodictyon*, and at least 10 others. These are absent from Africa.

In the Himalayas are a number of oceanic taxa that reoccur in such distant areas as Atlantic Europe, Pacific North America, and sometimes northern Japan. At least twelve species are found in Pacific North America, among them *Anastrophyllum donianum*, *Chandonanthus hirtellus*, *Dendrobazzania griffithiana*, *Herbertus sendtneri*, and *Radula auriculata*. Among those of Atlantic Europe are *Anastrophyllum joergensenii* and *Scapania nimbosa*, as well as other species found also in Pacific North America. These taxa suggest persistent remnants of an ancient flora of oceanic climates. Most of these taxa either have no diaspores suitable for long-distance dispersal or produce sporophytes infrequently.

4.4 Database

The foregoing discussion rests upon information extracted from the papers cited in the text, from those cited below, and from a database of the mosses assembled from *Index Muscorum* in collaboration with R. J. Belland. The database is somewhat flawed, because it has not been possible to ferret out all instances of synonymy since the cited publications appeared. Many floristic lists and manuals have been consulted, but are not cited. Unfortunately, many manuals have presented vague indications of distribution of the taxa treated. There is a great need of a modern generic flora of all bryophytes in which world distributions are outlined.

The following publications were sources of essential information for each of the floristic kingdoms and regions treated.

Holantarctic kingdom: Catcheside (1980), Fife (1986), Fulford (1951), Greene (1986), Martin (1958), Matteri (1986), Ramsay (1984), Ramsay *et al.* (1986), Sainsbury and Allison (1939), Schofield (1974), Schofield and Crum (1972), Schuster (1969, 1979, 1982, 1983*a*, *b*), Scott (1988), Scott and Bradshaw (1986), Stoneburner *et al.* (1990), Streimann and Curnow (1989), Vitt (1979).

South African kingdom: Arnell (1963), Magill (1981, 1987), Magill and Schelpe (1979), Schofield and Crum (1972).

Australian kingdom: Catcheside (1980), Schuster (1969, 1976, 1982, 1983*a*), Scott (1988), Scott and Bradshaw (1986), Stoneburner *et al.* (1990), Streimann and Curnow (1989).

Neotropical kingdom: Crosby (1969), Delgadillo (1971, 1984, 1987*a*, *b*, 1989), Fulford (1951), Gradstein and Pócs (1989), Gradstein *et al*, (1983), Grolle (1969), Miller (1982), Pursell and Reese (1970), Robinson (1986), Schuster (1969, 1982, 1983*a*, *b*).

Palaeotropical kingdom: Bischler and Jovet-Ast (1986), Bizot and Pócs (1974, 1976, 1982), Bizot *et al.* (1979), Frahm (1978), Frey and Kürschner (1983, 1988), Gradstein and Pócs (1989), Gradstein *et al.* (1983), Grolle (1969, 1978), Hyvonen (1989), Jovet-Ast (1948), Kis (1985), Long (1987), Miller (1982), Pócs (1975, 1976), Pursell and Reese (1982), Schultze-Motel (1975*a*, *b*).

Holarctic kingdom

Arctic region: Gams (1955), Johansson (1983), Mogensen (1985–1987), Schofield (1972, 1980), Steere (1953, 1977, 1979).

Boreal region: Abramova and Abramov (1969), Bishler and Jovet-Ast (1986), Frey and Kürschner (1983, 1988), Schofield (1969, 1972, 1980), Schuster (1983*a*), Steere (1965, 1979), Vana and Soldau (1985).

Pacific North American region: Hong (1987), Iwatsuki (1972), Schofield (1965, 1969, 1972, 1980, 1984, 1988, 1989), Schuster (1983*a*), Steere (1969).

Eastern North American region: Anderson (1971), Anderson and Zander (1973), Billings and Anderson (1986), Crum (1972), Delgadillo (1979), Iwatsuki (1958, 1972), Pursell and Reese (1970), Redfearn (1986), Schofield (1980, 1988), Schuster (1983*a*), Sharp (1938, 1939, 1941, 1972).

Atlantic European region: Duell (1984), Greig-Smith (1969), Nicholson (1930), Petit and Szmajda (1981), Ratcliffe (1968), Schofield (1969, 1980, 1988, 1989), Schofield and Crum (1972), Schuster (1983*a*), Størmer (1969, 1983).

Central European region: Duell (1983, 1984), Størmer (1983).

Mediterranean region: Bischler and Jovet-Ast (1986), Duell (1984), Jovet-Ast *et al.* (1976), Martincic (1965), Ratcliffe (1968), Schofield (1988), Størmer (1969).

Macaronesian region: Duell (1983), Eggers (1982), Sergio (1984).

South-east Asian region: Abramova and Abramov (1969), Deguchi and Iwatsuki (1984), Horikawa (1955), Iwatsuki (1958, 1972), Pócs (1976), Schofield (1965, 1984), Schofield and Crum (1972), Schultze-Motel (1975*b*), Sharp (1972), Steere (1969).

Himalayan region: Schofield (1985), Sharp (1974).

Acknowledgements

I am indebted to M. I. Schofield for deciphering a handwritten manuscript and providing a clean typed copy. René J. Belland has generously retrieved information from a database for moss genera and has offered useful suggestions concerning the manuscript. Research has been supported by the Natural Sciences and Engineering Research Council of Canada. To these I offer grateful thanks.

References

Abramova, A. L. and Abramov, L. I. (1969). Eastern-Asiatic affinities of the Caucasian bryoflora. *Journal of the Hattori Botanical Laboratory* **32**, 151–4.

Anderson, L. E. (1971). Geographical relationships of the mosses of the Southern Appalachian Mountains. *Research Monograph* 2, pp. 101–15. *Virginia Polytechnic Institute*, Blacksburg.

Anderson, L. E. and Zander R. H. (1973). The mosses of the southern Blue Ridge Province and their phytogeographic relationship. *Journal of the Elisha Mitchell Scientific Society*, **89**, 15–60.

Arnell, S. (1963). *Hepaticae of South Africa*. Swedish Natural Science Research Council, Stockholm.

Billings, W. D. and Anderson, L. E. (1966). Some microclimatic characteristics of habitats of endemic and disjunct bryophytes in the southern Blue Ridge. *The Bryologist* **69**, 79–95.

Bischler, H. and Jovet-Ast, S. (1986). The hepatic flora of south-west Asia: a survey. *Proceedings of the Royal Society, Edinburgh*, **89b**, 229–41.

Bizot, M. and Pócs T. (1974). East African bryophytes I. *Acta Academiae Paedagogicae Argriensis nova series* **12**, 383–449.

Bizot, M. and Pócs T. (1976). East African bryophytes II. *Acta Botanica Academiae Scientiae Hungaricae*, **22**, 1–8.

Bizot, M. and Pócs T. (1982). East African bryophytes V. *Acta Botanica Academiae Scientiae Hungaricae*, **28**, 15–24.

Bizot, M., Pócs T. and Sharp, A. J. (1979). Results of a bryogeographical expedition to East Africa in 1968 II. *Journal of the Hattori Botanical Laboratory*, **45**, 145–65.

Catcheside, D. G. (1980). *Mosses of South Australia*. Committee of South Australian Government, South Australia.

Crosby, M. R. (1969). Distribution patterns of West Indian mosses. *Annals of the Missouri Botanical Garden*, **56**, 409–16.

Crum, H. A. (1972). The geographic origins of the mosses of North America's eastern deciduous forest. *Journal of the Hattori Botanical Laboratory*, **35**, 269–98.

Deguchi, H. and Iwatsuki Z. (1984). Bryogeographical relationships in the moss flora of Japan. *Journal of the Hattori Botanical Laboratory*, **55**, 1–11.

Delgadillo, M. C. (1971). Phytogeographic studies on alpine mosses of Mexico. *The Bryologist*, **74**, 331–46.

Delgadillo, M. C. (1979). Mosses and phytogeography of the Liquidambar forest of Mexico. *The Bryologist*, **82**, 432–49.

Delgadillo, M. C. (1984). Mosses of the Yucatan Peninsula, Mexico. III: Phytogeography. *The Bryologist*, **87**, 12–16.

Delgadillo, M. C. (1987*a*). Moss distribution and the phytogeographical significance of the volcanic belt of Mexico. *Journal of Biogeography*, **14**, 69–78.

Delgadillo, M. C. (1987*b*). The Meso-American element in the moss flora of Mexico. *Lindbergia*, **12**, 121–4.

Delgadillo, M. C. (1989). Phytogeography of high elevation mosses from Chiapas, Mexico. *The Bryologist*, **92** 461–6.

Duell, R. (1983). Distribution of the European and Macaronesian liverworts. (Hepaticophytina). *Bryologische Beiträge*, **2**, 1–115.

Duell, R. (1984). Computerized evaluation of the distribution of European liverworts. *Journal of the Hattori Botanical Laboratory*, **56**, 1–5.

Eggers, J. (1982). Artenliste der Moose Makaronesiens. *Cryptogamie: Bryologie et Lichénologie*, **3**, 283–335.

Fife, A. J. (1986). The phytogeographic affinities of the alpine mosses of New Zealand. In *Flora and fauna of Alpine Australasia*, (ed. B. A. Barlow), pp. 301–35. CSIRO, Melbourne.

Frahm, J-P. (1978). Zur Moosflora der Sahara. *Nova Hedwigia*, **30**, 527–48.

Frey, W. and Kürschner, H. (1983). New records of bryophytes from Transjordan with remarks on phytogeography and endemism in SW Asiatic mosses. *Lindbergia*, **9**, 121–32.

Frey, W. and Kürschner, H. (1988). Bryophytes of the Arabian Peninsula and Socotra. *Nova Hewigia*, **46**, 37–120.

Fulford, M. (1951). Distribution patterns of the genera of leafy Hepaticae of South America. *Evolution*, **5**, 243–64.

Gams, H. (1955). Zur Arealgeschichte der Arktischen und arktisch-oreophytischen Moose. *Feddes Repertorium*, **58**, 80–92.

Gradstein, S. R. and Pócs T. (1989). Bryophytes. In *Tropical rain forest ecosystems*, (eds H. Leith and M. J. A. Werger), pp. 311–25. Elsevier, Amsterdam.

Gradstein, S. R., Pócs T. and Vana, J. (1983). Disjunct Hepaticae in tropical America and Africa. *Acta Botanica Hungarica*, **29**, 127–71.

Greene, D. M. (1986). *A conspectus of the mosses of Antarctica, South Georgia, the Falkland Islands and Southern South America*. British Antarctic Survey, Cambridge.

Greene, S. W. and Harrington, A. J. (1988). The conspectus of bryological taxonomic literature Part 1. *Bryophytorum Bibliotheca*, **35**, 1–272.

Greene, S. W. and Harrington, A. J. (1989). The conspectus of bryological taxonomic literature Part 2. *Bryophytorum Bibliotheca*, **37**, 1–321.

Greig-Smith, P. (1950). Evidence from hepatics on the history of the British flora. *Journal of Ecology*, **38**, 320–44.

Grolle, R. (1969). Grossdisjunktionen in Artenarealen Lateinamerikanischer Lebermoose. In *Biogeography and ecology in South America*, (eds E. J. Fittau, J. Illies, H. Klinge, G. H. Schwabe, and H. Sioli), pp. 562–82. Junk, The Hague.

Grolle, R. (1978). Die Lebermoose der Seychellen. *Wissentschaftliche Zeitsshrift Friedrich-Schiller Universität Jena, Mathematisch und Naturwissenschaftliche Reihe*, **27**, 1–17.

Grolle, R. (1981). On hepatics in Baltic amber. Present knowledge and promises. In *New Perspectives in bryotaxonomy and bryogeography*, (ed. J. Szweykowski), pp. 83–8. Adam Mieckiewicz University, Poznan.

Herzog T. (1926). *Geographie der Moose*. Fischer, Jena.

Hong, W. S. (1987). The distribution of western North American Hepaticae and taxa with a North Pacific arc distribution. *The Bryologist*, **90**, 344–61.

Horikawa, Y. (1955). Distributional studies in bryophytes in Japan and the adjacent regions. *Contributions to Phytotaxonomic and Geobotanical Laboratory, Hiroshima University n.s.* **27**, 1–152.

Hyvonen, J. (1989). On the bryogeography of western Melanesia. *Journal of the Hattori Botanical Laboratory*, **66**, 231–54.

Iwatsuki, Z. (1958). Correlations between the moss floras of Japan and the Southern Appalachians. *Journal of the Hattori Botanical Laboratory*, **20**, 304–52.

Iwatsuki, Z. (1972). Distribution of bryophytes common to Japan and the United States. In *Floristics and Palaeofloristics of Asia and eastern North America*, (ed. A. Graham), pp. 107–37. Elsevier, Amsterdam.

Johansson, B. (1983). A list of Icelandic bryophyte species. *Acta Naturelia Islandica*, **30**, 1–29.

Jovet-Ast, S. (1948). Les mousses et les sphaignes de Madagascar. *Memoires Institut Scientiques Madagascar Series B*, **9**, 43–56.

Jovet-Ast, S., Bischler, H. and Baudoin, P. (1976). Essai sur le peuplement hepatologique de la région Méditerranéenne. *Journal of the Hattori Botanical Laboratory*, **41**, 87–94.

Kis, G. (1985). *Mosses of South-East Africa. An annotated list with distributional data.* Vacratot, Hungary.

Long, D. G. (1987). Hepaticae and Anthocerotae of the Arabian Peninsula. *Nova Hedwigia*, **45**, 175–95.

Magill, R. E. (1981). *Bryophyta of South Africa Part 1. Mosses Fascicle*, **1**, 1–291. Botanical Research Institute, Department of Agriculture and Fisheries, Pretoria.

Magill, R. E. (1987). *Bryophyta of Southern Africa Part 2. Mosses Fascicle*, **2**, 293–443. Botanical Research Institute, Department of Agriculture and Fisheries, Pretoria.

Magill, R. E. and Schelpe, E. A. (1979). The bryophytes of southern Africa. An annotated checklist. *Memoirs of the Botanical Survey of South Africa*, **43**, 1–39.

Martin, W. (1958). Survey of moss distributions in New Zealand. *The Bryologist*, **61**, 105–15.

Martincic, A. (1965). Zum Arealkennzeichnung einiger 'Mediterranen' Moossippen. *Biologia Vestnik*, **13**, 41–52.

Matteri, C. (1986). Overview of the phytogeography of the moss flora from Southern Patagonia at 51° – 52° South latitude. *Journal of the Hattori Botanical Laboratory*, **60**, 171–4.

Miller, H. A. (1982). Bryophyte evolution and geography. *Biological Journal of the Linnaean Society*, **18**, 145–96.

Mogensen, G. S. (ed.) (1985–7). Illustrated Moss flora of Arctic North America and Greenland. *Meddelser om Grønland. Bioscience*, **17**, 1–57; **18**, 1–61; **23**, 1–36.

Nicholson, W. E. (1930). 'Atlantic' hepatics in Yunnan. *Annales Bryologici*, **3**, 151–3.

Petit, W. and Szmajda, P. (1981). Remarques sur la distribution des mousses 'Eu Atlantiques'. In *New perspectives in bryotaxonomy and bryogeography*, (ed. J. Szweykowski), pp. 89–104. Adam Mieckiewicza University, Poznan.

Pócs,T. (1975). Affinities between the bryoflora of East Africa and Madagascar. *Boissiera*, **24**, 125–8.

Pócs, T. (1976). Correlations between the tropical African and Asian Bryofloras I. *Journal of the Hattori Botanical Laboratory*, **41**, 95–106.

Pursell, R. A. and Reese, W. D. (1970). Phytogeographic affinities of the mosses of the Gulf Coastal Plain of the United States and Mexico. *Journal of the Hattori Botanical Laboratory*, **33**, 115–52.

Pursell, R. A. and Reese, W. D. (1982). The mosses reported from New Caledonia.

Journal of the Hattori Botanical Laboratory, **53**, 449–82.

Ramsay, H. P. (1984). Phytogeography of the mosses of New South Wales. *Telopea*, **2**, 535–48.

Ramsay, H. P., Streimann, H., Seppelt, R. and Fife, A. (1986). In *Flora and fauna of alpine Australia*, (ed. B. A. Barlow), pp. 301–35. CSIRO, Melbourne.

Ratcliffe, D. A. (1968). An ecological account of Atlantic bryophytes in the British Isles. *New Phytologist*, **67**, 365–439.

Redfearn, P. L. (1986). Bryogeography of the interior highlands of North America: taxa of critical importance. *The Bryologist*, **89**, 32–4.

Robinson, H. (1986). Notes on the bryogeography of Venezuela. *The Bryologist*, **89**, 8–12.

Sainsbury, G. O. K. and Allison, K. W. (1939). The relationship between Tasmanian and New Zealand mosses. *Proceedings of the 6th Pacific Science Congress*, **4**, 621–3.

Schofield, W. B. (1965). Correlations between the moss floras of Japan and British Columbia, Canada. *Journal of the Hattori Botancial Laboratory*, **28**, 17–42.

Schofield, W. B. (1969). Phytogeography of northwestern North America: Bryophytes and vascular plants. *Madroño*, **20**, 135–207.

Schofield, W. B. (1972). Bryology in arctic and boreal North America and Greenland. *Canadian Journal of Botany*, **50**, 1111–33.

Schofield, W. B. (1974). Bipolar disjunctive mosses in the Southern Hemisphere, with particular reference to New Zealand. *Journal of the Hattori Botanical Laboratory*, **38**, 13–32.

Schofield, W. B. (1980). Phytogeography of the mosses of North America (North of Mexico). In *The mosses of North America*, (eds R. A. Taylor and A. E. Leviton), pp. 131–70. Pacific Division, AAAS, San Francisco.

Schofield, W. B. (1984). Bryogeography of the Pacific coast of North America. *Journal of the Hattori Botanical Laboratory*, **55**, 35–43.

Schofield, W. B. (1985). *Introduction to bryology*. Macmillan, New York.

Schofield, W. B. (1988). Bryophyte disjunctions in the Northern Hemisphere: Europe and North America. *Botanical Journal of the Linnaean Society*, **98** 211–24.

Schofield, W. B. (1989). Structure and affinities of the bryoflora of the Queen Charlotte Islands. In *The outer shores*, (eds G. G. E. Scudder and N. Gessler), pp. 107–19. Queen Charlotte Islands Museum, Skidegate, British Columbia.

Schofield, W. B. and Crum, H. A. (1972). Disjunctions in bryophytes. *Annals of the Missouri Botanical Garden*, **59**, 174–202.

Schultze-Motel, W. (1975*a*). Die bryogeographische Stellung der Samoa-Inseln. *Bulletin du Société Botanique de France. Colloque Bryologie 1974*, **121**, 295–8.

Schultze-Motel, W. (1975*b*). Vergleichende Betrachtungen über die Artenzahlen von Laubmoosefloren in Südöst-Asien und Oceanien. *Lindbergia*, **3**, 57–9.

Schuster, R. M. (1969). Problems of Antipodal distribution in lower land plants. *Taxon*, **18**, 46–91.

Schuster, R. M. (1976). Plate tectonics and its bearing on the geographical origin and dispersal of angiosperms. In *Origin and evolution of angiosperms*, (ed. C. B. Beck), pp. 48–138. Columbia University Press, New York.

Schuster, R. M. (1979). On the persistence and dispersal of transantarctic Hepaticae. *Canadian Journal of Botany*, **57**, 2179–225.

Schuster, R. M. (1980). Paleoecology, distribution and evolution of the Hepaticae. In *Symposium on paleoecology*, (ed. K. J. Niklas), Praeger, New York.

Schuster, R. M. (1982). Generic and familial endemism in the hepatic flora of Gondwanaland: origin and causes. *Journal of the Hattori Botanical Laboratory*, **52**, 3–35.

Schuster, R. M. (1983*a*). Phytogeography of Bryophyta. In *New manual of bryology*, Vol. I, (ed. R. M. Schuster), pp. 463–626. Hattori Botanical Laboratory, Nichinan.

Schuster, R. M. (1983*b*). Reproductive biology, dispersal mechanisms and distribution patterns in Hepaticae and Anthocerotae. In *Dispersal and distribution*, (ed. V. Kubitzki), pp. 119–162. Verlag, Hamburg.

Scott, G. A. M. (1988). Australasian bryogeography: fact, fallacy and fantasy. *Biological Journal of the Linnaean Society*, **98**, 203–10.

Scott, G. A. M. and Bradshaw, J. S. (1986). Australian liverworts (Hepaticae): Annotated list of binomials and check-list of published species with bibliography. *Brunonia*, **8**, 1–171.

Sergio, C. (1984). The distribution and origin of Macaronesian bryophyte flora. *Journal of the Hattori Botanical Laboratory*, **56**, 7–13.

Sharp, A. J. (1938). Tropical bryophytes in the southern Appalachians. *Annales Bryologici*, **11**, 141–4.

Sharp, A. J. (1939). Taxonomic and ecological studies of eastern Tennessee bryophytes. *American Midland Naturalist*, **21**, 267–354.

Sharp, A. J. (1941). Some historical factors and the distribution of southern Appalachian bryophytes. *The Bryologist*, **44**, 16–18.

Sharp, A. J. (1972). Phytogeographical correlations between the bryophytes of Eastern Asia and North America. *Journal of the Hattori Botanical Laboratory*, **35**, 263–8.

Sharp, A. J. (1974). Some geographic relations in the Himalayan bryoflora. *Journal of the Hattori Botanical Laboratory*, **38**, 33–7.

Sjödin, A. (1980*a*). Index to distribution maps of bryophytes 1887–1975 I Musci. *Växtekologiska Studier*, **11**, 1–282. Svensk Växtgeografiska Sallskapet, Uppsala.

Sjödin, A. (1980*b*). Index to distribution maps of bryophytes 1887–1975 II Hepaticae. *Växtekologiska Studier* **12**, 1–143. Svensk Växtgeografiska Sallskapet, Uppsala.

Steere, W. C. (1953). On the geographical distribution of arctic bryophytes. *Stanford University Publications, Series Biological Sciences*, **11**, 30–47.

Steere, W. C. (1965). The boreal bryophyte flora as affected by Quaternary glaciation. In *The Quaternary of the United States*, (eds H. E. Wright, Jr and D. G. Frey), pp. 485–95. Princeton University Press, New Jersey.

Steere, W. C. (1969). Asiatic elements in the bryophyte flora of western North America. *The Bryologist*, **72**, 507–12.

Steere, W. C. (1977). Ecology, phytogeography and floristics of Arctic Alaskan bryophytes. *Journal of the Hattori Botanical Laboratory*, **11**, 47–72.

Steere, W. C. (1979). Taxonomy and phytogeography of bryophytes in boreal and arctic North America. In *Bryophyte systematics*, (eds G. C. S. Clarke and J. G. Duckett), pp. 123–57. Academic Press, London.

Steere, W. C. (1985). On the continental affiliations of the moss flora of Hispaniola. *Monographs in Systematic Botany, Missouri Botanical Garden*, **11**, 155–73.

Stoneburner, A., Wyatt, R., Catcheside, D. C. and Stone, I. G. (1990). Phytogeography of the mosses of Western Australia (abstract). *1990 ABLS Meeting, Wakulla Springs, Florida*, p. 9. American Bryological and Lichenological Society.

Størmer, P. (1969). *Mosses with a Western and Southern distribution in Norway.* Universitats forlaget, Oslo.

Størmer, P. (1983). *Characteristic features of the moss flora of the various parts of Europe.* Erling Sem Offsettrykkeri, A. S. 1–91.

Streimann, H. and Curnow, J. (1989). *Catalogue of mosses of Australia and its adjacent territories.* Australian Government Publishing Service, Canberra.

Takhtajan, A. (1986). *Floristic regions of the world.* University of California Press, Berkeley.

Vana, J. and Soldan, Z. (1985). Some new and phytogeographically interesting bryophytes from central Siberia. *Abstracta Botanica*, **9**, Suppl. 2,123–44.

Vitt, D. H. (1979). The moss flora of the Auckland Islands, New Zealand, with a consideration of habitats, origins and adaptations. *Canadian Journal of Botany*, **57**, 2226–63.

Zanten, B. O. van and Pócs T. (1981). Distribution and dispersal of bryophytes. *Advances in Bryology*, **1**, 479–562.

5

Invasions and range expansions and contractions of bryophytes

LARS SÖDERSTRÖM

5.1 Introduction

Changes in bryoflora have occurred all over the world, but the documentation varies between regions. The changes are often induced by mankind, and many species, e.g. *Bryum argenteum* and *Tortula muralis*, have proliferated in man-made environments. Most parts of the world have received additions of species from other regions, e.g. the boreal hepatics *Diplophyllum obtusatum*, *Jungermannia sphaerocarpa*, and *Lophozia incisa* in the tropics (Gradstein and Vána 1987) and the European *Scleropodium purum* and *Thuidium tamariscinum* in North America (Schofield 1988). Changes in the native flora involve, besides increases or decreases of abundance, both range expansions and range contractions. In this chapter, I give examples of changes in distribution ranges, mainly from Europe, and discuss the underlying mechanisms.

5.2 Invasion of exotic species

There are several examples of species which are recent immigrants to an area. An attempt to list them meets several problems. Range expansions and contractions occur naturally, but the number of species involved has increased due to human activity. It is often difficult to assess whether a species is new to an area or native. The time from which all new species have to be treated as immigrants must be arbitrarily set. Species which have reached an area afterwards are defined as immigrants or neophytes, irrespective of whether by human activity or not. Introduced species are those which have reached the area as result of human activity. Crundwell (1985) set the date from which new species have to be treated as immigrants to AD 1500 for Europe since the traffic from North America and South Africa become significant then.

Since bryophytes are never deliberately introduced, except in moss gardens, only indirect evidence can be used to detect if a species is native or introduced. Crundwell (1985) provides six criteria, of which none is absolute.

1. *Absence of subfossil records.* A species with subfossil records can be ruled out as an immigrant.

2. *Evidence of change in geographic distribution.* A species may be persistent in one or two localities for a few years and then disappear, or it may be found in well explored areas where it was absent earlier. Increase in number of localities, especially when radiating from a single point, may also be a good criterion.

3. *Anomalous geographical distributions.* This can be either on a world scale, e.g. occurring in the Southern Hemisphere and in Britain, or on a local scale, e.g. occurring in one field but not in others with similar environments.

4. *Association with some means of introduction,* e.g. botanical gardens or ports. Many species are introduced in glasshouses, but most of them do not survive outdoors. These species are potential introductions to areas with suitable habitats and climates. One example is *Splachnobryum obtusum*, common in many glasshouses in Europe, but found in only one locality outdoors, around a thermal bath in Eger, Hungary (Corley *et al.* 1981). With climate change the survival of such escapes may become more frequent.

5. *Genetically depauperate populations.* If the supposed introduction has considerably less genetic variation than native populations its population has probably arisen from only a few individuals. The extreme case is when a single diaspore is the source of the population. Low genetic variability may be difficult to detect, but in some dioecious species only one sex occurs, e.g. in *Atrichum crispum, Scopelophila cataractae, Tortula amplexa, T. rhizophylla,* and *Trichostomopsis umbrosa* in Europe, which indicates establishment from a limited number of diaspores.

6. *Association with open, disturbed, or temporary sites.* Ecological communities may be more or less susceptible to new invaders. Natural communities, especially stable ones, are less susceptible and disturbance facilitates invasion, since a larger proportion of the propagules survive in habitats where the mature vegetation is broken (Rejmánek 1989; Orians 1986).

Table 5.1 Species supposed to be immigrants to Europe, their origin, European occurrence, and means of reproduction in Europe

Species	Distribution as immigrant	Original distribution	Year of first record in Europe	Abundance in introduced areas	Sexual reproduction	Asexual reproduction	Reference
Widespread immigrants							
Campylopus introflexus	West and central Europe	South Africa, South America, Australia, New Zealand	1941	Frequent to very common	Dioecious Sporophyte common Spores 10–14 µm	Deciduous leaves	Gradstein and Sipman (1978), Meulen et al. (1987), Casas et al. (1988)
Orthodontium lineare	West and central Europe	Southern Hemisphere (circumpolar)	1910	Common	Autoecious Sporophyte common Spores 16–20 µm		Burell (1940), Ochyra (1982)
Ricca rhenana	Widespread in Europe	Widespread in the tropics	1903		Very rare	Propaguliferous shoots	Smith (1990), Crundwell (1958)
Immigrants with restricted distribution							
Atrichum crispum	Britain, Ireland, Spain, ?Belgium	Eastern North America	1848	Frequent and locally abundant	Dioecious Females unknown in Europe	Rhizoidal tubers	Smith (1978)
Calyptrochaeta apiculata	Britain	South Africa, South America, Australia, New Zealand, Antarctica	1967	Very rare	Only females in Britain		Paton (1968), Smith (1978)
Fossombronia crispa	Portugal	South Africa	1983	Rare	Dioecious Spores 45–65 µm		Sérgio (1985)

Species	Location	Distribution	Year	Frequency	Sexuality / spores	Reproduction	Reference
Fossombronia zeyheri	Portugal	South Africa	1982	Rare	Dioecious Spores 40–50 μm		Sérgio (1985)
Hypopterygium muelleri	Portugal	Australia, New Zealand, Lord Howe Island	1929	1 locality Rare	Archegonia frequent		Allorge (1974)
Lophocolea bispinosa	Britain	Australia, New Zealand, SW Asia	1962	Restricted (2 localities)	Dioecious Sporophytes occasional	Deciduous flagelliform branches	Paton (1974), Smith (1990)
Lophocolea semiteres	Britain	South Africa, South America, Australia, Tasmania, New Zealand, Vanuatu	1955	Restricted (2 localities) Locally abundant	Dioecious Sporophytes rare	Gemmae	Paton (1965), Long (1982), Smith (1990)
Lophozia herzogiana	Britain	New Zealand	1986	Rare	Dioecious Sterile in Europe Males unknown	Gemmae (abundant)	Crundwell and Smith (1989)
Racomitrium lamprocarpum	Portugal	South America, South Africa, East African Mountains	1878	Rare	Dioecious Spores 18–28 μm		Ochyra et al. (1988)
Scopelophila cataractae	Western Europe	South America, North America, SE Asia	1967	Scattered localities over a large area Rare	Only males in Europe	Protonemal gemmae	Sotiaux et al. (1987)
Sphaerocarpos stipitatus	Portugal	South America, South Africa	c. 1870	Rare	Dioecious Spores 90–110 μm		Sérgio and Sim-Sim (1991)
Splachnobryum obtusum	Hungary (also common in greenhouses in other parts of Europe)	South and Central America	?	Rare One locality	?		Corley et al. (1981)

Table 5.1 *(cont.)*

Species	Distribution as immigrant	Original distribution	Year of first record in Europe	Abundance in introduced areas	Sexual reproduction	Asexual reproduction	Reference
Telaranea murphyae	Britain	Unknown	1962	Rare Only in 2 localities	Dioecious Male in Tresco, female in Surrey		Paton (1965, 1971)
Tortula amplexa	Britain	North America	1973	Rare	Dioecious Only females in Britain and Ireland	Rhizoidal gemmae	Side and Whitehouse (1974), Smith (1978)
Tortula brevis (*Hennediella macrophylla*)	Britain	New Zealand	1965	Very rare	Autoecious Capsules frequent but seta short and spores large (20–22 μm)	Rhizoidal gemmae	Whitehouse and Newton (1988), Blockeel (1990)
Tortula rhizophylla	Britain, Italy, Spain, (Canary Islands, Azores, Japan?, Bolivia, Mexico, USA, Hawaii)	Unknown, possibly Japan	1964	Rare	Only females known	Rhizoidal gemmae	Warburg and Crundwell (1965), Martinez *et al.* (1989)
Tortula stanfordensis (*Hennediella stanfordensis*)	Britain, Ireland, (North America)	Unknown	1958	Scattered and rare	Dioecious Fruit very rare	Rhizoidal gemmae	Whitehouse and Newton (1988), Blockeel (1990)
Trichostomopsis umbrosa	British Isles, Iberian peninsula, (Canary islands)	California, Mexico, South America	1958	Scattered. Locally abundant	Only females in Europe	Tubers	Synnott and Robinson (1990), Guerra and Ros (1987)

Trichostomopsis trivialis	Spain	South Africa	1977	Rare	Sterile in Europe	Rhizoid tubers	Guerra and Ros (1987)
Species which may be native in Europe							
Bryoerythrophyllum inaequalifolium	Spain	Canary Islands, North America	1955	Rare	Sterile in Europe		Lloret (1987)
Campylopus pyriformis	Widespread in Europe	South America, South Africa, Australia, New Zealand	18th century	Widespread and locally common	Dioecious, fruit common	Deciduous leaves	Corley and Frahm (1982)
Oedopodiella australis	Spain	South Africa	1957	Rare	Sterile in Europe	Gemmae	Casas *et al.* (1981)
Orthodontium pellucens	France, Spain, (Canary Islands, Madeira)	South and Central America)	1931	Rare?	Autoecious		Allorge (1934), Meijer (1952)
Tortula bolanderi	France, Macaronesia	North America	1974	Rare	Sterile in Europe	Tubers	Crundwell and Whitehouse (1976)
Trichostomopsis aaronis	Spain	E. Mediterranean, Central Asia	?	Rare	Unknown	Rhizoidal bulbils	Guerra and Ros (1987)

5.2.1 Immigrants to Europe

Bryophyte immigrants in Europe are rather few (Table 5.1). I regard 22 species as clear or very likely immigrants, but at least 7 other species could probably be added. I have omitted doubtful data such as *Bryoerythrophyllum jamesonii* on Crete (Düll 1984) and *Didymodon reedii* in Britain (Appleyard 1985) since they may be misidentified (D. G. Long and T. Blockeel, personal communication). Reports of *Leskea obscura* in L. Maggiore (Corley *et al.* 1981) and *Vesicularia galerulata* on Malta (Reimers 1934) are suspect. A record of *Bryum apiculatum* agg. (a complex of tropical species) from the Isles of Scilly in Britain is not yet verified but must certainly be added if correct (A. C. Crundwell, personal communication). Doubts have also arisen about the neophytic status of some other species in the list since they grow in less well explored habitats and some may prove to be native.

5.2.2 Means of introduction among European neophytes

It may be difficult to distinguish between natural long-distance dispersal and accidental introduction by human activity, especially among older records. One example is *Atrichum crispum*, a species common in eastern North America. It was first detected in Wales in 1848 and grows on sandy or gravelly soil near water (Smith 1978), a substrate characterized by a natural disturbance regime. All British populations are males, which indicate a single successful introduction, but whether from a wind-borne spore or a spore hitch-hiking with human transports is unknown.

There are, however, species associated with some means of human introduction. Two hepatics, *Lophocolea semiteres* and *L. bispinosa* (Paton 1965, 1974), and a moss, *Calyptrochaeta apiculata* (Paton 1968), were all found in gardens on the Isles of Scilly off south-west England and have later been found in one additional locality each. All three species are native in the Southern Hemisphere and thought to have been introduced with garden plants about 1907–9 (Paton 1965). Since both hepatics are dioecious and both sexes (and capsules) are found, introductions by single spores are ruled out. It is more likely that fragments or small patches were introduced with garden plants from Australia.

Scopelophila cataractae grows in the vicinity of old works for processing minerals (Melick 1987; Sotiaux *et al.* 1987) and was probably introduced with ore from South America (Rumsay and Newton 1989). The widely separated localities (Fig. 5.1) may be a result of natural occurrences and a narrow habitat requirement. However, all fertile European specimens are males and dispersal is effected by protonemal gemmae.

Other methods of introduction are also known, e.g. with wool, as in the

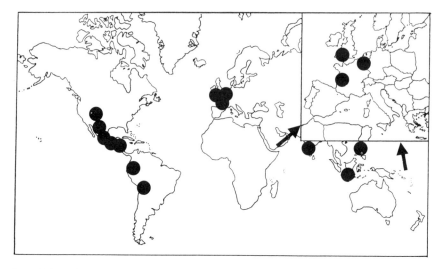

Fig. 5.1 The European and world distribution of *Scopelophila cataractae*. Also found in Japan, Korea, and northern India.

case of *Lophozia herzogiana* from New Zealand (Crundwell and Smith 1989). However, the mode of introduction is, in most cases, unknown.

5.2.3 Which areas receive immigrants and from where

Britain is probably the part of Europe where the bryoflora, both past and present, is best documented. This, together with the worldwide traffic to and from Britain during almost 500 years, and the degree that the natural habitats are disturbed, enhances immigration and detection of the immigrants. The Iberian Peninsula and France have been trading with other parts of the world for the same time, but the past and present bryoflora is less well known. However, some recent immigrants occur only on the Iberian Peninsula, e.g. *Fossombronia crispa*, *F. zeyheri* (Sérgio 1985), *Sphaerocarpos stipitatus* (Sérgio and Sim-Sim 1991), *Racomitrium lamprocarpum* (Ochyra *et al.* 1988), and *Trichostomopsis trivialis* (Guerra and Ros 1987). The only immigrants in Europe not occurring in Britain or the Iberian peninsula are *Tortula bolanderi* in France (Crundwell and Whitehouse 1976) and *Splachnobryum obtusum* in Hungary. The former may be native in Europe, but its neophytic status cannot be ruled out.

If a species is native just outside Europe, e.g. in the eastern Mediterranean, it may be difficult to prove neophytic status in Europe. Several regions are insufficiently known bryologically and species found as new there may be overlooked natives or introductions. For example, the recent increase in bryological activity on the Iberian peninsula has resulted in the discovery of many new species, some of them new to science (e.g. *Acaulon fontiquerianum*, Casas and Sérgio 1990) and others new for the area (e.g. *Bryoerythrophyllum inaequalifolium*, Lloret 1987). The more arid parts of Spain support many apparently relict species. One example is *Trichostomopsis aaronis* in south-east Spain. Outside this area it occurs in Egypt, Israel, Jordan, Iraq, Iran, and Central Asia (Guerra and Ros 1987). Is the Iberian population within the natural range or is it immigrant in origin? If native, it should be possible to find it in other places in North Africa. Certainly, some species have arrived by human activity from nearby areas, but undoubted neophytes in Europe originate almost exclusively from the Southern Hemisphere or North America. In three cases, however, the origin is unknown.

Tortula stanfordensis was discovered in Britain for the first time in 1958 in an earlier well explored area. It is described from California (Steere 1951) where it is abundant in suburban gardens and is now known also from other places (Whitehouse and Newton 1988; Fig. 5.2). *T. stanfordensis* is introduced, but the area where it is native is still unknown. It is dioecious and female plants are common, while male plants and juvenile capsules are very rare. The capsules do not reach maturity in Britain before the summer drought starts and spores are therefore not formed. Much of the dispersal is by vegetative means, such as by tubers carried by pedestrians or garden plants, with the result that some areas are colonized by only one sex. In its country of origin, when this is discovered, mature capsules may be expected to occur regularly. It is suggested that the species may be adapted to a

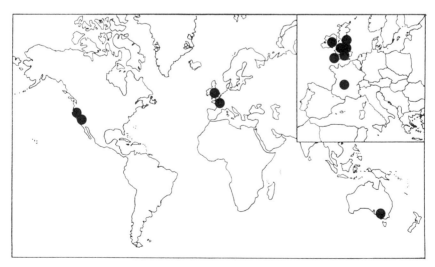

Fig. 5.2   The European and world distribution of *Tortula stanfordensis*.

climate with a warm wet season alternating with a dry one. If the wet season was warmer than the English winter, capsules might reach maturity before the onset of the dry season. The plants should therefore be looked for in areas with such a climate, for example around latitude 15° (Whitehouse 1971).

A similar life-history is shown by *Tortula rhizophylla*. Capsules are unknown, only female plants have been observed, and propagation is by subterranean tubers. It was discovered in an arable field in the Isle of Wight in 1964. Soon afterwards it was found to be conspecific with a plant known since 1960 from Louisiana where it is now widely distributed but only in man-made habitats. *T. rhizophylla* is presently known from many countries (Fig. 5.3) but must be introduced in most of the known area (Whitehouse and Newton 1988). There are, however, indications that it may be native in Japan.

5.2.4 Introduced species which have become widespread

Only 3 out of the 22 immigrants have spread successfully over a large part of Europe (Table 5.1). This is probably a function of spore production capacity. Of the neophytes with restricted distribution, a few are reported with sporophytes occasionally while two out of three widespread immigrants, *Orthodontium lineare* and *Campylopus introflexus*, produce sporophytes frequently. The third widespread species, *Riccia rhenana*, a mainly tropical aquatic plant attractive for use in aquaria (A. C. Crundwell, personal

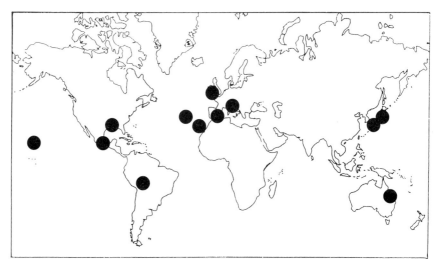

Fig. 5.3 The world distribution of *Tortula rhizophylla* (after Martinez *et al.* (1989) with permission from *Folia Botanica Miscellanea*).

communication), is an exception. This is probably the only bryophyte able to grow outdoors in Europe that is widely cultivated. It is transported around Europe as an aquarium plant and has escaped at several places (Crundwell 1958). A fourth species, *Campylopus pyriformis*, is also widespread and produces spores frequently, but its neophytic status is less clear.

The importance of sexual reproduction for the maintenance and spread of populations has been shown earlier (Gemmel 1950; Smith and Ramsay 1982). Longton (in press) showed that there is a lower proportion of rare species among British mosses with frequent sporophyte production than among those without or with rare sporophyte production. He concluded that a high frequency of genetic recombination increased the variability within the species. Low genetic diversity may be a factor causing a plant species to become rare by limiting both ecological tolerance and capacity to adapt to environmental changes, including those induced by human activity. Another advantage of reproduction is the production of spores, which are usually smaller than other types of propagule and more easily transported between localities. Species with copious spore production may colonize new localities more frequently than those producing fewer spores or only large diaspores (cf. Söderström 1989, 1990).

Orthodontium lineare is indigenous in the Southern Hemisphere where it has a pan-temperate distribution (Ochyra 1982; Fig. 5.4). It is introduced in Europe and occurs now over large parts of central and western Europe. The ecology and expansion of this species is rather well documented (e.g. Ochyra 1982; Herben 1987, 1990; Hedenäs *et al.* 1989*a*, *b*). In Sweden it

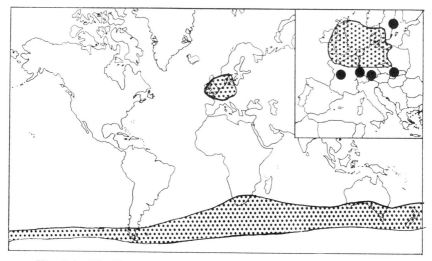

Fig. 5.4   The European and world distribution of *Orthodontium lineare*.

invades spruce and pine plantations, mostly in former broad-leaved forests (Hedenäs *et al.* 1989*a*). *O. lineare* grows mainly on temporary substrates like decaying wood, humus, and bare peat (Hedenäs *et al.* 1989*b*) which forces the species to 'move around' continuously to new substrate patches.

The ability of a competitively inferior species like *O. lineare* (Hedenäs *et al.* 1989*b*) to invade and persist in dynamic environments depends on the relationship between local population size and persistence, and between dispersal rate and establishment rate (Bazzaz 1986). A high dispersal capacity is therefore important. *O. lineare* produces large numbers of spores (up to 12 million per m^2 in favourable places) and can establish itself easily (Herben 1987). It may be expected that it can disperse freely, both within and between localities. There is a clear clumping in the occurrences of *O. lineare*, even in seemingly favourable habitats in south Sweden. With few exceptions, the clumping pattern is observed at all scales (between regions, between localities, and within localities) indicating some dispersal restrictions (Hedenäs *et al.* 1989*a*). Investigations of other species, e.g. *Ptilidium pulcherrimum* (Söderström and Jonsson 1989) and *Atrichum undulatum* (Miles and Longton 1987), show that most spores are deposited close to the source. The spores of *O. lineare* are released close to the ground with restricted opportunities for transfer into the turbulent air stream. Numbers of spores dispersed over longer distances may, therefore, be small.

The density of *O. lineare* localities differs between southern and western Sweden. Hedenäs *et al.* (1989*a*) showed that the most important factors for this were density of localities and amount of available substrate. In areas dense with suitable localities *O. lineare* may occupy a large proportion of them (Fig. 5.5) as in Skåne and southern Halland where natural deciduous forests are largely replaced by coniferous forests. In Småland and Väster-götland spruce is native, and larger parts of the natural deciduous forests remain. The density of available localities is less and offers few chances for colonization of *O. lineare*. In areas with only scattered suitable localities, e.g. in east and central Västergötland, *O. lineare* colonized only a few of them. Within these localities, it may, however, build up large local populations if the amount of available substrate is large.

Campylopus introflexus, also native in the Southern Hemisphere, has been spreading much like *Orthodontium lineare* (Frahm 1972*a*; Düll 1977) and is today present over a large part of Europe (Fig. 5.6). *C. introflexus* grows in more open places than *O. lineare* and colonizes dry, open sites with relatively acid and nutrient-poor soil and sparse, short herb vegetation (Meulen *et al.* 1987), or peatlands (Smith 1978). The distribution is always correlated with disturbance, either man-made or natural. It differs from *O. lineare* by relying more on vegetative propagules (tops of plants), at least in certain stages of the colonization. The production of spores enables, however, long distance dispersal, but the vegetative propagules are probably much more important for the formation of large local populations.

Meulen *et al.* (1987) studied the invasion of *C. introflexus* on open sites on sand dunes in The Netherlands. The native species are few, small, and more slow-growing than *C. introflexus* which rapidly outnumbers them.

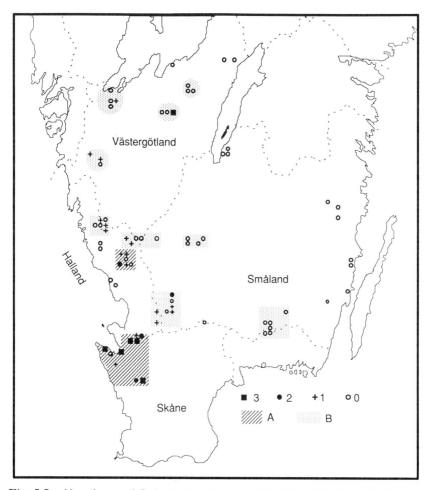

Fig. 5.5 Abundance of *Orthodontium lineare* at visited localities in south Sweden and the density of available localities in the region. The abundance of *O. lineare* within localities: 3 = several large or more than 10 small colonies, 2 = 1 large or 4–10 small colonies, 1 = 1–3 small colonies, 0 = absent. Density of suitable localities in the region: A = dense (>5 loc./10 km²), B = moderately dense (2–5 loc./10 km²). All other localities visited had less than 2 loc./10 km².

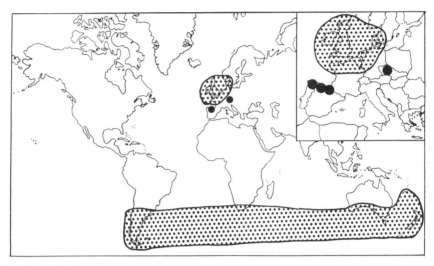

Fig. 5.6 European and world distribution of *Campylopus introflexus* (after Grad-stein and Sipman 1978 with permission from *The Bryologist*; large dots denote later records).

Meulen *et al.* deduce that the invasion begins at open, dry habitats, often after an open patch is created artificially by man. In this initial phase, *C. introflexus* has few competitors and adopts the 'colonist' strategy (*sensu* During 1979) with high rates of reproduction, both by spores and fragments. In the subsequent successional stages, *C. introflexus* shows characteristics of a 'perennial stayer'. Vegetative persistance predominates. Thick perennial cushions or mats develop, and they take all the space between the small and sparse native dune plants. In *C. introflexus* the individual cushions seem to survive better than in *O. lineare* and cushions did not change during five years in two populations followed by Meulen *et al.*

Campylopus pyriformis shows much the same pattern as *C. introflexus*. It is distributed in the Southern Hemisphere, Azores, and Europe (Corley and Frahm 1972) and occurs in Europe on open sand and peat. Frahm (1972*b*) regarded it as an introduction via the Azores, reaching Europe before the end of the eighteenth century. This may have been an introduction in 'prebotanical' times. During more than 200 years (from the start of overseas trade to the start of botanical exploration), this and perhaps other species may have become introduced without notice by botanists.

5.2.5 Immigrants with restricted distribution

Most introduced species have not successfully spread over Europe. A probable reason is restricted spore production, which decreases dispersal ability. With few exceptions, spore production is absent in immigrants with restricted distribution.

Tortula brevis is autoecious and produces sporophytes, but spores are rather large (20–22 μm) and the seta short. The spores may, therefore, be too large and released too close to the ground for effective wind dispersal. This species was described from Britain but has recently been shown to be identical with a New Zealand species (*Hennediella macrophylla*; Blockeel 1990). It occurs in two river basins in Britain and has been monitored since 1975 at one locality where it has decreased; it seems also to have decreased at the other locality (Whitehouse and Newton 1988).

Fossombronia crispa, F. zeyheri, and *Sphaerocarpos stipitatus* produce large spores better adapted to withstand regularly appearing unfavourable conditions than to long-distance dispersal. Although introduced around 1870 *S. stipitatus* remains rare (Sérgio and Sim-Sim 1991).

Several of the neophytic species not producing sporophytes in Europe are represented by only one sex. They all have more or less restricted distribution and disperse by vegetative propagules produced in, or close to, the ground. *Tortula amplexa* is a western American species found in Britain in 1973 with only females and tubers (Side and Whitehouse 1974). *Trichostomopsis umbrosa* is also from America and only females and tubers are known in Europe, where it is mainly found in man-made habitats (Synnott and Robinson 1990; Guerra and Ros 1987; Fig. 5.7), often in cities, as in all the Iberian localities.

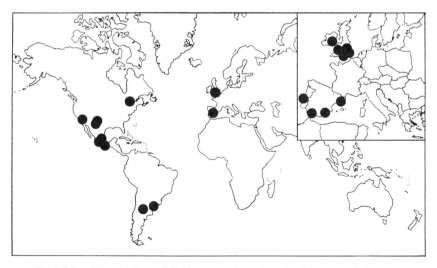

Fig. 5.7 Distribution of *Trichostomopsis umbrosa* in Europe and the world.

5.3 Native species in changing environments

Species may react to environmental changes either with changes in abundance, in the density of localities within existing distribution ranges, or with an expansion or decrease in distribution ranges. Changes of distribution may follow two different patterns. The *thinning* type of decrease starts with reduction in population sizes over all or part of an area. Next, extinction at several localities reduces the locality density in the area. This is eventually followed by extinction from a whole region. In the *retreat* type of decrease, the species disappears first from outpost and marginal localities and thereafter the range decreases progressively.

It is always difficult to separate between real changes in abundances or distribution of native species and those due to increased knowledge of a species. There are at least three important traps to avoid. First, when comparing earlier and recent occurrences one must be aware of the possibility that areas have been investigated with different intensity at different periods. Secondly, increased knowledge of the ecology and habitat requirements improves the search image among bryologists. Thirdly, the taxonomy of species may have changed.

The environmental changes most important for the bryoflora are creation, or increase, of potential new habitats, and fragmentation of existing habitats. Most important with creation of new habitat patches is the decrease of distances between available sites and an increase of the potential area for colonization, which ultimately leads to increased population size. This in turn increases the number of diaspores produced and so enhances further colonizations. Fragmentation of natural habitats has the reverse effect (Hansson *et al.*, in press).

Most studied bryophyte communities are dynamic (Kimmel 1962; During and ter Horst 1987; Herben 1987) and species growing there have to 'move around' to new spots, or new localities, regularly to survive, which means that dispersal is an important factor. Dispersal ability, i.e. how often a diaspore will succeed in a new locality, depends on (1) the number of spores produced, (2) how efficiently diaspores are transported, and (3) how easily diaspores will germinate in a new locality (Söderström 1990). Efficient transport of diaspores is less critical in spatially continuous communities where distances between suitable patches are small. Here, species with larger diaspores, such as gemmae and tubers, are favoured. In habitats where favourable conditions reappear at predictable intervals, dispersal over time will be more important than dispersal in space, and long-lived diaspores may be deposited in a diaspore bank. However, transport of diaspores between localities is of great importance in patchy, temporal habitats, where gaps appear unpredictably.

Herben *et al.* (1991) used a simulation model to investigate the most important factors for survival of species in patchy environments. There were far fewer species absent from a large part of the available substrate (i.e. dispersal-limited species) than species using most of the available substrate (i.e. substrate-limited species). However, two factors increased the proportion of dispersal-limited species, a variable establishment probability and a random arrangement of available substrate patches. Further analysis also showed that the ability of a diaspore to move between two localities, and the distances between localities, were important determinants of the stability of systems. Herben and Söderström (in press) used the same model to investigate which habitat parameters were most important for the persistence of species and found that an increased distance between the populations had the most serious effect, while decreased sizes and numbers of available localities had less effect.

5.4 Range expansions of native species

Many species are favoured by environmental changes. Tendencies to increase are, for example, shown for at least 60 species of mosses in Poland (Balcerkiewicz and Rusińska 1989) and in The Netherlands 80 moss species (20 per cent) increased while another 20 species (5 per cent) were regarded as new for the country after 1950 (Greven, in press).

The reclaimed polders of the former IJsselmeer in The Netherlands illustrate how new habitat patches may become colonized and provide examples of dispersal and expansion of bryophytes. Bremer and Ott (1991) studied the establishment and distribution of bryophytes in woods planted on these polders. They found several species that were not known from The Netherlands before or not recorded this century. Some must have arrived from areas 25 to more than 100 km away. They concluded that the large target area was important since several species were present only as a few large clones, each presumably representing one establishment event. There were also a correlation with stand age, indicating that new colonizations occur over a long period. Several species did not produce spores, and Bremer and Ott concluded that they had invaded from areas where spore production was common. Several species with large spores were present, but they were more common in older forests. This indicates that species with large spores may disperse over longer distances, but at a lower rate, than species with smaller spores.

The reason for an increase of a native species is almost always an alteration of habitats by man, e.g. deforestation, cultivation, or use of herbicides. The most obvious changes are the creation of urban environments with concrete, brick, and mortar. Epilithic species may find buildings suitable and their spread in an otherwise alien landscape may be enhanced. *Tortella tortuosa* is, for example, common in the uplands and mountains of central Europe, but occurs in the Polish lowlands only on buildings (Balcerkiewicz and Rusińska 1981).

Another habitat which has increased due to human acitivities is bare soil. *Tortula valenowskyi* is a Central Europe endemic which grows very well on steep slopes of loess ravines, formed by erosion along midfield roads. The number of localities is increasing within its small natural range (Balcerkiewicz and Rusińska, personal communication).

Pogonatum dentatum used to be a mountain plant in Scandinavia with a

Fig. 5.8 Past and present distribution of *Pogonatum dentatum* in Sweden. The main distribution area before 1950 (stippled) is in the mountains with a few outlying records (○). New records after 1960 (●) are scattered over most of northern Sweden. The dots reflect the pattern of recently investigated areas in the north.

few outlying localities along river banks in the north. During the second half of this century it has spread southwards, both in Sweden, where it has reached at least the south central parts (Hedenäs 1983, 1986; Fig. 5.8), and in Finland where it has reached the southernmost parts (Vaarama 1967). The species grows on disturbed soil, both on clear-felled areas, and on the sides of small forest roads. The number of new forest roads has increased rapidly during the last decades and they serve as excellent dispersal corridors with suitable substrate continuously distributed over long (but narrow) distances. Furthermore, since transports are canalized to these corridors, dispersal by vehicles is probably important. This is not the only mountain species now common along forest roads. *Anthelia juratzkana* is, for example, often found on moist, clayey soil near the larger rivers in the forest region of northern Sweden. This is probably a natural habitat and the river shores may serve as dispersal corridors. But it is also abundant now on moist, clayey roadsides all over northern Sweden and is spreading southwards.

River systems are often good dispersal routes for aquatic and shore plants. This was shown for a number of vascular plants (Nilsson 1986) but may apply also to bryophytes. With change in water quality or increase in suitable habitats a rapid response may be seen as exemplified by four aquatic mosses in The Netherlands which have increased during the last decades, in spite of (or due to) increased water pollution. *Fissidens crassipes* was formerly known from only three Dutch localities. At the beginning of the 1970s several new localities were discovered. This species is sensitive to frost and did not reach further north than SW Germany. But the increase of water temperature due to warm water effluents has increased the area where survival is possible (Florschütz *et al.* 1972) and it is now found in 27 Dutch localities (Touw and Rubers 1989).

Schistidium rivulare was found only once in the last century but has increased along the Dutch rivers since 1920, presumably because of the increased use of 'natural' stones and rocks when building dykes and embankments along the rivers (H. J. During, personal communication) and is now recorded from 31 localities (Touw and Rubers 1989). This may also explain the increase of *Cinclidotus danubicus*, first recorded in 1954 and now occurring in eight localities (Touw and Rubers 1989). The most successful of the increasing aquatic mosses is *Octodiceras fontanum*, first recorded in The Netherlands in 1976 (Melick 1985). This species can grow both on natural rocks and concrete and is found in both cold and warm water. The reason for the increase remains unclear.

Destruction of forests and replacement with semi-natural communities like meadows, pastures, and swards, has eliminated many forest species and allowed expansion of mosses with higher light requirements, e.g.

Brachythecium albicans and *Rhytidiadelphus squarrosus*. This change in land use significantly increased the number of localities and density of communities for over 30 moss species in Poland (Balcerkiewicz and Rusińska, personal communication). Also, at least 30 moss species benefited from the change of forest into arable fields, e.g. annual pioneers like *Barbula unguiculata*, *Phascum cuspidatum*, *Pterygoneuron ovatum*, and *Pseudephemerum nitidum*. Some of these species, however, have decreased in parts of Europe due to further changes in land use or increased use of fertilizers or herbicides.

Creation of new forests introduces possibilities for some species to increase. Bremer and Ott (1991) showed that several forest species have expanded their ranges to the planted forests reclaimed from the former IJsselmeer. *Distichium capillaceum*, *Plagiomnium medium*, *Pterigynandrum filiforme*, *Rhytidiadelphus subpinnatus*, and *Ulota drummondii* were new to The Netherlands and nine other species found there had only very few early records.

Air pollution is almost always considered to be devastating for bryophytes but some species are apparently favoured. *Brachythecium reflexum* and *Platygyrium repens* have both increased in The Netherlands during recent decades. They are mostly found on *Salix*, *Fraxinus*, and *Ulmus* stems, with relatively base rich bark, but the original habitat was usually rocks, more acid bark types, or rotten wood (H. J. During, personal communication). Localities with acid substrata must also have occurred earlier, but increased acidification has increased the number of suitable localities. Besides improvement of localities, decreased dispersal distances and increased locality sizes may have a positive effect on the population size of these species. This may be true also for other spreading acidophytic epiphytes, e.g. *Dicranum tauricum* and *D. montanum* in Britain (J. W. Bates, personal communication) and *D. tauricum*, *Herzogiella seligeri*, and *Plagiothecium latebricola* in The Netherlands (S. R. Gradstein and H. J. During, personal communications).

5.5 Range contractions of native species

Additions and increases of species are, as shown, common. The changes which most concern botanists today, however, are the decreases of native species. In Europe the number of decreasing species is largest in the centre and south while the number of decreasing species in boreal regions is still rather low. In Finland about 1.5 per cent of the species have disappeared, and in Sweden (with both boreal and temperate regions) about 10 per cent of bryophytes are decreasing and 3 per cent have already disappeared (Floravårdskommittén för mossor 1987). In The Netherlands 4 per cent have vanished and 22 per cent are declining (Greven, in press), while about

25 per cent of mosses are decreasing in Poland (Balcerkiewicz and Rusińska, personal communication).

The main reason for the decrease of so many species is the destruction or alteration of environment. Many habitats that were common earlier have decreased considerably. In boreal areas, the decrease of coniferous forests is the most obvious and severe reason for decreases, while exploitation of deciduous forests and rich fens, changes in land use, and increased chemical use are more severe in the temperate and Mediterranean parts of Europe.

Among the most rapidly disappearing habitats in south and central Europe are fens (especially rich fens) and deciduous and coniferous forests. Many rich fen species have become rare, except in boreal parts where fens are not so heavily exploited. In Finland only 1.8 per cent (1/56) of threatened species are confined to fens and mires (Rassi and Väisänen 1987). In Sweden, 19 per cent (13/68) of the threatened species are mire species, but most of them are from the southern, temperate part of the country (Floravårdskommittén för mossor 1987), and in Nordrhein-Westfalen 25 per cent (139/548) of the threatened bryophytes are mire species (Düll 1987).

Bryophytes, especially many short-lived acrocarps, have also decreased in Dutch chalk grasslands (During and Willems 1986) following termination of grazing. Grazing is sometimes replaced by mowing, but usually pastures are simply abandoned. This has increased the dense summer herb layer which decreases the light and, if the site is no longer mown, increases the herb litter. The only species increasing on these sites are litter species, e.g. *Amblystegium serpens*, *Brachythecium rutabulum*, *Eurhynchium striatum*, *Lophocolea bidentata*, and *L. heterophylla*. Destruction of sites due to mineral extraction, heavy fertilization, and invasion by shrubs and trees, is also important. All these factors contribute to the fragmentation of species ranges and the remaining sites become more isolated. Since many of the chalk grassland species reproduce mainly vegetatively (with relatively large diaspores) and a diaspore bank plays an important role (During and ter Horst 1983) it can be assumed that many of the species adapted to the environment are poorly dispersed and that immigration from other places is rare.

5.5.1 Fragmentation and decreasing density of localities

The decrease of species can be observed as range contractions or decrease in density of localities within a distribution area. In Europe the area originally covered by forests and peat bogs is now fragmented. Deforestation and juvenilization of forests has, for example, resulted in reduction or disappearance of habitats for epiphytic species such as *Neckera crispa*, *N. complanata*,

Ulota crispa, *Homalia trichomanoides*, *Antitrichia curtipendula*, and *Anomodon* spp. in Poland (Balcerkiewicz and Rusińska, personal communication).

The extinction rates increase due to smaller populations and less predictable substrate availability with increased fragmentation. In addition, successful diaspore dispersal becomes limited, due to lower diaspore production and increased distances to cross.

Old-growth spruce forests with decaying logs are rapidly decreasing in Sweden and the remaining stands are scattered and often well separated. Epixylic species are generally adversely affected (Söderström 1988) and many are rare and decreasing. This may depend both on shortage of habitats and on site isolation. The decrease of epixylic species in northern Sweden during this century is difficult to quantify due to insufficient knowledge about the earlier occurrence of forest bryophytes. The few old investigations (e.g. Tamm and Malmström 1926) indicate, however, that several species, e.g. *Anastrophyllum hellerianum* and *Lophozia ascendens*, were common. Both have disappeared from many localities during the last decade and examples of recent re-colonization are difficult to find.

5.5.2 Decreasing ranges

The disappearance of species from marginal localities reduces the general range. This is common for alpine and boreal fen mosses in the lowlands of central Europe. Glacial relics like *Meesia longiseta* and *Scorpidium turgescens*, for example, have vanished from the Polish lowlands. The latter was last observed at the end of the 1960s around Kraków in southern Poland. This locality was the last in the central European lowlands and was destroyed by drainage (Ochyra and Baryła 1988). *Trematodon ambiguus* occurs now only in one locality in Poland but was reported from 27 localities before 1945. These localities were situated on the eastern limit of its distribution range. *Meesia hexasticha*, a mainly central European species, had its north-eastern limit in Poland and was recorded from 9 localities before 1900 but never refound (Balcerkiewicz and Rusińska, personal communication).

Similar decreases indicate that many species will reduce their ranges in the future. In Poland, arctic–boreal species, e.g. *Calliergon trifarium*, *Helodium blandowii*, *Paludella squarrosa*, and *Scorpidium scorpioides*, are threatened in this way (Balcerkiewicz and Rusińska, personal communication), and in southern Sweden forest epixylics like *Anastrophyllum hellerianum* and *Lophozia ascendens* are at risk.

5.6 Conclusions

Mankind's impact on the environment has caused several bryophytes to change their distribution ranges. About 22 species have reached Europe in recent time, most of them through Britain or the Iberian Peninsula. Few of the immigrants have been spreading over larger parts of Europe. However, *Campylopus introflexus*, *C. pyriformis*, and *Orthodontium lineare* have been successful in this. It is believed that their ability to produce spores facilitates dispersal and invasion. Immigrants with restricted distribution either do not produce spores freely, or produce only large spores.

Many native species have reduced or expanded their ranges due to human activity. A major reason may be alteration of habitat density. Man-made habitats increase, so the dispersal distances between patches decreases and populations are more easily established in new localities. The reverse process, fragmentation of habitats, isolates localities and makes dispersal more difficult, with the result that the increased extinction rate is not balanced by new establishments. Increasing species are, for example, some Fennoscandian mountain species using forest roads as dispersal routes, some species along Dutch rivers taking advantage of increased use of natural stones in embankments or increased water temperature, and epilithic species using concrete buildings as a new substrate. Decreasing species are, for example, forest species living on decaying wood in northern Fennoscandia and rich fen species in central Europe. The habitats are heavily fragmented for both groups and their long-term survival is questioned for many regions.

Acknowledgements

When preparing this paper, I received valuable information from several persons. A. C. Crundwell gave me information about many neophytes and allowed me to use notes from a lecture he had given on the subject. Also I. Bisang, M. Brugues, R. Düll, H. During, S. R. Gradstein, T. Herben, D. G. Long, A. Rusińska, C. Sérgio, A. J. E. Smith, D. Synnott, E. Urmi, J. Váňa, and J. R. Wattez sent information about individual species. L. Hedenäs has helped me with several references. P. Bremer and E. C. J. Ott showed me an unpublished manuscript of their work. J. W. Bates, L. Ericson, A. M. Farmer, and C. Nilsson made valuable comments on the manuscript.

References

Allorge, P. (1934). Notes sur la flore bryologique de la Péninsule Ibérique. IX: Muscinées des provinces du Nord et du Centre d l'Espagne. *Revue Bryologique et Lichénologique*, **7**, 249–301.

Allorge, V. (1974). La Bryoflore de la Forêt de Bussaco (Portugal). *Revue Bryologique et Lichénologique*, **40**, 307–456.

Balcerkiewicz, S. and Rusińska, A. (1981). Interesujace mchy na ruinach umocnien Walu Pomorskiego w Strzalinach (woj. pilskie) [Interesting moss flora on ruins of fortifications 'Wal Pomorski' in Strzaliny (Pila Voivodship)]. *Badania Fizjograficzne nad Polska Zachodnia*, **33**, 190–1.

Balcerkiewicz, S. and Rusińska, A. (1989). Moss flora of Poland in the aspects of synanthropisation. In *Proceedings of the Sixth Central and East European Bryological Working Group*, (eds C. B. McQueen and T. Herben), pp. 95–102. Bot. Inst. CSAS, Pruhonice.

Bazzaz, F. A. (1986). Life history of colonizing plants: some demographic, genetic, and physiological features. In *Ecology of biological invasions of North America and Hawaii*, (eds. H. A. Mooney and J. A. Drake), pp. 96–110. Springer, New York.

Blockeel, T. (1990). The genus *Hennediella* Par.: a note on the affinities of *Tortula brevis* Whitehouse & Newton and *T. stanfordensis* Steere. *Journal of Bryology*, **16** 187–92.

Bremer, P. and Ott, E. C. J. (1991). The establishment and distribution of bryophytes in the woods of the IJsselmeerpolders, The Netherlands. *Lindbergia*, **16**, 3–18.

Burell, W. H. (1940). A field study of *Orthodontium gracile* (Wilson) Schwaegrishen and its variety *heterocarpum*. *Naturalist, London*, **785**, 295–302.

Casas, C. and Sérgio, C. (1990). *Acaulon fontiquerianum* sp. nov. de la Peninsula Ibèrica. *Cryptogamie, Bryologie et Lichénologie*, **11**, 57–62.

Casas, C., Brugués, M., and Cros, R. M. (1981). Contribució al coneixement de l'area geogràfica d'alguns briòfits. *Treballs de la Institució catalana d'historia natural*, **9** 169–78.

Casas, C., Heras, P., Resino, J., and Rodriguez-Oubiña, J. (1988). Consideraciones sobre la presencia en España de *Campylopus introflexus* (Hedw.) Brid. y *C. pilifer* Brid. *Orsis*, **3**, 21–6.

Corley, M. F. V., and Frahm, J.-P. (1982). Taxonomy and world distribution of *Campylopus pyriformis* (Schultz) Brid. *Journal of Bryology*, **12**, 187–90.

Corley, M. F. V., Crundwell, A. C., Düll, R., Hill, M. O., and Smith, A. J. E. (1981). Mosses of Europe and the Azores; an annotated list of species, with synonyms from recent literature. *Journal of Bryology*, **11**, 609–89.

Crundwell, A. C. (1958). *Riccia rhenana* Lorb. ex K. Müll. in Britian. *Transactions of the British Bryological Society*, **3**, 449–50.

Crundwell, A. C. (1985). The introduced bryophytes of the British Isles. *Bulletin of the British Bryological Society*, **45**, 8.

Crundwell, A. C. and Smith, A. J. E. (1989). *Lophozia herzogiana* Hodgson & Grolle in southern England, a liverwort new to Europe. *Journal of Bryology*, **15**, 653–7.

Crundwell, A. C. and Whitehouse, H. L. K. (1976). *Tortula bolanderi* (Lesq. & James) Howe in France, new to Europe. *Journal of Bryology*, **9**, 13–15.

Düll, R. (1977). Die vertbreitung der deutschen Laubmoose (Bryopsida). *Botanische Jahrbücher für Systematik (Stuttgart)*, **98**, 490–548.

Düll, R. (1984). Distribution of the European and Macaronesian mosses. *Bryologische Beiträge*, **4**, 1–113.

Düll, R. (1987). *Rote Liste der in Nordrhein-Westfalen gefährdeten Moose (Bryophyta).* Rote Liste der in Nordrhein-Westfalen gefährdeten Pflanzen und Tiere. 2. fassung. Landesanstalt für Ökologie, Landschaftsentwicklung und Forstplanung NW, Münster-Hiltrup.

During, H. J. (1979). Life strategies of Bryophytes: a preliminary review. *Lindbergia*, **5**, 2–18.

During, H. J. and ter Horst, B. (1983). The diaspore bank of bryophytes and ferns in chalk grassland. *Lindbergia*, **9**, 57–64.

During, H. J. and ter Horst, B. (1987). Diversity and dynamics in bryophyte communites on earth banks in a Dutch forest. *Symposia Biologica Hungarica*, **35**, 447–55.

During, H. J. and Willems, J. H. (1986). The impoverishment of the bryophyte and lichen flora of the Dutch Chalk Grasslands in the thirty years 1953–1983. *Biological Conservation*, **36**, 143–58.

Floravårdskommittén för mossor. (1987). Preliminär lista över hotade mossor i Sverige. *Svensk Botanisk Tidskrift*, **82**, 423–45.

Florschütz, P. A., Gradstein, S. R., and Rubers, W. V. (1972). The spreading of *Fissidens crassipes* Wils. (Musci) in the Netherlands. *Acta Botanica Neerlandica*, **21**, 174–9.

Frahm, J.-P. (1972*a*). Die Ausbreitung von *Campylopus introflexus* in Mitteleuropa. *Herzogia*, **2**, 317–30.

Frahm, J.-P. (1972*b*). Phytogeography of European *Campylopus* species. In *Proceedings of the Third meeting of the Bryologists from Central and East Europe*, (ed. J. Váňa), pp. 191–9. Univerzita Karlova, Prague.

Gemmel, A. R. (1950). Studies in the Bryophyta. I: The influence of sexual mechanism on varietal production and distribution of British Musci. *New Phytologist*, **49**, 64–71.

Gradstein, S. R. and Sipman, H. J. M. (1978). Taxonomy and world distribution of *Campylopus introflexus* and *C. pilifer* (= *C. polytrichoides*): a new synthesis. *The Bryologist*, **81**, 114–21.

Gradstein, S. R. and Váňa, J. (1987). On the occurrence of Laurasian Liverworts in the Tropics. *Memoirs of the New York Botanical Garden*, **45**, 388–425.

Greven, H. C. Recent changes in epilithic bryophytes in the Netherlands. *Biological Conservation* (in press).

Guerra, J. and Ros, R. M. (1987). Revision de la Seccion *Asteriscium* del genero *Didymodon* (Pottiaceae, Musci) (= *Trichostompsis*) en la Peninsula Iberica. *Cryptogamie, Bryologie et Lichénologie*, **8**, 47–68.

Hansson, L., Söderström, L., and Solbraeck, C. (In press). The ecology of dispersal in relation to conservation. In *Ecological principles of nature conservation*, (ed. L. Hansson), pp. 162–200. Elsevier, The Hague.

Hedenäs, L. (1983). *Pogonatum dentatum*—en norrlandsmossa på väg söderut. *Svensk Botanisk Tidskrift*, **77**, 147–50.

Hedenäs, L. (1986). Hur långt söderut i Sverige finns *Pogonatum dentatum* idag? *Mossornas Vänner*, **29**, 3–5.

Hedenäs, L., Herben, T., Rydin, H., and Söderström, L. (1989a). Ecology of the invading moss species *Orthodontium lineare* in Sweden: spatial distribution and population structure. *Holarctic Ecology*, **12** 163–72.

Hedenäs, L., Herben, T., Rydin, H., and Söderström, L. (1989b). Ecology of the invading moss species *Orthodontium lineare* in Sweden: substrate preference and interactions with other species. *Journal of Bryology*, **15**, 565–81.

Herben, T. (1987). The ecology of the invasion of *Orthodontium lineare* Schwaegr. in central Europe. *Symposia Biologica Hungarica*, **35**, 323–33.

Herben, T. (1990). Sociology of communities invaded by *Orthodontium lineare* (Bryophyta) in Europe (excl. the British Isles). *Preslia*, **62**, 215–20.

Herben, T. and Söderström, L. What habitat parameters are most important for the persistence of a bryophyte species on patchy, temporal substrates? *Biological Conservation* (in press)

Herben, T., Rydin, H., and Söderström, L. (1991). Spore establishment probability and the persistence of the fugitive invading moss, *Orthodontium lineare*: a spatial simulation model. *Oikos*, **60**, 215–21.

Kimmell, U. (1962). Entwicklung einiger Moose und Flechten auf Dauer-Untersuchungsflächen. *Oberheissische Gesellschaft für Natur- und Heilkunde, Berichte N. F. Naturwissenschaftliche Abteilung*, **32**, 151–60.

Lloret, F. (1987). *Bryoerythrophyllum inaequalifolium* (Tayl.) Zander, new to the European continent. *Lindbergia*, **13**, 127–9.

Long, D. G. (1982). *Lophocolea semiteres* (Lehm.) Mitt. established in Argyll, Scotland. *Journal of Bryology*, **12**, 113–15.

Longton, R. E. Reproduction and rarity in British mosses. *Biological Conservation* (in press).

Martinez Lacal, F., Mateo, F. D., and Varo, J. (1989). *Tortula rhizophylla* (Sak.) Iwats. & Saito, musgo nuevo para la península Ibérica. *Folia Botanica Miscellanea*, **6**, 81–4.

Meijer, W. (1952). The genus *Orthodontium*. *Acta Botanica Neerlandica*, **1**, 3–80.

Melick, H. van (1985). De verspreiding van *Octodiceras fontanum* (La Pyl.) Lindb. in Nederland. *Lindbergia*, **11**, 169–71.

Melick, H. van (1987). *Scopelophila cataractae* (Mitt.) Broth. ook in Nederland. *Lindbergia*, **12**, 163–5.

Meulen, F. van der, Hagen, H. van der, and Kruijsen, B. (1987). *Campylopus introflexus* invasion of a moss in Dutch coastal dunes. *Proceedings K. Nederlandse akademie van wetenschappen, Serie C Biological and Medical Sciences*, **90**, 73–80.

Miles, C. J. and Longton, R. E. (1987). Life history of the moss, *Atrichum undulatum* (Hedw.) P. Beauv. *Symposia Biologica Hungarica*, **35** 193–207.

Nilsson, C. (1986). Change in riparian plant community composition along rivers in northern Sweden. *Canadian Journal of Botany*, **64**, 589–92.

Ochyra, R. (1982). *Orthodontium lineare* Schwaegr.—a new species and genus in the moss flora of Poland. *Bryologische Beiträge*, **1**, 23–36.

Ochyra, R. and Baryła, J. 1988. Wyginiecie skorpionowca obłego *Scorpidium turgescens* (Musci) Poland. *Chrońmy przyrode ojczysta*, **44** 68–74.

Ochyra, R., Sérgio, C., and Schumacker, R. (1988). *Racomitrium lamprocarpum* (C.

Muell.) Jaeg., an austral moss disjunct in Portugal, with taxonomic notes. *Bulletin de Jardin Botanique National de Belge*, 58 225–258.

Orians, G. H. (1986). Site characteristics favouring invasions. In *Ecology of biological invasions of North America and Hawaii*, (eds H. A. Mooney and J. A. Drake), pp. 133–48. Springer, New York.

Paton, J. A. (1965. *Lophocolea semiteres* (Lehm.) Mitt. and *Telaranea murphyae* sp. nov. established on Tresco. *Transactions of the British Bryological Society*, 4, 775–9.

Paton, J. A. (1968). *Eriopus apiculatus* (Hook. f. & Wils.) Mitt. established on Tresco. *Transactions of the British Bryological Society*, 5, 460–2.

Paton, J. A. (1971). *Telaranea murphyae* Paton with female inflorescens in Surrey. *Transactions of the British Bryological Society*, 6, 228–9.

Paton, J. A. (1974). *Lophocolea bispinosa* (Hook. f. & Tayl.) Gottsche, Lindenb. & Nees established in the Isles of Scilly. *Journal of Bryology*, 8, 191–6.

Rassi, P. and Väisänen, R. (1987). *Threatened animals and plants in Finland*. Valtion painatuskeskus, Helsinki.

Reimers, H. (1934). Eine *Vesicularia* im Mediterraneangebiet. *Revue Bryologique et Lichénologique*, 7 306–7.

Rejmánek, M. (1989). Invasibility of plant communities. In *Biological invasions*, (eds J. A. Drake, H. A. Mooney, F. di Castri, R. H. Groves, F. J. Kruger, M. Rejmánek, and M. Williamson), pp. 369–88. Wiley, Chichester.

Rumsay, F. J. and Newton, M. E. (1989). *Scopelophila cataractae* (Mitt.) Broth. in Wales. *Journal of Bryology*, 15, 519–24.

Schofield, W. B. (1988). Bryophyte disjunctions in the Northern Hemisphere: Europe and North America. *Botanical Journal of the Linnean Society*, 98, 211–24.

Sérgio, C. (1985). Notulae Bryoflorae Lusitanicae. I. 5: Notas acerca do género *Fossombronia* Raddi em Portugal. *Portugaliae Acta Biologica*, 14, 181–98.

Sérgio, C. and Sim-Sim, M. (1991). *Sphaerocarpos stipitatus* Bisch. ex Lindenb. na Europa. Espécie introduzida em Portugal desde o século passado. *Portugalia Acta Biologica*, 15 (in press).

Side, A. G. and Whitehouse, H. L. K. (1974). *Tortula amplexa* (Lesq.) Steere in Britain. *Journal of Bryology*, 8 15–18.

Smith, A. J. E. (1978). *The moss flora of Britain and Ireland*. Cambridge University Press, Cambridge.

Smith, A. J. E. (1990). *The liverworts of Britain and Ireland*. Cambridge University Press, Cambridge.

Smith, A. J. E., and Ramsay, H. (1982). Sex, cytology and frequency of bryophytes in the British Isles. *Journal of the Hattori Botanical Laboratory*, 52, 275–81.

Söderström, L. (1988). The occurrence of epixylic bryophyte and lichen species in an old natural and a managed forest stand in northeast Sweden. *Biological Conservation*, 45, 169–78.

Söderström, L. (1989). Regional distribution patterns of bryophyte species on spruce logs in northern Sweden. *The Bryologist*, 92, 349–55.

Söderström, L. (1990). Dispersal and distribution patterns in patchy, temporary habitats. In *Spatial processes in plant communities*, (eds F. Krahulec, A. D. Q. Agnew, S. Agnew, and J. H. Willems), pp. 103–13. SPB, The Hague.

Söderström, L. and Jonsson, B. G. (1989). Spatial pattern and dispersal in the leafy hepatic *Ptilidium pulcherrimum*. *Journal of Bryology*, **15**, 793–802.

Sotiaux, A., De Zuttere, P. D., Schumacker, R., Pierrot, R. B., and Ulrich, C. (1987). Le genre *Scopelophila* (Mitt.) Lindb. en Europe. *Cryptogamie, Bryologie et Lichénologie*, **8**, 95–108.

Steere, W. C. (1951). *Tortula stanfordensis*, a new species from California. *The Bryologist*, **54**, 119–23.

Synnott, D. M. and Robinson, D. W. (1990). The moss *Trichostomopsis umbrosa* (C. Mueller) H. Robinson in Ireland. *Glasra*, **1**, 15–19.

Tamm, C. O. and Malmström, C. (1926). *The experimental forests of Kulbäcksliden and Svartberget in north Sweden*. Skogsförsöksanstaltens exkursionsledare, XI. Stockholm.

Touw, A. and Rubers, W. V. (1989). *De Nederlandse Bladmossen* KNNV, Utrecht.

Vaarama, A. (1967). A find of *Pogonatum capillare* (Michx.) Brid. in southern Finland and reflections on its bryo-geographical significance. *Aquilo, Serie Botanica*, **6**, 209–18.

Warburg, E. F. and Crundwell, A. C. (1965). *Tortula vectensis*, a new species from the Isle of Wight. *Transactions of the British Bryological Society*, **4**, 763–6.

Whitehouse, H. L. K. (1971). Some problems associated with the distribution and life-history of the moss *Tortula stanfordensis* Steere. *Lizard Field Club*, **4**, 17–22.

Whitehouse, H. L. K. and Newton, M. E. (1988). *Tortula brevis sp. nov.* and *T. stanfordensis* Steere: morphology, cytology and geographical distribution. *Journal of Bryology*, **15**, 83–99.

6

Lichen reinvasion with declining air pollution

OLIVER L. GILBERT

6.1 Introduction

The decline of lichen floras under the impact of air pollution has been a worldwide phenomenon. So sensational is the decline that a Scandinavian worker coined the term 'lichen desert' to describe the centre of towns where trees are devoid of foliose and fruticose lichens (Sernander 1926). In the UK the degree of impoverishment was shown to correspond closely with levels of sulphur dioxide, a correlation made possible using the results of a national network of nearly 1300 air pollution gauges (Clifton 1969). Though it was not fully appreciated at the time, stable SO_2 emission rates aided the calibration of semi-quantitative biological scales for estimating levels of air pollution that were constructed in the late 1960s (Gilbert 1970a; Hawksworth and Rose 1970). This was the period when ecological interactions between lichens and air pollution were most vigorously investigated. Species were listed in order of their sensitivity, the effect of environmental variables such as pH, nutrient availability, and microclimate were investigated, and a number of other air pollutants were screened to determine whether they too were having any effect. The only ones convincingly shown to be influencing lichen distribution were point source emissions of flourides, which are harmful (Martin and Jacquard 1968; Gilbert 1973b), and alkaline dust which is able to counteract SO_2 producing small oases of lichen-rich vegetation (Gilbert 1976).

In those days of consistently high levels of urban SO_2, reinvasion was never mentioned, and if encountered would probably have been misinterpreted as a relic or an anomalous phenomonon. Major reviews of the lichen-air pollution field up to 1976 did not mention it (Ferry et al. 1973; Gilbert 1973a; Hawksworth and Rose 1976). The only examples known at that time involved industrial localities. In the steel town of Port Talbot,

South Wales, Pyatt (1970) interpreted the appearance of numerous very small lichen thalli on trees in the north-west suburbs as indicating that the area was recovering from a previous period of more intense industrial atmospheric pollution, while Skye and Hallberg (1969) noted that one year after the closure of an oil shale works in southern Sweden, strong healthy lobe growth on foliose species was possibly a sign of recovery, but later doubted it (Skye 1980). Today opportunities to study the reinvasion on a large scale are still limited to a few countries in Europe. A recent multi-author volume reviewing the effect of air quality on lichens and bryophytes, with particular reference to North America, only gives examples of pollution abatement around point sources from that continent (Nash and Wirth 1988).

6.2 Air pollution trends in the United Kingdom

Total sulphur dioxide emissions in the UK were relatively stable throughout the 1960s at about 6 million tonnes a year. Thereafter, annual emissions declined, falling to about 5 million tonnes in 1976 and under 4 million in 1983. They have remained at around this level with a low of 3.66 million tonnes in 1988, the last year for which figures are available (Department of the Environment 1989). To achieve the limits set by EEC Directive 88/609, the UK will need to fit flue gas desulphurization equipment to six of its largest coal-fired power stations, so further reductions are to be expected. The Department of the Environment coordinates a 287 site network of gauges to monitor compliance with EEC directives on SO_2 and smoke levels. Of more relevance to lichen studies are ground level concentrations of SO_2. These have fallen by nearly 80 per cent since 1962, reflecting the reduction in emissions from low-level sources, particularly domestic properties (Department of the Environment 1989). The course of this decline is also strongly related to the increasing use of sulphur-free fuels such as natural gas, lower industrial energy demands, and energy conservation. Today, 91 per cent of the SO_2 is emitted from the tall stacks of industry and power stations (71 per cent). Currently, average urban ground level concentrations in the UK, as measured by the DOE network are $c.$ 33–36 μg m^{-3}. It is not possible to give a precise figure, as weather variations from year to year affect both the quantity of fuel consumed and dispersion of the pollutants.

Though it is useful to know about general trends, if correlations are to be made between changing lichen distributions and levels of SO_2 the information needs to be much more site-specific. A great deal of the air pollution data held by the government at the Warren Spring Laboratory has not been analysed in detail, but Laxen and Thompson (1987) have

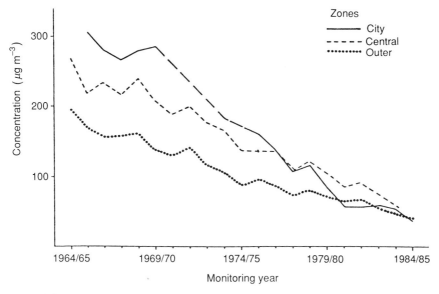

Fig. 6.1 London-wide trends in annual mean sulphur dioxide concentrations between 1964/65 and 1984/85. Present-day concentrations are the same in all zones, the steady downward trend showing no sign of abating (from Laxen and Thompson 1987).

used these statistics to construct a picture of SO_2 in Greater London between 1931 and 1985. In the capital there were no clear trends till 1964, when a slow decline set in which became particularly pronounced after 1970. If London is divided into a City, a Central, and an Outer Zone, SO_2 is seen to be declining in all of them, with an especially steep fall in the City zone, which was originally the most heavily polluted. By 1985 mean annual SO_2 concentrations were around 40 μg m^{-3} in all zones (Fig. 6.1.). Highest daily SO_2 levels also declined steadily between 1964 and 1985 (1500–2000 down to 200–250 μg m^{-3}) and the difference between winter and summer peaks has almost been eliminated. This could be very significant, but lichenologists have never understood the effect of high winter peaks on their organisms. The steady downward trend in annual concentrations shows no sign of abating.

It would be particularly valuable to know about trends in rural areas, but the national network only monitors compliance with EEC directives and background levels in remote coastal sites, so has little information on, for example, SO_2 concentrations in rural East Anglia, over the Pennine moorlands or in the Lake District. In the 1970s, when more extensive monitoring was being undertaken, rural gauges provided evidence that levels outside towns were falling much more slowly, if at all, as tall stack

emissions were spreading the pollution over a wide area. Today, lichenologists who have been monitoring lichens on rural trees probably know as
much as anyone about quantitative changes in air pollution in such areas.
Over much of Britain, sulphur dioxide concentrations are getting down to
a level where analytical methods which involve titrating acidified hydrogen
peroxide against a base to determine atmospheric acidity are not particularly
reliable (Roberts and Lane 1981).

6.3 Studies involving epiphytes

Most work on reinvasion has involved epiphytes. The majority take the
form of surveys carried out in the 1970s and repeated in the 1980s. All but
one demonstrate that in areas where there has been air pollution abatement
there is at least a small increase in lichen diversity or expansion of species
already present. In their detail the surveys differ considerably, as once SO_2
levels have fallen below a threshhold value other ecological factors start to
control the composition of the epiphytic communities. The subject is still
at the stage when nearly every paper throws some new light on the topic.

A study which can be used to set the scene involves a 125 km transect
stretching west from Liverpool to the clean air of North Wales (Cook *et al.*
1990). The epiphytic cover on mature, free-standing oaks (*Quercus* spp.)
was assessed at 53 sites in 1973 and again in 1986. As was to be expected
changes during that time interval were not uniform throughout the 125 km.
There was a marked increase in the cover and extent of the crustose lichen
Lecanora conizaeoides at the urban end of the transect, a small increase in
lichen diversity along the middle stretch (mean no. taxa per tree 3.09–4.36),
and a possible decline of large foliose species at the western, least polluted
end. Most of the new species in the middle stretch, e.g. *Evernia prunastri*,
Hypogymnia physodes, *Parmelia saxatilis*, *P. sulcata*, *Physcia tenella*, were
present as small juveniles. When sites were classified according to the
Hawksworth–Rose 10 point pollution scale it was apparent that over the
thirteen years the zones had moved inwards considerably at the eastern end
while at the western end they remained unchanged (Fig. 6.2). These results
are more or less consistent with estimated changes in pollution levels along
the transect. The most noteworthy points are the long implied time lag
involved between pollution levels falling sufficiently for foliose and fruticose
species to colonize the trees and their actual appearance and build-up into
communities, and the suggestion that long distance transport of atmospheric pollutants may be affecting lichens at the rural end.

Events at the urban end of the transect were confirmed by Alexander
(1982) who in 1980 repeated a huge mapping exercise first carried out by
Vick and Bevan (1976) during the 1972/74 period. One of his findings was

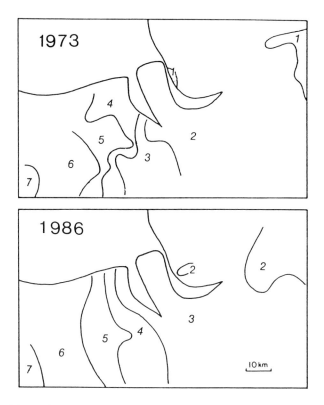

Fig. 6.2 Maps of the Liverpool–North Wales area showing how the Hawksworth–Rose (1970) lichen zones altered between 1973 and 1986. Changes reflect amelioration on the east side but no apparent improvement in the originally less polluted western areas (after Cook *et al.* 1990).

that *Lecanora conizaeoides*, which in 1972/74 was absent from trees over an area of 40 km², had spread throughout the built-up area and so was no longer useful for monitoring. He discovered that the reinvasion rates of the four indicator species mapped showed variation, and used this to explain the anomalous situation where the leaf fungus, tar-spot of sycamore (*Rhytisma acerinum*), which responds quickly to falling levels of air pollution, sometimes occurred further in towards the city centre than the less sensitive *Xanthoria parietina* which shows a time lag of several years before colonizing newly ameliorated areas.

Another long transect involving oaks (*Quercus robur*), this time extending 70 km SSW from central London was recorded annually between 1979 and 1990 (Bates *et al.* 1990). Their results differ markedly from those obtained by Cook *et al.* (1990). Despite SO₂ levels having declined greatly prior to and during the period of observation (Fig. 6.1) no firm evidence was obtained for the recolonization of oak boles or bases, even deep in Sussex. The method used to detect changes in the lichen cover was incapable of identifying small differences and could have missed widely scattered juvenile thalli, but it does appear that over the twelve years, reinvasion and expansion of existing species was negligible. The conclusion was that lichen communities on oak in the London area continue to predict the high SO_2 levels that prevailed thirty years ago. This is quite easily the longest time lag ever recorded and means that oaks in SE England are of little value for estimating SO_2 levels under ameliorating conditions.

The results of the London oak transect are all the more perplexing in the light of other reinvasion studies that have been undertaken in the capital. The decline of London's lichen flora has been documented by Laundon (1967, 1970) and the recovery of its epiphytes, which commenced around 1970, by Rose and Hawksworth (1981) and Hawksworth and McManus (1989). The first paper reports a general reappearance of sensitive lichens in the north-western suburbs and suggests, from observing the size of thalli, that many were established between 1973 and 1977. This reinvasion was a response to rapidly falling SO_2 levels, but the predictions of pollution control bodies were that this was a limited and temporary improvement before air pollution levels rose once again. However, pollution levels continued to fall and trees in London continued to be recolonized at an ever increasing rate. The second paper, reporting fieldwork up to 1988, gives details of 49 epiphytes, 25 of which had not been seen within 16 km of the centre of London this century. In addition, fruticose species such as *Evernia prunastri* and *Ramalina farinacea* were now in many central London parks, while the even more sensitive *Parmelia caperata* was expanding in the outer suburbs.

Their most significant finding was that under the prevailing pollution regime (mean winter $SO_2 < 50 \, \mu g \, m^{-3}$) the total number of species occurring at sites in the capital was no longer correlated with distance from the city centre, some central parks having similar numbers to those in the outer suburbs (Fig. 6.3). The recolonization process, instead of following an orderly sequence through the zones recognized by Hawksworth and Rose (1970) was being controlled by dispersal efficiency rather than SO_2 sensitivity, as central London was now available to all species from zones 1–7. This led to some unexpected discoveries such as a zone 7 species, *Candelaria concolor*, in Chelsea Physic Garden (site 5 in Fig. 6.3) and *Usnea subfloridana* in Dollis Hill (site 19). This phenomenon, first noticed by Alexander (1982), was christened 'zone skipping' (Hawksworth and McManus 1989) and defined as recolonization without the return of species progressively lost under conditions of gradually rising ambient air

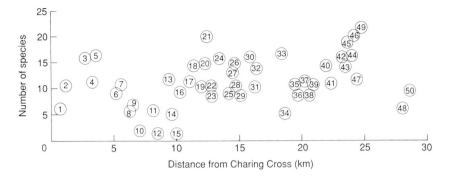

Fig. 6.3 Chart showing the lack of a relationship between the number of lichen species at sites lying different distances from Charing Cross, central London (from Hawksworth and McManus 1989).

pollution levels. Species which show a reluctance to zone skip were identified as *Chaenotheca ferruginea* and *Hypocenomyce scalaris*. From measuring the size of the largest thalli and interpolating with growth rate curves, widespread colonization of central London (9 km from Charing Cross) by foliose species was thought to have commenced in 1983. The authors do not indicate the time lag involved but from matching their data against Fig. 6.1 it would appear to be in the order of 4 to 5 years.

Studies in Cheshire have focused attention on a factor that is of central importance to an understanding of the ecology of reinvasion, it is habitat dependent. Fifteen years ago B. W. Fox discovered that epiphytes were returning to this polluted county, but only in one specialized habitat, willow carr containing large leaning specimens of crack and white willow (*Salix fragilis. S. alba*). In such sites he observed the return first of *Hypogymnia physodes*, *H. tubulosa*, and *Parmelia sulcata* soon followed by *Evernia prunastri*, *Parmelia glabratula*, *P. subaurifera*, *Platismatia glauca*, and *Ramalina farinacea*. As pollution abated further *Parmelia revoluta*, *P. subrudecta*, *Physcia aipolia*, and *Usnea subfloridana* joined in what were now quite rich lichen assemblages with a high cover. The latest species to start spreading through the carrs have been *Parmelia caperata* and *P. perlata*. Often the species dominate individual branches. In the last few years a number of the lichens have spread from the carrs to the base of field trees, especially ash and sycamore (Guest 1989). Ten years ago it would have seemed inconceivable that such sensitive species would appear in Cheshire; it raises the question of where the propagules originated.

Long series of observations at a single site are especially valuable as they enable direct comparisons to be made between the original lichen flora and those newly establishing following pollution abatement. Recently the Jardin de Luxembourg in Paris, where Nylander first demonstrated the link between lichens and air pollution (Nylander 1866, 1896), has been resurveyed a hundred years after it was declared devoid of lichens (Seaward and Letrouit-Galinou 1991). It appears that the trees were still bare in 1986 but by 1990 eleven lichens had recolonized a variety of species. Out of sixteen epiphytes recorded by Nylander (1866) only four had so far returned. It is too early to conclude that the re-establishing lichen flora differs from that formerly wiped out by air pollution. Another long series of observations has been made on a wood near the Hook of Holland (de Bakker 1986). In 1949 twelve species were recorded; this number fell to seven in 1973 and then recovered to twelve by 1984. A comparison of the three lists could be interpreted as showing that a different community is establishing, as only 25 per cent of the original species have returned. However, as each survey was carried out by a different person, recorder bias and the fact that recovery is still at an early stage complicate any analysis.

The suggestion that a new and different lichen flora may be filling up at least parts of certain lichen deserts receives support from studies in the Netherlands. De Bakker (1989) compared the lichen flora of oak trees in areas of intensive farming with similar trees in adjacent nature reserves. He found that the field trees carried 'nitrophytic' communities in which *Candelariella vitellina*, *Physcia* spp. and *Xanthoria* spp. were prominent, while those in the nature reserves supported acidophytic assemblages that included *Evernia prunastri*, *Hypogymnia physodes*, *Lecanora conizaeoides* and *Lepraria incana*. This difference was shown to be associated with high ammonia emissions from the area under modern intensive agriculture, this factor replacing earlier acidification by SO_2. Evidence that this may be a general phenomenon has been provided by van der Knaap and van Dobben (1987) who in 1986, repeated a survey of epiphytes, first carried out in 1972, at 689 sites near Rotterdam. During the 15 year time interval, mean species richness per site increased from 5.35 to 7.11 and the mean frequency of all species from 5.4 per cent to 7.0 per cent. This was correlated with declining SO_2 concentrations in the atmosphere. The other change noted was that in this area of mixed dune and polder, nitrophytic species had increased more than others. Kandler (1987) has reported a similar pattern of reinvasion around Munich.

6.4 Epiphytes: current state of knowledge

Epiphyte reinvasion in response to falling levels of SO_2 air pollution, has been well documented in a range of European countries but several features of the ecology of this event are not fully understood. Central to this is the time lag required between pollution abatement and the establishment of lichen thalli sufficiently large to be detected. Several variables appear to be involved and even the onset of colonization is hard to define, as a number of workers (Lindsay 1981; Guest 1989) have reported juvenile thalli disappearing after a year or two, presumably casualties of peaks in fluctuating pollution levels.

The time lag is partly controlled by dispersal capacity. As a consequence, reinvasion offers a rare opportunity to study the relative mobility of species. Many appear to have a far greater dispersal capacity than expected: for example the most likely source of the *Usnea subfloridana* that is colonizing willows in Lincolnshire is North Wales, over 150 km away. There is a high proportion of sorediate species among the early invaders with *Physcia aipolia*, *P. stellaris* (de Bakker 1987), and *Xanthoria polycarpa* being exceptions. In Table 6.1 a selection of species is presented, which, from the literature and the author's observations, appear to be either unusually fast or slow colonizers of lichen deserts in the UK. Rapid colonizers have been christened 'zone skippers' (Hawksworth and McManus 1989); the name 'zone dawdlers' is now proposed for the latter group.

Table 6.1 Lists of epiphytic lichens that are either 'zone skippers' or 'zone dawdlers'

Zone skippers	Zone dawdlers
Candelaria concolor	*Calicium* spp.
Evernia prunastri	*Chaenotheca* spp.
Parmelia caperata	*Chrysothrix candelaris*
P. perlata	*Diploicia canescens*
P. revoluta	*Hypocenomyce scalaris*
P. subrudecta	*Lecanactis abietina*
Physcia aipolia	*Opegrapha vulgata*
Ramalina farinacea	*Parmelia saxatilis*
Usnea subfloridana	*Parmeliopsis ambigua*
Xanthoria polycarpa	*Pertusaria amara*

All zone skippers so far identified have been foliose or fruticose species, though *Dimerella pineti* might be considered for inclusion as it is spreading throughout the Midlands on willow trees. The majority of the species (Table 6.1) are members of the *Parmelion perlatae* alliance which occurs on well-lit mature deciduous trees in sites subject to event wetting. By contrast, most zone dawdlers are crustaceous lichens characteristic of dry bark that rarely gets wetted by the rain. Such sites retain acidity acquired in the past. The groups therefore appear to be ecologically distinct, so habitat is as important in determining the detail of the order in which species return to lichen deserts as factors such as dispersal efficiency.

Most workers agree that pH is of fundamental importance in controlling re-establishment in formerly polluted areas. This is not unexpected, as it was identified as a key factor in the expansion of lichen deserts (Gilbert 1970*b*; Puckett *et al.* 1973; Turk *et al.* 1974). Low pH can be toxic to lichens *per se* but its most damaging role is in determining the form in which the sulphurous acid is present. As pH falls from 7.0 down to 3.0, relatively innocuous SO_3^{--} is replaced by mildly toxic HSO_3^-. Then, below pH 4.0, highly damaging undissociated 'H_2SO_3' molecules start to appear, their concentration increasing as the pH falls further (Vass and Ingram 1949). Currently this is the best explanation we have as to why lichens return first to high pH barks. Most apparent anomalies in the field survey data can be explained by this mechanism. Leaning trunks of white and crack willows have among the highest bark pH of any common tree, now that elms have declined, and are the niche to which epiphytes return first. Bark throughout the London oak transect was extremely acid (pH 2.9 to 4.0). This acidity, acquired during years of SO_2 exposure (Grodzinska 1977, 1979), is probably due to the leaching of cations by sulphurous acid and persists following amelioration, so the trunks remain poorly colonized. Often the smooth bark of young oaks which have not experienced the conditions which acidified the parent trees are the first to be recolonized.

De Bakker (1989) explained the positive relationship between oak epiphytes and ammonia emissions from farmland as an effect not of nitrogen but of pH; the alkaline ammonia gas raised bark pH from a background of 3.5–5.0 to 5.0–6.1. The number of species on a tree correlated strongly with bark pH, but hardly at all with its NH_4^+ concentration. A further example of the same general effect has been noted by most field workers; it involves the 'canine zone' at the base of town trees which, whatever the species, has an enhanced pH (Fig. 6.4). This microhabitat is greatly favoured by reinvading lichens, particularly an entity within the *Lecanora dispersa* agg. that has a conspicuous white thallus.

Reinvasion is also speeded up by dust contamination of tree trunks. On a macro-scale this can be observed around cement factories and quarries producing roadstone (Gilbert 1976). Towns are also dusty places with a 'traffic film' developing on trees near roads, and fine materials being released on demolition sites and by the construction industry. At a meso-scale, dust contamination is probably responsible for the rich invading lichen flora of sites like Chelsea Physic Garden which is completely surrounded by buildings; its corticolous flora includes *Lecanora muralis*, *Candelariella vitellina*, and other species typical of dust enriched bark (Hawksworth and McManus 1988). An example of beneficial dust (soil) contamination at a micro-scale is the preferential invasion of roots projecting through grassland that has been reported by various workers (e.g.

Fig. 6.4 The 'canine zone' at the base of town trees is a microhabitat favoured by reinvading lichens. The upper *Lecanora conizaeoides* covered area has a pH *c*.4.0 while the darker, strongly eutrophicated basal zone of pH *c*.5.5 is being colonized by *Xanthorion* species (from Gilbert 1989).

Seaward and Letrouit-Galinou 1991) and which can be observed in many town parks.

Recording beyond the stage at which a scatter of juvenile thalli are present will be important. Apart from the special case of willow carrs and asbestos roofs, where diverse communities with a high cover have developed quite rapidly, the build up of a lichen coat on trees has been slow to follow on from the initial colonization by foliose and fruticose species. Ten years of monitoring over 200 trees in the Sheffield area (by the author) and long-term observations in West Yorkshire (Seaward and Henderson 1991) suggest that the persistence of the first colonizers is low, despite pollution levels continuing to fall. There appears to be a link between persistence and the buffering capacity of the substratum.

6.5 Studies involving saxicolous species

The most original and detailed work on reinvasion has been the long-term study by Mark Seaward of *Lecanora muralis* in Leeds, West Yorkshire. It involved a baseline survey of the distribution of this lichen in a range of habitats, the establishment of permament quadrats, and transplant work. Although the survey was originally set up to follow decline under increasing pollution stress, by 1970 it became apparent that reinvasion was occurring. The first papers, covering the period 1969–74, appeared in 1976, since when there have been a number of updates (Seaward 1976*a*, 1976*b*, 1980, 1982, Henderson-Sellers and Seaward 1979). During the period 1970 to 1980 *Lecanora muralis* spread towards the centre of Leeds at an average rate of 150 m per annum by colonizing asbestos cement sheeting and asbestos roof tiles. The density and variety of habitats in urban Leeds enabled detailed maps to be prepared which showed that three advancing waves were involved. The front runners were on asbestos cement roofs, 550 m behind this vanguard there was a wave advancing at a similar speed on cement, concrete, and mortar, then 1900 m further out there was a more slowly advancing (but later to speed up) wave on the capstones of siliceous walls (Fig. 6.5). Invading fronts are characterized by a dense splatter of very small colonies which suggests that the inoculum pressure is

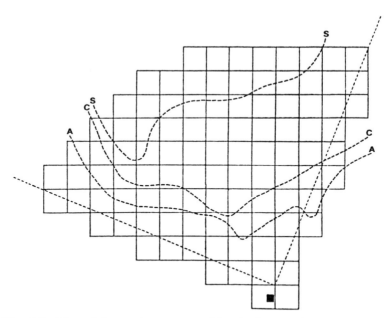

Fig. 6.5 Map of the northern sector of Leeds showing the major inner limit of *Lecanora muralis* on asbestos cement (A), on cement, concrete and mortar (C), and on siliceous wall-tops (S), in 1970. By 1975 fronts A and C had each advanced *c.* 750 m in response to pollution abatement. The grid lines are 1 km apart (from Seaward 1976*a*).

high. The time lag between SO_2 falling below the threshold limit for establishment on asbestos roofs (c. 200 μg m^{-3} daily mean) and juvenile thalli appearing was about five years, by which time air pollution levels alone were not the major factor affecting their distribution at the site. Germination and establishment were affected by a complex of factors, the most influential of which was substrate chemistry, especially pH. The mean pH values of the different substrates were old asbestos cement, 9.80; cement, concrete and mortar, 8.35; and siliceous wall capstones, 5.25.

The permanent quadrat results highlighted a phenomenon that is still not fully explained. The first three annual cohorts to invade asbestos roofs as pollution abated differed in a number of respects from later cohorts. In addition to colonizing earlier, they displayed a fast growth rate, averaging 3.87 mm per year, and apparently established during the winter months; their subsequent spread on the roofs showed no aggregation of colonies. Thalli establishing subsequently had a significantly slower mean annual growth rate of 2.50 mm, they first appeared in May, and their ensuing spread showed aggregation, suggesting that colonies were exhibiting local dispersal. The 'urban super-race' may be the first example of a resistant lichen variety evolving under the impact of SO_2 stress. *Lecanora muralis* is morphologically highly variable, it fruits abundantly, and probably spreads by spores. For further details readers are referred to the original papers (e.g. Seaward 1976a, 1982).

Zone skippers and zone dawdlers have hardly been identified among saxicolous assemblages, as their order of sensitivity to SO_2 is only poorly known. There is some evidence that *Parmelia saxatilis* behaves as a dawdler, the reason being that the quality of the isidia produced under pollution stress is adversely affected (Gilbert 1988); the species has no other regular means of spread. *Lecanora campestris* and *Ochrolechia parella* have also been slow to spread in response to declining levels of air pollution. Saxicolous zone skippers include *Lecanora polytropa* agg. and *Trapelia coarctata*, which have become widespread in many towns over the last few years. Other candidates such as *Scoliciosporum umbrinum* and the opportunists mentioned in Gilbert (1990) are difficult to evaluate, having been seriously overlooked in the past.

6.6 Discussion

An understanding of the ecology of reinvasion has been hampered by SO_2 levels falling more rapidly than the rate at which many lichens could reinvade the ameliorated areas. For this reason workers have been reluctant to attach SO_2 levels to the recolonization gradients which have developed as, at best, they will be four to five years out of date and at worst meaningless. While lichens can provide a general picture of pollution abatement, organisms which invade annually, such as tar-spot (*Rhytisma acerinum*), a leaf pathogen of sycamore, are more appropriate indicators of rapidly ameliorating conditions. Another difficulty to be overcome is deciding which pollution parameter controls establishment and subsequent growth. Lichenologists have never been very certain over this, but in a rough and ready way, workers in the UK have linked fatal injury near urban areas to mean winter levels, as this is a figure that has often been available. Showman (1981) demonstrated that around a point source that switched to taller stacks the pollution void was colonized by *Parmelia caperata* despite annual average levels of SO_2 remaining the same. He considered that a decrease in peak levels was sufficient to trigger off the reinvasion. It is unrealistic to assume that lichen responses can all be correlated with the same SO_2 criterion, but it is difficult to do otherwise, as in practice there is always a shortage of air pollution data.

The persistence of the first colonizers is often low despite progressively falling pollution levels. This was revealed by the long-term observations in West Yorkshire by Seaward and Henderson (1991) and by the author's studies of epiphytes around Sheffield over ten years. Diverse lichen communities with high cover have so far only been reported on willow bark and asbestos. On other tree species recovery has been slow following initial establishment of foliose and fruticose species. Data collected so far suggest

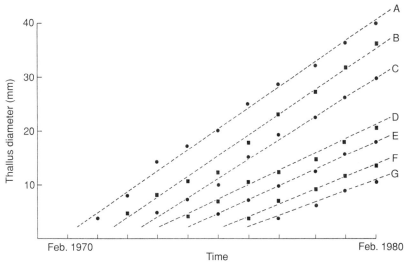

Fig. 6.6 Mean growth of successive cohorts of *Lecanora muralis* thalli colonizing asbestos roofs in Leeds during the period 1970–80. Cohorts A, B, and C have been designated the 'urban super-race' (from Seaward 1982).

a positive link between expansion and the pH and buffering capacity of the substratum. In the same way that successive waves of colonization have been reported on different stone surfaces (Fig. 6.6) it should be possible to demonstrate a similar effect on trees, which can have very distinctive bark chemistries (Barkman 1958; Skye 1968).

Sufficient floristic studies have been undertaken to provide a perspective on the way lichen deserts are filling up. Reinvasion is not an orderly event like the closing of an iris-diaphragm; instead, it is rather patchy, being controlled by habitat conditions and to a lesser extent by dispersal efficiency. Table 6.2 shows some of the factors which influence it. Regarding epiphytes, lichens return first to tree willows, followed by species

Table 6.2 Some factors which influence the way lichen deserts are reinvaded following pollution abatement

Most favourable habitats/ conditions	⟶	Least favourable habitats/ conditions
Willow, poplar, sycamore, ash, elm	beech, oak, lime	alder, pine, birch
Alkaline dust contamination, canine zone	ammonia emissions uncontaminated bark	acidified bark
Old asbestos cement	cement, concrete, mortar	siliceous

with basic, neutral, and acidic bark, in that order. The inherent properties of the bark can be modified by extraneous sources of 'pollution' that can enhance or decrease its receptivity to lichen propagules. On an individual tree, conditions vary considerably with the bole, the upperside of slanting surfaces, and nutrient streaks below wounds, all of which have a slightly higher pH than other niches, being picked out by reinvading lichens. The position with regard to saxicolous habitats is less clear, but work by Seaward has highlighted the importance of quite small differences in pH.

Considerable work has been directed towards screening the urban environment for other pollutants that might start restricting lichen distributions once SO_2 ceased to be a restraint. The small amount of work done on lead, nitrogen oxides, ozone, and peroxyacetyl nitrate suggests they are not toxic at field concentrations (Nash and Wirth 1988), while flouride levels in towns have remained low. Ammonia and alkaline dust are the only influential pollutants and they facilitate reinvasion by raising pH, which affects the speciation of SO_2. It would be premature to forecast the effect acid rain might have as its action on lichens under natural conditions is still uncertain (Richardson 1988).

An aspect which requires further study is the performance under declining SO_2 of those species that were originally favoured by it. For decades *Lecanora conizaeoides* has formed a monotonous grey–green cover on tree trunks and siliceous habitats in urban areas. It has been too common to bother with, an accepted background against which other changes occurred. A study on Tyneside in the 1960s (Gilbert 1969), when it was probably near its maximum extent, showed that this species differed from all others in reaching its greatest cover in the outer suburbs. Since that time, according to studies on Merseyside (Alexander 1982; Cook *et al.* 1990), the peak is likely to have moved in towards the city centre and there will have been a withdrawal from rural areas. An alternative scenario is that instead of behaving in a predictable manner, *L. conizaeoides* swards have experienced a general disintegration over large areas. Little is known about the detailed response of this enigmatic lichen which was first recorded c. 1860 from Leicestershire, expanded widely in Europe as air pollution spread, and is now declining in step with pollution abatement. Populations of other species believed to have been favoured by the spread of SO_2, e.g. *Parmelia incurva* and *Parmeliopsis ambigua*, also require monitoring if an understanding of the legacy of high air pollution is to be gained. Adjustments occurring within communities that are only mildly influenced by SO_2 will be harder to determine, but baseline studies are in place, e.g. Cook *et al.* (1990).

So far only the initial stages of reinvasion have been observed, and these predominantly on trees in the more polluted areas. Early indications are that a narrow range of mobile, competitive species with a high reproductive capacity and a wide ecological amplitude are the first invaders of the shrinking lichen deserts. Infiltration is not uniform, high pH and eutrophicated habitats being favoured. It seems that since the rise of air pollution in the middle of the last century a general eutrophication of much of lowland Europe has taken place. Now that the masking effect of SO_2 is being removed it is becoming apparent that one form of pollution has been replaced by another, with the consequence that the original pre-industrial revolution lichen flora is likely to return only locally. Reinvasion of lichen deserts is not their expansion in reverse. A hard lesson for lichenologists to grasp is that due to differential time lag phenomena their organisms can only indicate pollution abatement, not monitor it.

References

Alexander, R. W. (1982). The interpretation of lichen and fungal response to decreasing sulphur dioxide levels on Merseyside. *Environmental Education and Information*, **2**, 193–202.

Barkman, J. J. (1958). *Phytosociology and ecology of the cryptogamic epiphytes*. Assen, Netherlands.

Bates., J. W., Bell, J. N. B., and Farmer, A. M. (1990). Epiphyte recolonisation of oaks along a gradient of air pollution in south-east England, 1979–1990. *Environmental Pollution*, **68**, 81–99.

Clifton, M. (1969). The national survey of air pollution in the United Kingdom. In *Air Pollution. Proceedings of the First European Congress on the Influence of Air Pollution on Plants and Animals*, pp. 303–13. Centre for Agricultural Publishing and Documentation, Wageningen, Netherlands.

Cook, L. M., Rigby, K. D., and Seaward, M. R. D. (1990). Melanic moths and changes in epiphytic vegetation in north-west England and north Wales. *Biological Journal of the Linnean Society*, **39** 343–54.

de Bakker, A. J. (1986). Veranderingen in de epifytische korstmos-flora van het Staelduinse bos in de periode 1949–1984. *Gorteria*, **13**, 70–4.

de Bakker, A. J. (1987). Physcia stellaris (L.) Ach. in Nederland. *Gorteria*, **13**, 210–15.

de Bakker, A. J. (1989). Effects of ammonia emission on epiphytic lichen vegetation. *Acta Botanica Neerlandica*, **38**, 337–42.

Department of the Environment (1989). *Digest of Environmental Protection and Water Statistics*, (no. 12). HMSO, London.

Ferry, B. W., Baddeley, M. S., and Hawksworth, D. L. (1973). *Air pollution and lichens*. Athlone Press, London.

Gilbert, O. L. (1969). The effect of sulphur dioxide on lichens and bryophytes around Newcastle upon Tyne. In *Air Pollution, Proceedings of the First European Congress on the Influence of Air pollution on Plants and Animals*, pp. 223–33. Centre for Agricultural Publishing and Documentation, Wageningen, Netherlands.

Gilbert, O. L. (1970a). A biological scale for the estimation of sulphur dioxide air pollution. *New Phytologist*, **69**, 629–34.

Gilbert, O. L. (1970b). Further studies on the effect of sulphur dioxide on lichens and bryophytes. *New Phytologist*, 69, 605–27.

Gilbert, O. L. (1973a). Lichens and air pollution. In *The lichens*, (eds V. Ahmadjian and M. E. Hale), pp. 443–72. Academic Press, New York.

Gilbert, O. L. (1973b). The effect of airborne flourides. In *Air pollution and lichens*, (eds B. Ferry, M. S. Baddeley, and D. L. Hawksworth) pp. 176–91. Athlone Press, London.

Gilbert, O. L. (1976). An alkaline dust effect on epiphytic lichens. *Lichenologist*, **8**, 173–8.

Gilbert, O. L. (1988). Colonisation by *Parmelia saxatilis* transplanted onto a suburban wall during declining SO_2 air pollution. *Lichenologist*, **20**, 197–8.

Gilbert, O. L. (1989). *The ecology of urban habitats*. Chapman and Hall, London.

Gilbert, O. L. (1990). The lichen flora of urban wasteland. *Lichenologist*, **22**, 87–101.

Grodzinska, K. (1977). Acidity of tree bark as a bioindicator of forest pollution in southern Poland. *Water, Air, Soil Pollution*, **8** 3–7.

Grodzinska, K. (1979). Tree bark sensitive biotest for environmental acidification.

Environment International, **2**, 173–6.

Guest, J. (1989). Further recolonisation of Cheshire by epiphytic lichens. *British Lichen Society Bulletin*, **64**, 29–31.

Hawksworth, D. L. and McManus, P. M. (1988). Yet more lichens on trees in the Chelsea Physic Garden. *British Lichen Society Bulletin*, **63**, 24–25.

Hawksworth, D. L. and McManus, P. M. (1989). Lichen recolonization of London under conditions of rapidly falling sulphur dioxide levels, and the concept of zone skipping. *Botanical Journal of the Linnean Society*, **109**, 99–109.

Hawksworth, D. L. and Rose, F. (1970). Qualitative scale for estimating sulphur dioxide air pollution in England and Wales using epiphytic lichens. *Nature*, **227**, 145–8.

Hawksworth, D. L. and Rose, F. (1976). *Lichens as pollution monitors*. Edward Arnold, London.

Henderson-Sellers, A. and Seaward, M. R. D. (1979). Monitoring lichen reinvasion of ameliorating environments. *Environmental Pollution*, **19**, 207–13.

Kandler, O. (1987). Lichen and conifer recolonisation in Munich's cleaner air. In *Air Pollution and Ecosystems. Proceedings of an International Symposium held in Grenoble, France, 18–24 May, 1987*, (ed. P. Mathy), pp. 784–90. Reidel, Dordrecht.

Laundon, J. R. (1967). A study of the lichen flora of London. *Lichenologist*, **3**, 277–327.

Laundon, J. R. (1970). London's lichens. *London Naturalist*, **49**, 20–69.

Laxen, D. P. H. and Thompson, M. A. (1987). Sulphur dioxide in Greater London, 1931–1985. *Environmental Pollution*, **43**, 103–14.

Lindsay, D. C. (1981). The lichens of the Birmingham region. *Proceedings of the Birmingham Natural history Society*, **24**, 125–52.

Martin, J.-F. and Jacquard, F. (1968). Influence des fumees d' usines sur la distribution des lichens dans la vallee de la Romanche (Isere). *Pollution Atmospherique*, **38**, 95–9.

Nash, T. H. and Wirth, V. (1988). *Lichens, bryophytes and air quality*. J. Cramer, Berlin and Stuttgart.

Nylander, W. (1866). Les lichens du Jardin du Luxembourg. *Bulletin Society Botanique de France*, **13**, 364–72.

Nylander, W. (1896). *Les Lichens des Environs de Paris*. Schmidt, Paris.

Puckett, K. J., Nieboer, E., Flora, B., and Richardson, D. H. S. (1973). Sulphur dioxide: its effect on photosynthetic ^{14}C fixation in lichens and suggested mechanisms of phytotoxicity. *New Phytologist*, **72**, 114–54.

Pyatt, F. B. (1970). Lichens as indicators of air pollution in a steel producing town in south Wales. *Environmental Pollution*, **1**, 45–56.

Richardson, D. H. S. (1988). Understanding the pollution sensitivity of lichens. *Botanical Journal of the Linnean Society*, **96**, 31–43.

Roberts, T. M. and Lane, P. (1981). *Trends in sulphur dioxide concentrations, (1969–1979) in relation to effects on crops*. Central Electricity Research Laboratory, RD/L/N 208/90 Leatherhead, UK.

Rose, C. I. and Hawksworth, D. L. (1981). Lichen recolonisation in London's cleaner air. *Nature*, **289** 289–92.

Seaward, M. R. D. (1976*a*). Performance of *Lecanora muralis* in an urban environment. In *Lichenology: progress and problems* (eds D. H. Brown, D. I. Hawksworth, and R. H. Bailey), pp. 323–357. Academic Press, London.

Seaward, M. R. D. (1976*b*). Lichens in air polluted environments: Multi variate analysis of the factors involved. In *Proceedings of the Kuopio Meeting on Plant Damages Caused by Air Pollution*, (ed. L. Karenlampi), pp. 57–63. University of Kuopio, Finland.

Seaward, M. R. D. (1980). The use of lichens as bioindicators of ameliorating environments. In *Bioindikation auf der Ebene der Indivduen*, (eds R. Schubert and J. Schuh), pp. 17–23. Martin-Luther-Universität Halle-Wittenberg, Germany.

Seaward, M. R. D. (1982). Lichen ecology of changing urban environments. In *Urban Ecology: 2nd European Ecological Symposium*, (eds R. Bornkamm, J. A. Lee, and M. R. D. Seaward), pp. 181–9. Blackwell Scientific Publications, Oxford.

Seaward, M. R. D. and Henderson, A. (1991). Lichen flora of the West Yorkshire Conurbation—Supplement IV (1984-90). *Naturalist*, **116**, 17–21.

Seaward, M. R. D. and Letrouit-Galinou, M. A. (1991). Lichens return to the Jardin du Luxembourg after an absence of almost a century. *Lichenologist*, **23**, 118–6.

Sernander, J. R. (1926). *Stockholms Natur.* Uppsala.

Showman, R. E. (1981). Lichen recolonisation following air quality improvement. *The Bryologist*, **84**, 492–7.

Skye, E. (1968). Lichens and air pollution. *Acta Phytogeograghica Suecica*, **52**, 1–123.

Skye, E. (1980). Continued investigations of epiphytic lichen flora around Kvarntorp in Karke. *Acta Phytogeographica Suecica*, **68**, 141–52.

Skye, E. and Hallberg, I. (1969). Changes in the lichen flora following air pollution. *Oikos*, **20**, 547–52.

Turk, R., Wirth, V., and Lange, O. L. (1974). CO_2-Gaswechseluntersuchungen zur SO_2-resistenz von flechten. *Oecologia*, **15**, 33–64.

van der Knapp, W. O. and van Dobben, H. F. (1987). *Veranderingen in de Epifytenflora van Rijnmond sinds 1972*. RIN- rapport 87/1. Rijkinstituut voor Naturbeheer, Arnhem.

Vass, K. and Ingram, M. (1949). Preservation of fruit juices with less sulphur dioxide. *Food Manufacture*, **24**, 414–16.

Vick, C. M. and Bevan R. J. (1976). Lichens and tar spot fungus (*Rhytisma acerinum*) as indicators of sulphur dioxide pollution on Merseyside. *Environmental Pollution*, **11**, 203–16.

7

Changes in moss-dominated wetland ecosystems

DALE H. VITT AND PETER KUHRY

7.1 Introduction

The Bryopsida, or mosses, are a diverse assemblage of species that form the second largest group of green land plants (Vitt 1982). In general, mosses are poikilohydric and drought tolerant (Proctor 1972; Levitt 1980). As a result of these two characteristics, species can maintain active photosynthesis only when water is available. The ability of this group of plants to tolerate frequent dry periods by coming to thermodynamic equilibrium with their surroundings has enabled them to occupy highly stressed habitats. The abundance and species richness of mosses in tropical montane rain forests; on montane cliff faces and canyons; in boreal forests; in subarctic, boreal, and temperate peatlands; and in arctic meadows, indicates that this group of plants can exist under water- and nutrient-stressed conditions, and also be a dominant component of the ecosystem. Recent studies of tropical montane rain forests (Coxson and Vogel 1989), boreal forests (Oechel and Lawrence 1985), peatlands (Bayley et al. 1987), and arctic meadows (Vitt and Pakarinen 1977) have all suggested that the moss component can have a great influence on the nutrient dynamics and water retention of the ecosystem. However, it is only in peatlands in which continuous records of past species occurrences are found.

7.2 Definition and classification of wetlands

Wetlands can be classified into five classes (Zoltai 1988), including two that form significant deposits of peat. Areas of shallow open water are truly aquatic systems. Swamps dominated by a well developed tree layer, and marshes dominated by Cyperaceae, *Typha*, and/or other monocots, are eutrophic ecosystems. Production is high; however, decomposition is also high, with little peat formation. Production–decomposition processes are influenced by relatively high water flow, eutrophic nutrient regime, and high seasonal water level fluctuations (Fig. 7.1). Mosses are usually a minor component in marshes and swamps, probably because they are unable to compete with the larger, rapidly growing vascular plants due to seasonally variable water levels.

Peatlands are ecosystems that develop when plant production is greater than plant decomposition. The partially decomposed material that results gradually accumulates and forms deposits of peat. In many peatlands the

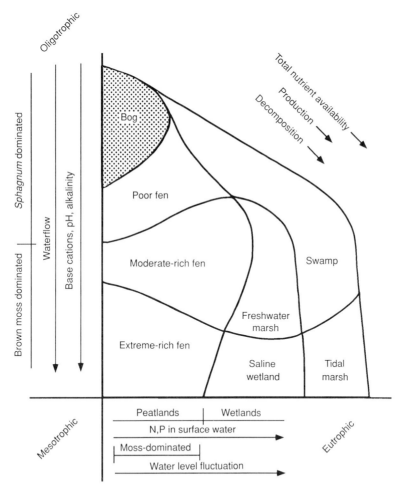

Fig. 7.1 Relationships of bryophyte-dominated peatland types to important physical, chemical, and ecological gradients of wetlands. The shaded area is ombrogenous; white areas are geogenous.

peat deposit is composed largely of moss stems and leaves. In other peatlands, partially decomposed remains of vascular plants form parts of the peat. Fens (minerotrophic systems with geogenous vegetation) and bogs (ombrotrophic systems with ombrogenous vegetation) are mesotrophic or oligotrophic wetlands that develop significant peat deposits; thus they are peatlands. These systems have relatively little seasonal water level fluctuation and are almost always dominated by bryophytes (Fig. 7.1). Fens are systems in contact with groundwater. Based on vegetation, DuRietz (1949) and Sjörs (1950) suggested that fens can be divided into two main types, both of which have correlated chemical characteristics. Rich fens have pHs above 5.5 and are dominated by brown mosses. Sjörs (1952) and more recently Malmer (1986) and Vitt (1990) have stressed the importance of the relationship of pH to electrical conductivity, the latter used as an indicator of total cation availability. Vitt and Chee (1990) have suggested that brown moss dominated mires are developed in alkaline waters where the dominant anion is bicarbonate. Brown mosses, including species of *Scorpidium, Calliergon, Drepanocladus* (*sensu lato*), *Meesia, Campylium, Brachythecium, Tomenthypnum,* and *Catoscopium,* all tolerate alkaline waters. Poor fens have pHs between 4.0 and about 5.5 and are dominated by species of the genus *Sphagnum* (peat mosses). The bicarbonate ion is absent, replaced by higher concentrations of either chloride in oceanic conditions or sulphate in continental conditions (Malmer *et al.* 1992). Cation exchange, taking place on the cell walls of *Sphagnum* (Richter and Dainty 1989) is an important process in poor fens, releasing hydrogen ions in exchange for base cations. This natural acidification is a process seemingly controlled by species of *Sphagnum* (Clymo 1963). With increasing acidity, decomposition is retarded and rates of peat accumulation may increase. Bogs are unique ecosystems that occur only when ombrotrophic conditions are present, which means that all available water and nutrients are derived from precipitation. Ombrotrophy results in extremely low nutrient availability and relatively little water flow; these in turn result in decreased productivity and decreased decomposition. Acidity is increased owing to humic and fulvic acid production during decomposition (Hemond 1980). As a result, pHs of 3.0–4.0 are achieved and the dominant anion becomes organic in nature. *Sphagnum* dominates, also contributing to the acidity by cation exchange. Thus, in both fens and bogs, mosses play a dominant role in nutrient retention and release and in acidification of the ecosystem. The change from an ecosystem dominated by brown mosses to one dominated by *Sphagnum* is directly related to the change from a system influenced by alkaline waters to one influenced by acidic waters.

7.3 Development of peatlands

Peatlands can be formed through the lateral expansion of peat over upland areas (paludification) or through infilling of lakes (terrestrialization). Paludification occurs as a result of a rise of the regional water table induced by climatic change or mediated by local peat build-up itself. Terrestrialization results from sediment and peat infilling of a water-filled depression, with aquatic habitats gradually becoming drier. Eventually the original lake can be completely covered over with peat and terrestrial vegetation. In both scenarios (paludification and terrestrialization), large vegetation changes are evident. At the same time, chemical changes are occurring in the peatland due to peat build-up, oligotrophication, and acidification. Acidification produces wholesale changes in species. Nutrient stress also causes many species to be replaced by others more tolerant to oligotrophy. Secondary internal developmental processes in both paludified and terrestrialized peatlands result in patterning, pool development, and the differentiation of hummocks and hollows.

The development of a peatland is influenced by factors of two types. Allogenic factors are those that influence the peatland through the regional environment, in particular the climate. Also important are edaphic, topographic, and geological features. Autogenic influences are those mediated by the peatland itself. Especially important are peat build-up, acidification by *Sphagnum*, increased humic acid production under acidic conditions, oligotrophication, surface pattern development, and shade produced by trees adapted to particular peatland conditions. Whereas some authors have stressed the importance of allogenic factors (Granlund 1932; Damman 1979; Barber 1981), others have suggested that autogenic processes exert the overriding influences (Frenzel 1983; Glaser and Janssens 1986; Foster *et al.* 1988). In general, it is clear that oceanic peatland complexes are quite different from continental ones (Malmer *et al.* 1992), and that along a climatic gradient, peatlands change so that ombrotrophic bogs in continental areas are distinctly different from those in oceanic regions (Damman 1979; Glaser and Janssens 1986). Likewise geological and bedrock differences, along with climate, influence the type of fens found (Gignac and Vitt 1990). However, it is also clear that the internal development of a peatland is greatly influenced by processes mediated by the peatland itself. The successional patterns that are present from lake to fen to bog occur regardless of allogenic factors. Bogs can develop directly from extreme-rich fens as well as from poor fens, and succession from brown mosses to *Sphagnum*, with the associated chemical changes, takes place under a variety of edaphic, climatic, and geological conditions.

7.4 Macrofossil analysis of peat deposits

Plant macrofossils accumulated in a peat sequence allow the reconstruction of the successive plant communities that formed the organic deposit. They have been deposited where they occurred as living organisms. Macrofossils from outside the peat system are rarely incorporated into the peat because of the lack of appreciable water flow into and through a peat deposit. Microfossils (pollen and spores) contained in the peat are derived from the plants living both on the peatland and in the area surrounding the peatland and can be utilized in reconstructing past peatland and regional vegetation types.

Several aspects have to be taken into consideration in the analysis of peat macrofossils; firstly, not all species occurring on the peatland are uniformly decomposed, and some species may be either over- or under-represented in the fossil record; secondly, vascular plants occurring on the peatland are relatively large plants that may send roots and underground stems far into the peat deposit; thirdly, larger macrofossils (twigs, leaves, seeds) may settle at a different rate than smaller ones; finally, identification of many of the vegetative plant parts, especially of the vascular plants, is often difficult, and different sized macrofossils make quantification difficult. The moss component in the macrofossil record is less affected by these problems. Although selective preservation of bryophytes clearly occurs (as hepatics are almost never found), within brown mosses and *Sphagnum* preservation appears to be relatively uniform (Janssens 1988). Mosses do not have roots, and are deposited exactly where they occurred as living organisms. Bryophytes can mostly be identified on the basis of one or a few leaves, and they are mostly of similar size.

In Europe, detailed studies on peatland development are often based on the analysis of both microfossils and macrofossils, the latter including vascular plant and moss remains. Traditionally, histories of peatland development are based on a comparison of fossil assemblages of vascular plants and mosses with present-day syntaxonomical units from the same region (Rybnícek and Rybnícova 1968; Bakker and Van Smeerdijk 1982; Kuhry 1985). This has resulted in detailed physiognomic reconstructions of successive plant communities that occurred on peatlands through time. As the synecology of the modern counterparts is well known (Ruuhijärvi 1960; Eurola 1962; Westhoff and den Held 1969), inferences regarding environmental changes in peatlands, including water level, nutrient status, and pH have been made. At times, moisture indices based on fossil moss assemblages have been used as an additional tool (Dupont 1986; Van der Molen and Hoekstra 1988). Moreover, an integration of results of the microfossil and the macrofossil analysis has often resulted in a comprehensive view of peatland development as related to climatic, hydrologic, anthropogenic, and autogenic factors (Van Geel 1978; Van Geel and Middeldorp 1988).

In North America, paleoecological studies of peatland development are mostly based on the moss component of the macrofossil record. Because of the relatively uniform decomposition and size of mosses, quantification of results can be achieved. Bryophytes have well-known habitat tolerances (Janssens 1988; Gignac *et al.* 1991*b*) and are quite sensitive indicators of surface water chemistry, nutrient status, height above water level and amount of shade occurring in peatland microhabitats (Horton *et al.* 1979; Gauthier 1980; Andrus *et al.* 1983; Vitt and Slack 1984). The occurrence and abundance patterns of moss species across the flark–string pattern in patterned peatlands has been described in detail by, among others, Slack *et al.* (1980) for extreme-rich fens and by Vitt *et al.* (1975) for poor fens. Species assemblages are often correlated with edaphic and microclimatic situations (Miller 1980; Dickson 1986). As a result, species of bryophytes can be used to infer past chemistry. Reconstructions of peatland development have been expressed in the form of curves indicating inferred changes in pH and moisture (height above water table) through time (Janssens 1988; Nicholson and Vitt 1990; Janssens *et al.* 1992). In addition, bryophyte species have regional distribution patterns controlled largely by climate (Gignac and Vitt 1990), from which it may be possible to infer regional climate.

7.5 Species changes in the Tertiary and Pleistocene

The modern day distributions of bryophyte species are discussed in Chapter 4. Modern day distribution patterns of many of the moss species that are abundant in wetlands are well documented (Janssens 1983; Ochyra and Wojterski 1983; Daniels and Eddy 1985; Crum 1986; Vitt *et al.* 1988). Numerous papers have proposed a historical rationale in order to explain modern day ranges. Wetland species of the northern hemisphere were greatly affected by the Pleistocene glaciations; however, few studies are available that actually document changes in the distribution patterns of species. Although not a wetland species, Janssens *et al.*, (1979) study of *Aulacomnium heterostichum* illustrates the possible extent of changes in the distribution pattern of a species with time. They found this species in Eocene sediments from southern British Columbia, whereas its present day distribution is disjunct between southern China and south-eastern United States. Also, Janssens and Zander (1980) documented *Leptodontium flexifolium* and *Pseudocrossidium revolutum* from 60 000 year old sediments from the Yukon Territory, whereas their present distribution lies either much to the south or to the east. Studies by Dickson (1973), Miller (1980), and Janssens and Glaser (1986) all describe localities for wetland species that are outside their present ranges. These studies all document the extensive changes in the past distributions of wetland species relative to their current ranges and it seems likely that these changes were due to changes in past climate. Kuc and Hill's (1971) report of 13 species from the Pliocene of Banks Island in the western Canadian Arctic Archipelago indicated the presence of rich fen vegetation associated with the coniferous forests present at the locality. Included in these species were *Drepanocladus exannulatus*, *Scorpidium scorpioides*, and *Meesia triquetra*, species that remain common on Banks Island today. Mogensen (1984) reported a similar assemblage of species from the Pliocene of northernmost Greenland. Among 13 species, nine occur in northern Greenland at present, and the remaining four are known from farther south on the island. These deposits, characterized by brown mosses that are typical of boreal rich fens and calcareous arctic fens, have many similarities with a 42 000 yr BP deposit in The Netherlands, where *Scorpidium scorpioides* dominates the moss component of the macrofossil record. Also common in this Pleistocene deposit are *Drepanocladus aduncus*, *D. lycopodioides*, *Calliergon giganteum*, and *Hygrohypnum luridum* (Brinkkemper *et al.* 1987). The Usselo deposits from the east of The Netherlands provide an example of changes in peatland moss assemblages during the latest Pleistocene. Between about 13 000 and 11 000 yr BP a eutrophic pool dominated by aquatic vascular and algal taxa gradually terrestrialized. A rich fen developed with numerous brown mosses. This was followed by a *Sphagnum* dominated poor fen. A subsequent rise in the water table resulted in the reappearance of aquatic taxa (Van Geel *et al.* 1989).

7.6 Documented species changes during the Holocene

7.6.1 Peatland development, the Grenzhorizont, cyclic succession and the decline of *Sphagnum imbricatum* in Europe

As early as 1811, the occurrence of depressions (hollows) and small hills (hummocks) over the peatland surface was noticed and described by William Aiton. In Eville Gorham's (1957) historical review of peatland ecology, Aiton was recognized as one of the first researchers to suggest that hummocks and hollows may alternate with one another. As well, he recognized the general plant succession from lakes to fens and bogs and the importance of topography and chemistry in determining pattern in peatlands. Later von Post and Sernander (1910) proposed that hummock–hollow sequences might be important in bog development and in 1916 Clements published his fully developed theory on plant succession. At this time there was much interest in bog succession and how the patterns observed were related to climate changes. Much of this debate followed from Blytt's (1876) divisions of post-glacial time into climatic periods, and also from Weber's (1900) observation that the upper peat in many German bogs often had a sharp division between a lower, humified, dark peat and an upper, non-humified, light peat. Weber called this boundary the Grenzhorizont. This transition was thought by many to have been climate mediated and was thought to have been associated with the warmth or dryness of the late Sub-Boreal at about 3000–2500 yr BP.

These early papers recognized that the peat deposit contained a variety of plant and animal parts that could be used to reconstruct past plant communities. However, few researchers examined the fossil contents of the peat in order to distinguish individual moss species. Instead, the early paleofloristic studies used seeds, scales, or gross physionomic changes to characterize zones within the peat deposit. This is not to say that the importance of different peat zones, each dominated by a different species of *Sphagnum*, was not considered important. For example, Kulczynski (1949) described a 'Bryales peat' and also a peat profile of *Sphagnum recurvum* peat followed by a *S. medium* peat. Oswald (1923) closely followed *Sphagnum* succession from a plant sociological point of view in an attempt to document his 'cyclic succession' theory, but there are no data from peat stratigraphy or details of moss species composition in the peat column. In 1932, Granlund published his work on the stratigraphy of Swedish bogs in which he reported observations on different types of *Sphagnum* peat, but without detailed data on specific species changes. In 1961, Walker published a paper on post-glacial hydroseres, again without examination of changes in the moss species. However, he stated that *Sphagnum*, once established, expands very rapidly indeed and quickly imposes its dominating influence on any further development. Walker and Walker (1961) presented a semi-quantified expression of the past occurrence of several species of *Sphagnum*. They showed that in peat sections between 30 and 50 cm in length, a definite succession of species occurred. Most of these authors were interested in bog development and how autogenic and allogenic factors (especially climatic changes) have influenced the complicated surface patterns found on oceanic bogs. Little work was done on fen development (Tallis 1983), other than the early and gross stratigraphic studies of Weber (1902) which discussed terrestrialization (Verlandunghypothese).

Since the 1960s, there have been numerous publications documenting

Table 7.1 Important peat-forming moss species found in deposits that correspond to Pollen zones III–VIII in eastern Finland. Extracted from Tolonen *pers. comm.* and (1967).

Pollen zone	Approximate age limits (yr BP)	Brown mosses	Mesotrophic sphagna	Oligotrophic sphagna
VIII	2500	*Drepanocladus badius* *D. exannulatus* *D. fluitans* *Scorpidium scorpioides*		*S. cuspidatum* *S. fuscum* *S. magellanicum* *S. tenellum*
	3850			*S. balticum* *S. magellanicum* *S. parvifolium*
VII	4700	*Calliergon stramineum* *Scorpidium scorpioides*	*S. pulchrum* *S. subsecundum*	*S. fallax* *S. magellanicum* *S. majus* *S. papillosum* *S. parvifolium*
	6700			
VI	6700	*Calliergon richardsonii* *C. stramineum* *Drepanocladus exannulatus* *D. fluitans* *Scorpidium scorpioides*	*S. obtusum* *S. subsecundum* *S. teres*	
	8000			
V	8000	*Calliergon richardsonii* *C. stramineum* *C. trifarium* *Catoscopium nigritum* *Drepanocladus exannulatus* *Paludella squarrosa* *Scorpidium scorpioides*	*S. aongstroemii* *S. flexuosum* *S. subsecundum* *S. teres* *S. warnstorfii*	*S. compactum* *S. fallax* *S. lindbergii* *S. majus* *S. papillosum*
	8800			

Table 7.1 (*Cont.*)

Pollen zone	Approximate age limits (yr BP)	Brown mosses	Mesotrophic sphagna	Oligotrophic sphagna
IV	8750	*Calliergon giganteum* *C. richardsonii* *C. stramineum* *Cinclidium stygium* *Drepanocladus exannulatus* *D. fluitans* *D. tundrae* *Paludella squarrosa* *Scorpidium scorpioides*	*S. flexuosum* *S. subsecundum* *S. teres* *S. warnstorfii*	*S. fallax* *S. jensenii* *S. majus*
	9700			
III	9150	*Aulacomnium palustre* *Campylium stellatum* *Calliergon stramineum* *C. trifarium* *Drepanocladus exannulatus* *D. vernicosus* *Paludella squarrosa* *Scorpidium scorpioides*		
	10150			

wetland development inferred from either the pollen record or from gross analysis of peat. However, few studies have documented in detail the changes occurring in bryophyte communities. A summary of the vegetation and moss species changes that can be documented from more detailed studies on European peatland development carried out in the last three decades follows.

Ruuhijärvi (1963) presented a study on the developmental history of four peatlands in northern Finland, with detailed information on the related moss successions. Sonesson (1968) used species occurrence of mosses to build the sequence of strata for his study of a peatland at Abisko, Sweden. Tolonen (1967) made extensive use of bryophytes to characterize his peatland developmental stratigraphy in eastern Finland. His qualitative study of these macrofossil associations (Table 7.1) showed clearly that on a regional landscape, the peatlands in this area were gradually developing from Braunmoor (rich fen) to aapamoor (patterned fen) to Hochmoor (bog), and that species of *Sphagnum* became more important components of the regional landscape in the late Holocene. The development of peatlands and associated species changes in central Europe is documented by Rybníček (1973). Mosses were dominant components of these ecosystems, with *Scorpidium scorpioides* common only during the early Holocene, but nearly disappearing by the middle Holocene. *Sphagnum subsecundum* was abundant during the early and middle Holocene becoming rare in the late Holocene. More oligotrophic species of *Sphagnum* are common only after the early Holocene, when according to Rybníček (1973) and Tolonen (1967), oligotrophic communities developed throughout central and northern Europe during the Atlantic period (about 7500–5000 yr BP). A detailed paleoecological study including microfossils (pollen and spores) and macrofossils (vascular plant remains and mosses) from the 'Bláto' mire in southeastern Bohemia, Czechoslovakia, has been published by Rybníček and Rybníčková (1968) and included specific notes on the occurrence of fossil bryophytes in Czechoslovakia.

These analyses led to important questions about habitat availability for species of *Sphagnum* occurring in ombrotrophic habitats. If large-scale development of ombrotrophy occurred only after the early Holocene in central and northern Europe, then in what habitats and in what geographic localities did these species occur before that period? The data of Tolonen (1967) and Rybníček (1973) indicate significant changes in the habitat availability for bryophyte species, with rich fen habitats being rare at present, and hence populations of species of *Scorpidium*, *Calliergon*, and *Drepanocladus* are small and widely scattered. In Europe, it is within this group of species that our rarest species are found and for which habitats need to be secured.

Weber's original concept of a climate mediated Grenzhorizont in German peatlands (1900) has been the subject of extensive discussion. Averdieck (1957) renamed it as the Schwarztorf-Weisstorf Kontakt (SWK). The most recent reviews include those by Overbeck *et al.* (1957), Casparie (1969), Overbeck (1975), and Frenzel (1983). Overbeck (1975), among others, observed that the Grenzhorizont (or SWK), although recorded in Germany, The Netherlands, England, Ireland, and Scandinavia, was not synchronous, even in peatlands within a small area and within individual peatlands themselves. This renders a climatic interpretation of this event less probable. More recent detailed studies on raised bogs in Germany and The Netherlands, however, seem to support a climatic cause for this transition at least in the Dutch–German region. An example is the study of a peatland in the east of The Netherlands by Van Geel (1978). Until around 3200 yr BP, *Sphagnum* sect. Acutifolia (*S.* cf. *rubellum*) was the

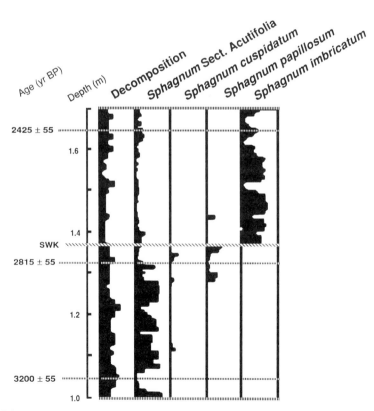

Fig. 7.2   Abundance changes of four important *Sphagnum* species associated with the Grenzhorizont (or SWK) transition, at Engbertsdijksveen, eastern part of The Netherlands (after Van Geel 1978).

dominant moss on the dry hummocks of an ombrotrophic raised bog (Fig. 7.2). This was followed by a transitional phase in which first *S. cuspidatum* and later *S. papillosum* became dominant, indicating the local development of pools. This was interrupted by several periods in which *S.* sect. Acutifolia was dominant. At around 2815 yr BP, *S. cuspidatum* and *S. papillosum* were generally dominant. Shortly after this period, *S. imbricatum* began to dominate, indicating the development of low hummocks. It is at this last transition that the SWK can be observed. According to Van Geel, these changes in moss composition were mediated by climatic change. He observed a clear correlation of events in the peatland with changes in forests surrounding the raised bog area. At the same time, pollen indicators of human influence were markedly out of phase, and important human influences began only after *S. imbricatum* was well established in the peatland. The stepwise changes in both the peatland and surrounding forest were interpreted to be associated with climatic instability at the end of the Sub-Boreal (between 3400 and 2600 yr BP) leading to the more oceanic climate of the Sub-Atlantic period.

Several other detailed studies fit well into this model, although differences between peatlands have resulted in somewhat different local development. In some cores, *S. papillosum* is missing (Klinger 1968; Van Geel 1972). At other sites both *S. cuspidatum* and *S. papillosum* are missing, and *S.* sect. Acutifolia is immediately replaced by *S. imbricatum* (Dupont 1986, hummock sequence). Still in other cases the sequence as described above is present, but the SWK is located at a slightly lower level, in the *S. cuspidatum–S. papillosum* dominated phase (Middeldorp 1986), or is unclear (Dupont 1986, hollow sequence). In one case, the establishment of *S. imbricatum* was delayed until about 2000 yr BP (Dupont and Brenninkmeijer 1984) with the SWK located at the transition of the *S.* sect. Acutifolia peat to the *S. papillosum–S. cuspidatum* assemblages. However, in all these studies, marked changes in the moss composition started after 3400 yr BP, corresponding with the beginning of climatic instability at the end of the Sub-Boreal. With the exception of the study by Dupont and Brenninkmeijer (1984), *S. imbricatum* was consistently established by 2600 yr BP. As a result of local differences, the beginning of uninterrupted growth of regenerative peat (SWK) occurred at slightly different times; however, this almost always occurred within this period of climatic instability. As indicated by several authors (Van Geel 1972; Dupont and Brenninkmeijer 1984; Dupont 1986; Middeldorp 1986), *Sphagnum* successions were associated with changes in surrounding forests; however, they were out of phase with periods of human influence. Stepwise climatic change towards more oceanic regimes is the most probable cause of these changes. Moreover, similar successions starting at about the same period in peat-

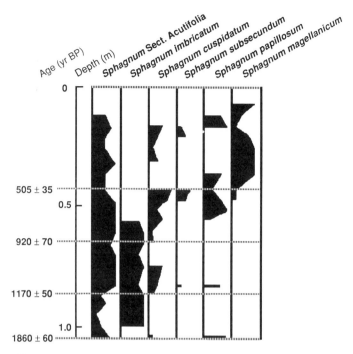

Fig. 7.3   Abundance changes of six important species of *Sphagnum* at Bolton Fell Moss, north-western England (after Barber 1981).

lands of different ages and having different developmental histories can hardly be assigned to purely autogenic processes.

An overview of paludification pathways in four peatlands located in Germany and The Netherlands, including several of the sites mentioned above, was presented by Dupont (1987). Paludification at these sites began in the late Atlantic and early Sub-Boreal periods (between 7000 and 4500 yr BP), whereas initiation of peat development is normally associated with wetter climatic regimes during the Atlantic period (about 7500–5000 yr BP). In the original studies on these peatlands, the moss component of the macrofossil record was analysed. At first, mosses characteristic of the lagg-zone of peatlands were common and included *Aulacomnium palustre*, *Polytrichum strictum*, *Pohlia nutans*, and *Sphagnum palustre*. After a relatively short time span of 250 to 1500 years, depending on the site, poor fen and bog mosses of the genus *Sphagnum* began to dominate. In the British Isles, the formation of extensive blanket bogs has been associated with forest clearance by early man. Deforestation caused an increase in the ground-water table resulting in extensive paludification (Moore 1973, 1975).

Since Oswald's (1923) ideas on 'cyclic succession' were published, much effort has been devoted to determining how the topography of a bog surface develops and is maintained. Barber (1981) presented a detailed analysis of peat profiles from north-western England. He showed clearly (Fig. 7.3) that over the past 2000 years, a gradual succession of *Sphagnum* species occurred. He interpreted the surface topography from one profile to have varied between pool, wet lawn, and dry lawn. The dominant species changes were from *S. imbricatum* to *S. cuspidatum* to *S. papillosum* to *S. magellanicum*. *Sphagnum imbricatum* remained dominant for about 1200 years, followed by a 300 year period of *S. cuspidatum* and *S. papillosum* dominance. During the last 500 years *S. magellanicum* was dominant. Barber concluded that these bog, pool, and lawn communities were characterized by associations of species, not by monospecific replacements, and that the time-scale of species interplay is markedly out of step with any theory of cyclic regeneration. His overall conclusion, based on both macrofossil and microfossil data, is that autogenically controlled cyclic regeneration (*sensu* Oswald 1923) does not occur in British bogs. Rather the surface topography and species composition has 'been very sensitive to climatic changes over the last two millenia or more, reacting swiftly to changes in temperature and/or rainfall in forming pools or else broad dry surfaces which are revealed to us by their characteristic stratigraphy and macrofossil content'. For a wide geographic area, the overall validity of this statement can be questioned in view of more recent data that have become available. Bryophyte successions similar to those observed by Barber (1981) took place between 2000 and 1600 yr BP in a peatland in The Netherlands (Dupont 1986, hollow sequence). In an Irish bog (Van Geel and Middeldorp 1988) a similar moss succession occurred within the last 700 years (Fig. 7.4).

Overbeck (1975) reviewed the possible causes for the decline of *S. imbricatum* in north-western Europe. Barber (1981) explained the marked present day absence of *S. imbricatum* on British peatlands in the light of climatically induced cycles, whereas Pearsall (1956) thought that burning and/or grazing was responsible for its disappearance. Tallis (1964) stated that the disappearance of *S. imbricatum* is undoubtedly due to drying out of the peat. A number of species sequences from several widespread localities in the British Isles and mainland Europe all show the replacement of *S. imbricatum* by *S. magellanicum* (Barber 1981—north-western England; Tallis 1964—southern England; Van Geel and Middeldorp 1988—Ireland; Dupont 1986—The Netherlands). However, the timing and water level changes accompanying this species change differ between sites. At the Irish site (Fig. 7.4), an *S. imbricatum* dominated moss assemblage was replaced by *S.* sect. Acutifolia with some *S. imbricatum* at c. 700 yr BP. This transition coincided with an increase in human activities in the area

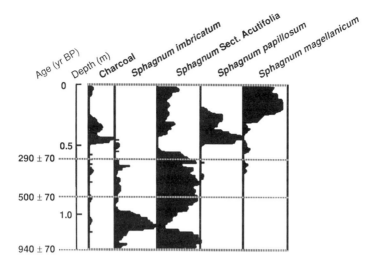

Fig. 7.4 Abundance changes of the dominant *Sphagnum* species and occurrence of charcoal at Carbury, eastern Ireland (after van Geel and Middeldorp 1988).

surrounding the bog, also indicated by the presence of several thin charcoal layers in the peat column. Then, at *c.* 300 yr BP *S. papillosum* became the dominant peat moss following an important rise in the local water table that coincided with massive deforestation in the surrounding area. A concentration of charcoal is present in the peat sequence. The water table rise could be associated with either increased regional precipitation or with changes in rainfall interception and evapotranspiration related to loss of forest cover (Bosch and Hewlett 1985; see also Moore 1973, 1975). Van Geel and Middeldorp (1988) concluded that the disappearance of *S. imbricatum* is due to the increased influx of anthropogenic dust and charcoal.

Thus in oceanic climates, where ombrotrophic bogs predominate, changes in species dominance patterns can be documented. These species changes have been attributed to both autogenic and allogenic influences. In the case of changes in moss assemblages associated with the Grenzhorizont (or SWK) new evidence suggests a climatic cause for these successions, although more detailed analyses are needed in western, central, and northern Europe to confirm the general validity of this interpretation. At least in the British Isles, the influence of early man may be a primary factor in the initiation of some extensive blanket mires (Moore 1973, 1975); man may also be the overriding cause of at least some species changes. Barber's work (1981) indicated that climatic changes during the last 2000 years may have had an important effect on abundance of *Sphagnum* species in oceanic climates. However, recent studies show that the species changes reported by Barber occur at markedly different times in western Europe, while at one site a clear relation of these successions to human influence has been observed. A complexity of climatic, anthropogenic, and autogenic factors seem to be responsible for the varying spatial and temporal development of ombrotrophic bogs during the last 2000 years in western Europe. More recently, atmospheric pollution since the Industrial Revolution is almost certainly the cause of the demise of *Sphagnum* in the Southern Pennines of England (Tallis 1964; Lee *et al.* 1987; see also Chapter 12).

7.6.2 Terrestrialization, paludification, and patterning in North America

Peatland development has been postulated to proceed either through terrestrialization or paludification (Tallis 1983). Chemical changes within either of these developmental sequences are generally from alkaline to acid surface waters, from flowing to stagnant water, and from mesotrophy to oligotrophy (Fig. 7.1). Bogs can exist within patterned peatland complexes that vary from extreme-rich fens to poor fens. Foster *et al.* (1983) and Foster and King (1984) described the development of patterning in poor fens from eastern North America and Europe. They suggested that the flark–string pattern is secondary, with these landforms developing some time after the initiation of the peatland.

In North America, several studies have shown that both terrestrialization and paludification can be important processes in the development of an individual peatland (Futyma and Miller 1986; Kubiw *et al.* 1989; Nicholson and Vitt 1990). Figures 7.5 and 7.6, modified from Kubiw *et al.* (1989), illustrate the changes in moss species in relation to these two important peatland processes at Muskiki Lake, Alberta, Canada. At the site closest to the lake, terrestrialization started at about 9000 yr BP (Fig. 7.5). The core contains a sequence of species beginning with *Chara*, then *Scorpidium scorpioides*, *Drepanocladus revolvens*, and *Calliergon richardsonii*. Kubiw *et al.* interpret this developmental sequence to be strongly autogenically influenced. At about 7300 yr BP, an abrupt change to dry habitat species (*Paludella squarrosa*, *Tomenthypnum nitens*, *Sphagnum warnstorfii*, and other species of *Sphagnum* sect. Acutifolia) might reflect a change in the regional climate to drier conditions that is also reflected in local lake histories (Schweger and Hickman 1989), or these dry habitat species may have become more abundant due to a continuation of the terrestrialization process. Between about 4700 and 2800 yr BP, the area reverted to wet fen vegetation dominated by *Drepanocladus revolvens* and *Scorpidium scorpioides*. At present, the site is dominated by *Tomenthypnum nitens* and *Sphagnum warnstorfii* and no patterning is present at this terrestrialized site. In the paludified area of this peatland, peat formation was studied at two sites

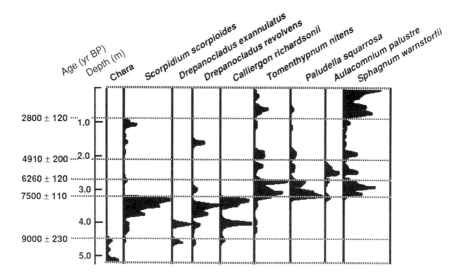

Fig. 7.5   Abundance changes of the dominant moss species and the genus *Chara* from the terrestrialized area at Muskiki Lake, Alberta, Canada (after Kubiw *et al*. 1989).

(Fig. 7.6). Peatland development was initiated between 8500–7500 yr BP, in a wooded environment dominated by *Sphagnum warnstorfii*, along with *S. magellanicum*, *S. subsecundum*, and *Calliergon richardsonii*. Mesotrophic *Sphagnum* species remained dominant until gradually replaced by brown mosses. Based on the macrofossil analyses, both sites began as relatively dry sites, gradually over a 2500–3000 year period becoming considerably wetter and more alkaline. Subsequently, pattern formation took place. This study shows clear evidence of the secondary nature of flark–string patterning at these sites. Flarks, dominated by *Scorpidium scorpioides*, differed from strings, dominated by *Tomenthypnum nitens*, *Paludella squarrosa*, and *Sphagnum warnstorfii*. The cores show a clear spatial and temporal alternation of flark and string vegetation in the developmental history of the peatland.

Patterns of change associated with terrestrialization and paludification are remarkably similar across the boreal region of the world. The species involved in successive seral stages generally have well known habitat preferences and often can be used to predict either chemical, hydrologic, or even climatic conditions associated with the seral stages (Gignac *et al*. 1991*b*). When the abundance of peatlands in a given area is related to pH of the surface water, a bimodal curve results, with rich fens dominated by brown mosses dominating one peak, and bogs and poor fens dominating the second (Gorham *et al*. 1987; Gorham and Janssens 1992). Relatively

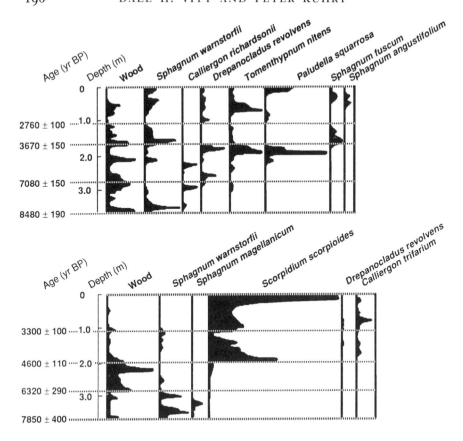

Fig. 7.6 Abundance changes of the dominant moss species and occurrence of wood from the two paludified sites at Muskiki Lake, Alberta, Canada (after Kubiw *et al.* 1989). *Top*: paludified site with present day string vegetation; *Bottom*: paludified site with present day flark vegetation.

few peatlands can be found at present that have pHs of about 5.5–6.0. Chemically this can easily be explained as the region along the acidity–alkalinity gradient where HCO_3^- alkalinity becomes zero (Wetzel 1975). When surface vegetation patterns are studied the vegetation change from brown mosses to *Sphagnum* is clearly correlated with this chemical change. However, the general rarity of peatlands having pHs near this chemical and vegetational change seems to have a time component as well. The development of poor fens and bogs from rich fens must pass through this change over pH. We suggest that the rarity of peatlands of this pH is largely due to the short period of time that peatlands of this pH exist in the developmental sequence. Figure 7.7 supports this hypothesis. Rich fen vegetation dominated by the brown moss *Drepanocladus lapponicus* existed

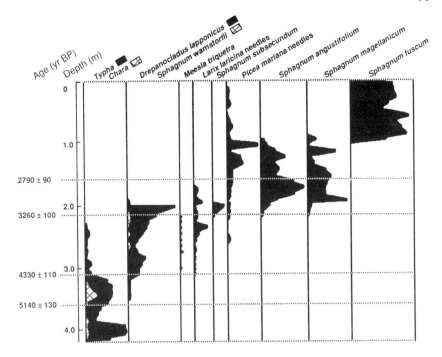

Fig. 7.7 Abundance changes of the dominant moss species and other important indicators of local conditions at Beauval, western Saskatchewan, Canada (after Kuhry *et al.* 1992*b*).

for a considerable length of time; likewise poor fen and bog vegetation dominated by three species of *Sphagnum* was present for several thousand years. However, the replacement of the brown mosses by *Sphagnum* species was extremely rapid, occurring over approximately 150 years and in this case was associated with the presence of *S. subsecundum*. Thus, the most significant changes in the chemistry, vegetation, and development of this and several other fen–bog ecosystems can be documented in the macrofossil record by the occurrence of *Sphagnum subsecundum sensu lato* (Kuhry *et al.* 1992*b*; Janssens *et al.* 1992). In other cases, the rich fen-poor fen/bog transition occurred directly from brown mosses (Kuhry *et al.* 1992*a*, *b*). Inferred pH reconstructions (Janssens 1988; Nicholson and Vitt 1989; Janssens and Engstrom 1992) strongly suggest that pH regimes in the 5.5–6.0 range are highly unstable and are rapidly influenced by the change from brown moss to *Sphagnum*-dominated species assemblages.

In the southern boreal region of western Canada, Zoltai and Vitt (1990) demonstrated the marked influence of regional climatic change on wetland development, with peat initiation and the subsequent seral species changes

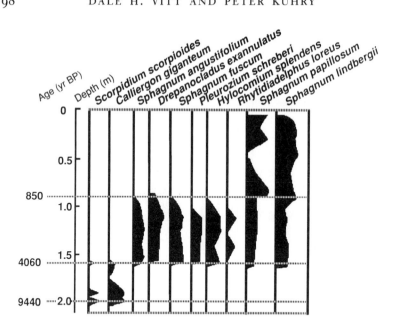

Fig. 7.8 Abundance changes of the dominant moss species at Pleasant Island, south-eastern Alaska (after Janssens and Engstrom 1992).

restricted to the past 6000 yr BP. This is associated with prevailing warm and dry climatic conditions during the early and middle Holocene. When conditions gradually became wetter peatland development began. However, at the beginning the influence of still drier conditions is reflected in the abundance of *Typha* and other eutrophic indicators and the scarcity of mosses, indicating marked fluctuations in the water table associated with drier climatic regimes (Fig. 7.7). Farther north, the influence of this warmer and drier period was less dramatic. Although peatland development began earlier at these middle boreal sites and a *Typha*-dominated phase was not present, extensive paludification was restricted to approximately the last 6000 years (Nicholson and Vitt 1990).

In south-eastern Alaska, Janssens and Engstrom (1992) described a profile that began with *Calliergon giganteum* and *Scorpidium scorpioides* about 9500 yr BP (Fig. 7.8). At present, *Scorpidium scorpioides* is known from only two sites in oceanic western North America (Miller 1980; Janssens and Engstrom 1992), while other continental species are rare in the area. This rich fen vegetation changed abruptly to a *Sphagnum*-dominated association at around 4100 yr BP. Interestingly, this association consisted of *S. angustifolium*, *S. fuscum*, and *S. centrale*, all widespread sub-continental species that are not abundant on the coast at present, mixed with more oceanic species such as *S. papillosum, S. lindbergii* and *Rhytidiadelphus loreus.* At 850 yr BP, these species of drier habitats disappeared leaving an association dominated by *S. papillosum, S. lindbergii,* and *Rhytidiadelphus loreus.* Janssens and Engstrom attribute the change at 4100 yr BP to autogenic influences, and the change at 850 yr BP to a rise of the associated lake level.

7.6.3 Species changes in polar regions

Whereas in most boreal forest sites, a distinct directed succession of moss species is evident, in the Arctic, where permafrost dominates the landscape, species changes associated with chemical gradients are much less prevalent. Instead species changes are directed by changes in water availability. The availability of water was considered by Vitt and Pakarinen (1977) and Longton (1988) as the most important environmental factor controlling occurrence and abundance of mosses in studies of present day community pattern. Recently, LaFarge-England (1989) presented results from Elles- mere Island that corroborated these earlier results. Whereas these studies were concerned with present day spatial species abundance patterns, LaFarge-England *et al.* (1991), determined similar patterns through a time sequence. Over a 1000 year period (3450–2430 yr BP), associations fluctuated between those dominated by *Tortula ruralis* and those dominated by *Philonotis fontana* and *Aulacomnium palustre* (Fig. 7.9). Over the 1000

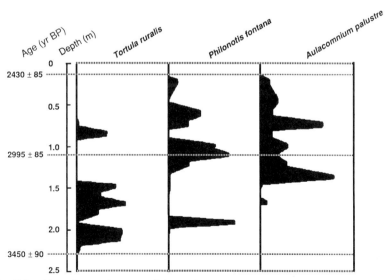

Fig. 7.9   Abundance changes of three dominant moss species at Southern Piper Pass, northern Ellesmere Island, Canada (after LaFarge-England *et al.* 1991).

year period, *Tortula ruralis* was the dominant moss on three separate occasions, each time being succeeded by *Philonotis fontana*. Using modern day habitat preferences, the authors calculated moisture indices for each of the successive species associations, and inferred significant changes in the moisture regime of the microhabitats. This study illustrates the highly dynamic nature of species occurrences in a soligenous fen, without an overall change in community composition. It contrasts sharply with a report from the eastern Canadian Arctic (Carey Island) of *Aplodon wormskjoldii* existing continuously for 2000 years (Brassard and Blake 1978). In the Antarctic, Baker's (1972) long profile of a single, unchanging species, *Chorisodontium aciphyllum*, also stands out in marked contrast.

7.6.4 Species changes in the northern Andes of South America

Detailed studies from the Colombian Andes reveal dynamics similar to those present in the boreal zone, albeit with different species. At the 'Páramo de Laguna Verde' site shown in Fig. 7.10, located at 3650 m elevation near Bogota, peat formation began via terrestrialization about

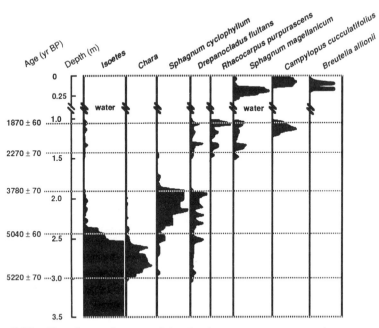

Fig. 7.10 Abundance changes of the dominant moss species and two important indicator species at Páramo de Laguna Verde, Eastern Cordillera, Colombia (after Kuhry 1988*a*).

5000 yr BP (Kuhry 1988a). Abundant remains of *Chara* and *Isoetes* indicate a shallow pool that became dominated by *Drepanocladus fluitans* and *Sphagnum cyclophyllum* for some 1200 years. About 2300 yr BP, an association developed dominated by *Rhacocarpus purpurascens*, with *Sphagnum magellanicum* and *Campylopus cucculatifolius*. More recently, *Sphagnum magellanicum*, *Breutelia allionii* and *Campylopus cucculatifolius* have dominated. At the 'El Bosque' site, located at an altitude of 3650 m in a volcanic part of the Colombian Central Cordillera, a rich wet fen with *Drepanocladus aduncus* dominated between about 5800 and 4000 yr BP. It was replaced by a fen with *Campylopus*, suggesting drier conditions. On several occasions, there were sudden increases in the local water table related to disturbances in the forested catchment of the area caused by volcanic activity. These events resulted in complicated patterns of moss succession (Kuhry 1988b).

7.7   Conclusions

The evolution of regional landscapes has greatly affected the distribution of peatland types and the species that are restricted to them. During the early Holocene, fens were abundant on deglaciated temperate and boreal landscapes. In continental boreal North America, peatlands were restricted in occurrence and *Sphagnum* was not common in peat deposited during this time. During the middle and late Holocene, *Sphagnum* dominated ecosystems became more prevalent. Although there are few definitive studies, it appears that peatland development and associated changes in moss species assemblages were influenced by both autogenic peatland developmental processes and by regional climatic changes. In polar regions, water availability appears to be the controlling factor in species changes.

Studies that document changes in wetland moss species during the Holocene are not numerous, but several generalizations can be made based on the available literature. In Figure 7.11, documented changes in moss assemblages from representative peatland sites in temperate and boreal regions, oceanic and continental areas, in north-western Europe and North America are schematically illustrated. Regional climatic change may have been critical in determining site specific changes in oceanic areas, while in continental areas it may have been more influential in restricting the occurrence and development of peatlands in general. In oceanic Europe, paludification occurred over a wide area during the Atlantic period as a result of wetter climatic conditions. Because of the moist conditions, paludification often led quickly to ombrotrophic conditions. Subsequent successional phenomena, as in the case of the Grenzhorizont (or SWK), were most probably influenced by climatic change. In Europe, especially in the late Holocene, patterns in peatland development may have been largely

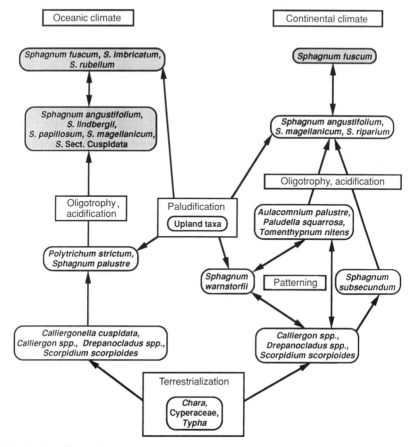

Fig. 7.11 Generalized species associations, seral pathways, and peatland processes based on representative documented Holocene bryophyte macrofossil records from Europe and North America under oceanic and continental climatic regimes. Shaded areas are ombrogenous; white areas are geogenous. Data taken from Van Geel (1989), Bakker and van Smeerdijk (1982), Dupont (1986), Kubiw *et al.* (1989), Kuhry *et al.* (1992*b*), Janssens and Engstrom (1992), and Kuhry (unpublished data).

mediated by man. In continental Canada, peatland development was at first restricted to the northern boreal region because of dry conditions to the south. In the southern boreal region, peatland development began only after 6000 yr BP and at the present boundary with the parkland zone as recent as 200 yr BP. In continental areas, peat accumulation may be inhibited by seasonal drought, especially in the late summer, that reduces the effective growing season for the major peat-forming bryophyte species.

This has resulted in a slower and more infrequent development of truly ombrotrophic conditions. Alkaline surface waters also may possibly influence slow developmental patterns. As a result, many paludified and terrestrialized sites in continental western Canada have not yet developed oligotrophic conditions, much less ombrotrophy. The differential development of peatlands in continental and oceanic areas is also reflected in secondary developmental patterns, with ombrotrophic pool formation restricted to oceanic bogs and flark–string development restricted to more continental fens.

Clearly, peatland formation is in general restricted to certain ecoclimatic regions. While peatland development under oceanic climatic regimes can be quite different from that under continental conditions, under both regimes peatlands develop acidity, oligotrophy, and ombrotrophy as a result of peat build-up and species changes (Fig. 7.11). These processes are autogenic in nature. However, the rate of change is influenced by climate.

Mosses provide excellent evidence for the occurrence of paludification, terrestrialization, and patterning in peatlands. Moss species assemblages provide clear evidence of the rapidity with which chemical and developmental change can occur in peatlands. *Sphagnum subsecundum* (*sensu lato*) is a species characteristic of the transition from brown moss dominated vegetation to that dominated by other *Sphagnum* species of more oligotrophic habitats, although this transition can also occur directly from brown mosses. These mosses of mesotrophic habitats serve to initiate acidification events that are further mediated by other species of *Sphagnum*.

Future studies of recurrence horizons (including the Grenzhorizont or SWK), peatland developmental sequences, and peatland vegetation responses to climatic change must include the analysis of the bryophyte components. Wetlands, especially peatlands, are ecosystems that are particularly sensitive to disturbance by man and to regional climatic change. Both will play a dominating role in the occurrence, much less the existence, of wetland moss species of the future.

Acknowledgements

This chapter was made possible through funding from the Natural Sciences and Engineering Research Council of Canada; especially through operating grant A-6390 to D. H. Vitt and a Strategic Grant to S. E. Bayley, D. H. Vitt, and D. W. Schindler. The computer graphics were done by Sandra F. Vitt, to whom we are grateful. Our continued enthusiasm for peatlands, bryophytes, and paleoecology has been much enhanced through discussions and interactions with Jan Janssens and Bas van Geel.

References

Aiton, W. (1811). *Treatise on the origin, qualities, and cultivation of moss-earth, with directions for converting it into manure.* Wilson and Paul, Air.

Andrus, R. E., Wagner, D. J., and Titus, J. E. (1983). Vertical zonation of *Sphagnum* mosses along hummock–hollow gradients. *Canadian Journal of Botany*, **61**, 3128–39.

Averdieck, F. R. (1957). Zur geschichte der Mooren und Wälder Holsteins. Ein Beitrag zur Frage der Rekurrenz-Flächen. *Nova Acta Leopold*, **19**, 130.

Baker, J. H. (1972). The rate of production and decomposition of *Chorisodontium aciphyllum* (Hook. f. & Wils.) Broth. *British Antarctic Survey Bulletin*, **27**, 123–9.

Bakker, M. and van Smeerdijk, D. G. (1982). A palaeoecological study of a late Holocene section from 'Het Ilperveld', western Netherlands. *Review of Palaeobotany and Palynology*, **36**, 95–163.

Barber, K. E. (1981). *Peat stratigraphy and climatic change.* A. A. Balkema, Rotterdam.

Bayley, S. E., Vitt, D. H., Newbury, R. W., Beaty, K. G., Behr, R., and Miller C. (1987). Experimental acidification of a *Sphagnum*-dominated peatland. *Canadian Journal of Fisheries and Aquatic Sciences (Suppl.)*, **44**, 194–205.

Blytt, A. (1876). *Essay on the immigration of the Norwegian flora during alternating rainy and dry periods.* Cammermeye, Christiana.

Bosch, J. M. and Hewlett. J. D. (1982). A review of catchment experiments to determine the effect of vegetation changes on water yield and evapotranspiration. *Journal of Hydrology*, **55**, 3–23.

Brassard, G. R. and Blake, Jr, W. (1978). An extensive subfossil deposit of the arctic moss *Aplodon wormskjoldii. Canadian Journal of Botany*, **56**, 1852–9.

Brinkkemper, O., van Geel, B., and Wiegers, J. (1987). Paleoecological study of a middle-pleniglacial deposit from Tilligte, The Netherlands. *Review of Paleobotany and Palynology*, **51**, 235–69.

Casparie, W. A. (1969). Bult- und Schlenkenbildung in Hochmoortorf. *Vegetatio*, **19**, 146–80.

Clements, F. E. (1916). Plant succession: an analysis of the development of vegetation. *Carnegie Institution of Washington Publication*, **242**, 1–512.

Clymo, R. S. (1963). Ion exchange in *Sphagnum* and its relation to bog ecology. *Annals of Botany, N.S.*, **27**, 309–24.

Clymo, R. S. (1983). Peat. In *Mires: swamp, bog, fen and moor, (Ecosystems of the world: 4A)*, (ed. A. J. P. Gore), pp. 159–224. Elsevier, Amsterdam.

Coxson, D. S. and Vogel, J. H. (1989). Drought exposure and nutrient cycling— The role of canopy epiphytes in a tropical cloud mist forest. (Abstract). *Supplement to American Journal of Botany*, **76**, (6), 5.

Crum, H. A. (1986). Illustrated moss flora of arctic North America and Greenland. 2: Sphagnaceae. *Meddelelser om Groenland Bioscience*, **18**, 3–61.

Damman, A. W. (1979). Geographic patterns in peatland development in eastern North America. In *Classification of peat and peatlands*, Proceedings of the International Peat Society Symposium, Hyytiälä, Finland, (eds E. Kivinen, L.

Heikurainen, and P. Pakarinen), pp. 42–57. International Peat Society, Helsinki.

Daniels, R. E. and Eddy, A. (1985). *Handbook of European Sphagna*. Institute of Terrestrial Ecology, Abbots Ripton, UK.

Dickson, J. H. (1973). *Bryophytes of the Pleistocene. The British record and its chronological and ecological implications*. Cambridge University Press, London.

Dickson, J. H. (1986). Bryophyte analysis. In *Handbook of Holocene paleoecology and paleohydrology*, (ed. B. E. Berglund), pp. 627–43. Wiley, New York.

Dupont, L. M. (1986). Temperature and rainfall variation in the Holocene based on comparative palaeoecology and isotope geology of a hummock and a hollow (Bourtangerveen, The Netherlands). *Review of Paleobotany and Palynology*, **48**, 71–159.

Dupont, L. M. (1987). Paleoecological reconstruction of the successive stands of vegetation leading to a raised bog in the Meerstalblok area (The Netherlands). *Review of Paleobotany and Palynology*, **51**, 271–87.

Dupont, L. M. and Brenninkmeijer, C. A. M. (1984). Palaeobotanic and isotopic analysis of late Subboreal and early Subatlantic peat from Engbertsdijksveen VII, The Netherlands. *Review of Paleobotany and Palynology*, **41**, 241–71.

DuRietz, G. E. (1949). Huvudenheter och huvudgränser i svensk myrvegetation. *Svensk Botanisk Tidskrift*, **43**, 274–309.

Eurola, S. (1962). Über die regionale Einteilung der südfinnischen Moore. *Annales Botanici Societatis Zoologicae Botanicae Fennicae 'Vanamo'*, **33**, (2), 1–243.

Foster, D. R., King, G. A., Glaser, P. H., and Wright, Jr, H. E. (1983). Origin of string patterns in boreal peatlands. *Nature*, **306**, 256–8.

Foster, D. R. and King, G. A. (1984). Landscape features, vegetation and developmental history of a patterned fen in south-eastern Labrador, Canada. *Journal of Ecology*, **72**, 115–43.

Foster, D. R., Wright, Jr, H. E., Thekaus, M., and King, G. A. (1988). Bog development and landform dynamics in central Sweden and south-eastern Labrador, Canada. *Journal of Ecology*, **76**, 1164–85.

Frenzel, B. (1983). Mires—repositories of climatic information or self-perpetuating ecosystems? In *Mires: swamp, bog, fen and moor (Ecosystems of the world: 4A)*, (ed. A. J. P. Gore). Elsevier, Amsterdam.

Futyma, R. P. and Miller, N. G. (1986). Stratigraphy and genesis of the Lake Sixteen peatland, northern Michigan. *Canadian Journal of Botany*, **64**, 3008–19.

Gauthier, R. (1980). *La végétation des tourbières et les sphaignes du Parc des Laurentides, Québec*. Etudes écologique No. 3. Laboratoire d'écologie forestière, Université Laval, Québec.

Gignac, L. D. and Vitt, D. H. (1990). Habitat limitations of *Sphagnum* along climatic, chemical, and physical gradients in mires of western Canada. *The Bryologist*, **93**, 7–22.

Gignac, L. D., Vitt, D. H., and Bayley, S. E. (1991*a*). Bryophyte response surfaces along ecological and climatic gradients. *Vegetatio*, **93**, 29–45.

Gignac, L. D., Vitt, D. H., Zoltai, S. C., and Bayley, S. E. (1991*b*). Bryophyte response surfaces along climatic, chemical, and physical gradients in peatlands of western Canada. *Nova Hedwigia*, **53**, 27–71.

Glaser, P. H. and Janssens, J. A. (1986). Raised bogs in central North America:

transitions in landforms and gross stratigraphy. *Canadian Journal of Botany*, **64**, 395–414.

Gorham, E. (1957). The development of peatlands. *Quarterly Review of Biology*, **32**, 145–66.

Gorham, E., Janssens, J. A., Wheeler, G. A., and Glaser, P. H. (1987). The natural and anthropogenic acidification of peatlands. In *Effects of atmospheric pollutants on forests, wetlands and agricultural ecosystems*, (eds T. C. Hutchinson and K. M. Meema), pp. 493–512. Springer, Berlin.

Gorham. E. and Janssens, J. A. (1992). Concepts of fen and bog re-examined in relation to bryophyte cover and the acidity of surface waters. *Acta Societatis Botanicum Polandiae* (in press).

Granlund, E. (1932). De svenska högmossarnas geologi. *Sveriges Geologiska Undersoekning Afhandlingar och Uppsatser*, **26**, 1–193.

Hemond, H. F. (1980). Biogeochemistry of Thoreau's bog, Concord, Massachusetts. *Ecological Monographs*, **50**, 507–26.

Horton, D. G., Vitt, D. H., and Slack, N. G. (1979). Habitats of circumpolar-subarctic Sphagna. I: A quantitative analysis and review of species in the Caribou Mountains, northern Alberta. *Canadian Journal of Botany*, **57**, 2283–317.

Janssens, J. A. (1983). Past and extant distribution of *Drepanocladus* in North America, with notes on the differentiation of fossil fragments. *Journal of the Hattori Botanical Laboratory*, **54**, 251–98.

Janssens, J. A. (1988). Fossil bryophytes and paleoenvironmental reconstruction of peatlands. In *Methods in Bryology*, Proceedings of the Bryological Methods Workshop, Satellite Conference of the XIV International Botanical Congress under the auspices of the International Association of Bryologists 17–23 July 1987, Mainz, Federal Republic of Germany, (ed. J. M. Glime), pp. 299–306. The Hattori Botanical Laboratory, Nichinan.

Janssens, J. A. and Engstrom, D. R. (1992). Paleobryological reconstruction of Holocene peatland development, Pleasant Island, southeastern Alaska. (submitted to *Journal of Ecology*).

Janssens, J. A. and Glaser, P. H. (1986). The bryophyte flora and major peat-forming mosses at Red Lake peatland, Minnesota. *Canadian Journal of Botany*, **64**, 427–42.

Janssens, J. A. and Zander, R. H. (1980). *Leptodontium flexifolium* and *Pseudocrossidium revolutum* as 60 000 year-old subfossils from the Yukon Territory, Canada. *The Bryologist*, **83**, 486–96.

Janssens, J. A., Hansen, B. C. S., Glaser, P. H., and Whitlock, C. (1992). Development of a raised bog complex in northern Minnesota. In *Patterned peatlands in northern Minnesota*, (eds H. W. Wright Jr, B. Coffin, and N. Aasing) University of Minnesota Press, Minneapolis.

Janssens, J. A., Horton, D. G., and Basinger, J. F. (1979). *Aulacomnium heterostichoides* sp. nov., an Eocene moss from south central British Columbia. *Canadian Journal of Botany*, **57**, 2150–61.

Klinger, P. U. (1968). Feinstratigraphische untersuchugen an Hochmooren. Mit Hinweisen zur Bestimmung der wichtigsten grossreste in nordwestdeutschen

Hochmoortorfen und einer gesonderten Bearbeitung der mitteleuropäischen Sphagna Cuspidata. Unpublished dissertation, University of Kiel.

Kubiw, H., Hickman, M., and Vitt, D. H. (1989). The developmental history of peatlands at Muskiki and Marguerite lakes, Alberta. *Canadian Journal of Botany*, **67**, 3534–44.

Kuc, M. and Hills, L. V. (1971). Fossil mosses, Beaufort Formation (Tertiary), northwestern Banks Island, western Canadian Arctic. *Canadian Journal of Botany*, **49**, 1089–94.

Kuhry, P. (1985). Transgressions of a raised bog across a coversand ridge originally covered with an oak-lime forest. *Review of Paleobotany and Palynology*, **44**, 303–53.

Kuhry, P. (1988*a*). Palaeobotanical–palaeoecological studies of tropical high Andean peatbog sections (Cordillera Oriental, Colombia). *Dissertationes Botanicae*, Band 116. Cramer, Belin.

Kuhry, P. (1988*b*). A palaeobotanical and palynological study of Holocene peat from the El Bosque Mire, located in a volcanic area of the Cordillera Central of Colombia. *Review of Palaeobotany and Palynology*, **55**, 19–72.

Kuhry, P., Halsey, L., Bayley, S. E., and Vitt, D. H. (1992*a*). Peatland development in relation to Holocene climatic change in Manitoba and Saskatchewan (Canada). (Manuscript).

Kuhry, P., Nicholson, B. J., Gignac, L. D., Vitt, D. H., and Bayley, S. E. (1992*b*). Development of *Sphagnum* dominated peatlands in boreal continental Canada. (Manuscript).

Kulczynski, M. St (1949). Peat bogs of Polesie. In *Mémoires de l'Académie Polonaise des Sciences et des Lettres, Classe des Sciences Mathématiques et Naturelles, Serie B: Sciences Naturelles*, **15**, 1–356.

LaFarge-England, C. (1989). The contemporary moss assemblages of a high arctic upland, northern Ellesmere Island, N.W.T., Canada. *Canadian Journal of Botany*, **67**, 491–504.

LaFarge-England, C., Vitt, D. H., and England, J. (1991). Holocene soligenous fens on a high arctic fault block, northern Ellesmere Island (82° N), Canada. *Arctic and Alpine Research*, **23**, 80–98.

Lee, J. A., Press, M. C., Woodin, S., and Ferguson, P. (1987). Responses to acidic deposition in ombrotrophic mires in the UK. In *Effects of atmospheric pollutants on forests, wetlands and agricultural ecosystems*, (eds T. C. Hutchinson and K. M. Meema), pp. 549–60. Springer, Berlin.

Levitt, J. (1980). *Response of plants to environmental stresses. Vol. 2: Water, radiation, salt, and other stresses*. Academic Press, New York.

Longton, R. E. (1988). *The biology of polar bryophytes and lichens*. Cambridge University Press, Cambridge.

Malmer, N. (1986). Vegetational gradients in relation to environmental conditions in northwestern European mires. *Canadian Journal of Botany*, **64**, 375–83.

Malmer, N., Horton, D. G., and Vitt, D. H. (1992). Elemental concentrations in mosses and surface waters of western Canadian mires relative to precipitation chemistry and hydrology. *Holarctic Ecology* (in press).

Middeldorp, A. A. (1986). Functional palaeoecology of the Hahnenmoor raised bog ecosystem—a study of vegetation history, production and decomposition by means of pollen density dating. *Review of Palaeobotany and Palynology*, **49**, 1–73.

Miller, N. G. (1980). Mosses as paleoecological indicators of late glacial terrestrial environments: some North American studies. *Bulletin of the Torrey Botanical Club*, **107**, 373–91.

Mogensen, G. S. (1984). Pliocene or early Pleistocene mosses from Kap Koben-havn, north Greenland. *Lindbergia*, **10**, 19–26.

Moore, P. D. (1973). The influence of prehistoric cultures upon the initiation and spread of blanket bog in upland Wales. *Nature*, **241**, 350–3.

Moore, P. D. (1975). Origin of blanket mires. *Nature*, **256**, 267–9.

Nicholson, B. J. and Vitt, D. H. (1990). The paleoecology of a peatland complex in continental western Canada. *Canadian Journal of Botany*, **68**, 121–38.

Ochyra, R. and Wojterski, T. (1983). Series V. Mosses (Musci), part I. In *Atlas of geographical distribution of spore-plants in Poland*, (eds J. Szweykowski and T. Wojterski), pp. 3–31. Polska Akademia Nauk, Poland.

Oechel, W. G. and Lawrence, W. T. (1985). Taiga. In *Physiological ecology of North American plant communities*, (eds B. F. Chabot and H. A. Mooney), pp. 66–94. Chapman and Hall, New York.

Oswald, H. (1923). Die Vegetation des Hochmoores Komosse. *Svenska Vaxsociolo-giska Sallskapets Handlingar*, **1**.

Overbeck, F. (1975). *Botanisch-Geologische Moorkunde*. Wachholtz, Neumünster.

Overbeck, F., Munnich, K. O., Aletsee, L., and Averdieck, F. R. (1957). Das alter des 'Grenzhorizonts' norddeutscher Hochmoore nach Radiocarbon-Datierun-gen. *Flora*, **145**, 37–71.

Pearsall, W. H. (1956). Two blanket-bogs in Sutherland. *Journal of Ecology*, **44**, 493–516.

Proctor, M. C. F. (1972). An experiment on intermittent dessication with *Anomodon viticulosus* (Hedw.) Hook and Tayl. *Journal of Bryology*, **8**, 337–65.

Richter, C. and Dainty, J. (1989). Ion behaviour in plant cell walls. I: Characteri-zation of the *Sphagnum russowii* cell wall ion exchanger. *Canadian Journal of Botany*, **67**, 451–9.

Ruuhijärvi, R. (1960). Über die regionale Einteilung der nordfinnischen Moore. *Annales Botanici Societatis Zoologicae Botanicae Fennicae 'Vanamo'*, **31** (1), 1–360.

Ruuhijärvi, R. (1963). Zur Entwicklungsgeschichte der nordfinnischen Hoch-moore. *Annales Botanici Societatis Zoologicae Botanicae Fennicae 'Vanamo'*, **34** (2), 1–39.

Rybnícek, K. (1973). A comparison of the present and past mire communities of Central Europe. In *Quaternary plant ecology*, The 14th Symposium of The British Ecological Society, University of Cambridge, 28–30 March 1972, (ed. H. J. B. Birks and R. G. West), pp. 237–61. Blackwell Scientific Publications, London.

Rybnícek, K. and Rybnícková, E. (1968). The history of flora and vegetation of the Bláto mire in southeastern Bohemia, Czechoslovakia, (palaeoecological study). *Folia Geobotanica and Phytotaxonomica*, **3**, 117–42.

Schweger, C. E. and Hickman, M. (1989). Holocene paleohydrology of central Alberta: testing the general-circulation-model climate simulations. *Canadian*

Journal of Earth Sciences, **26**, 1826–33.

Sjörs, H. (1950). Regional studies in north Swedish mire vegetation. *Botaniska Notiser*, **1950**, 173–222.

Sjörs, H. (1952). On the relation between vegetation and electrolytes in north Swedish mire waters. *Oikos*, **1950** (2), 241–58.

Slack, N. G., Vitt, D. H., and Horton, D. G. (1980).Vegetation gradients of minerotrophically rich fens in western Alberta. *Canadian Journal of Botany*, **58**, 330–50,

Sonesson, M. (1968). Pollen zones at Abisko, Torne Lappmark, Sweden. *Botaniska Notiser*, **121**, 491–500.

Tallis, J. H. (1964). Studies on southern Pennine peats. III: The Behaviour of *Sphagnum*. *Journal of Ecology*, **52**, 345–53.

Tallis, J. H. (1983). Changes in wetland communities. In *Mires: swamps, bog, fen and moor (Ecosystems of the world; 4A)*, (ed. A. J. P. Gore), pp. 311–47. Elsevier, Amsterdam.

Tolonen, K. (1967). Über die Entwicklung der Moore im finnischen Nordkarelien. *Annales Botanici Fennici*, **4**, 219–416.

Van der Molen, P. C. and Hoekstra, S. P. (1988). A palaeoecological study of a hummock-hollow complex from Engbertsdijksveen, in The Netherlands. *Review of Palaeobotany and Palynology*, **55**, 213–74.

Van Geel, B. (1972). Palynology of a section from the raised peat bog 'Wietmarscher Moor', with a special reference to fungal remains. *Acta Botanica Neerlandica*, **21**, 261–84.

Van Geel, B. (1978). A palaeoecological study of Holocene peat bog sections in Germany and The Netherlands. *Review of Palaeobotany and Palynology*, **25**, 1–120.

Van Geel, B. and Middeldorp, A. A. (1988). Vegetational history of the Carbury Bog (Co. Kildare, Ireland) during the last 850 years and a test of the temperature indicator value of $2H/1H$ measurements of peat samples in relation to historical sources and meteorological data. *New Phytologist*, **109**, 377–92.

Van Geel, B., Coope, G. R., and van der Hammen, T. (1989). Palaeoecology and stratigraphy of the lateglacial type section at Usselo (The Netherlands). *Review of Palaeobotany and Palynology*, **60**, 25–129.

Van Post, L. and Sernander, R. (1910). Pflanzen-physiognomische Studien auf Torfmooren in Narke. In *XI International Geological Congress: Excursion Guide No. 14 (A7)*, Stockholm.

Vitt, D. H. (1982). Sphagnopsida and Bryopsida. In *Synopsis and classification of living organisms*, (ed. S. P. Parker), pp. 305, 307–36. McGraw-Hill, New York.

Vitt, D. H. (1990). Production–decomposition dynamics of boreal mosses over climatic, topographic and chemical gradients. *Botanical Journal of the Linnean Society*, **104**, 35–59.

Vitt, D. H. and Chee, W.-L. (1990). The relationships of vegetation to surface water chemistry and peat chemistry in fens of Alberta, Canada. *Vegetatio*, **89**, 87–106.

Vitt, D. H. and Pakarinen, P. (1977). The bryophyte vegetation, production, and organic components of Truelove Lowland. In *Truelove Lowland, Devon Island,*

Canada: A high arctic ecosystem, (ed. L. C. Bliss), pp. 225–44. University of Alberta Press, Edmonton.

Vitt, D. H. and Slack, N. G. (1984). Niche diversification of *Sphagnum*, relative to environmental factors in northern Minnesota peatlands. *Canadian Journal of Botany*, **62**, 1409–30.

Vitt, D. H., Achuff, P., and Andrus, R. E. (1975). The vegetation and chemical properties of patterned fens in the Swan Hills, north central Alberta. *Canadian Journal of Botany*, **53**, 2776–95.

Vitt, D. H., Marsh, J. E., and Bovey, R. B. (1988). *Mosses, lichens and ferns of northwest North America*. Lone Pine Publishing, Edmonton.

Walker, D. (1961). Peat stratigraphy and bog regeneration. *Proceedings of the Linnean Society of London*, **172**, 29–33.

Walker, D. and Walker, P. M. (1961). Stratigraphic evidence of regeneration in some Irish bogs. *Journal of Ecology*, **49**, 169–85.

Weber, C. A. (1900). Über die Moore, mit besondere Berücksichtigung der zwischen Unterweser und Unterelbe liegenden. *Jahresbericht der Manner von Morgenstern*, **3**, 3–23.

Weber, C. A. (1902). *Über die Vegetation und Entstehung des Hochmoores von Augstumal im Memeldelta mit vergleichenden Ausblicken auf andere Hochmoore der Erde*. Paul Parey, Berlin.

Westhoff, V. and Den Held, A. J. (1969). *Plantengemeenschappen in Nederland*. Thieme, Zutphen.

Wetzel, R. G. (1975). *Limnology*. Saunders, Philadelphia.

Zoltai, S. C. (1988). Wetland environments and classification. In *Wetlands of Canada*, (coordinator C. D. A. Rubec), pp. 1–26. Ecological Land Classification Series, No. 24. Sustainable Development Branch, Environment Canada, Ottawa, Ontario, and Polyscience Publications, Montreal, Quebec.

Zoltai, S. C. and Vitt, D. H. (1990). Holocene climatic change and the distribution of peatlands in western interior Canada. *Quaternary Research*, **33**, 231–40.

8

Temperate forest management: its effects on bryophyte and lichen floras and habitats

FRANCIS ROSE

8.1 Historical introduction

It is now clear that most of Europe, until at least Neolithic times, was covered with woodland of some kind. Palynological data from many sites has made this evident (e.g. Godwin 1975). It has also become clear recently that these early Flandrian forests would have been more open in character than was previously thought, at least in lowland areas. Numerous glades and open 'lawns' would have existed due to the presence of considerable numbers of large grazing and browsing herbivores such as the red deer and other deer species, the wild ox, the wild swine, and probably a species of wild horse. Bison were certainly also widespread in some parts of Europe. Grigson (1978) reviewed the evidence for the presence of numerous large herbivores in the Flandrian up to Atlantic times at least. The situation would have been even more spectacular in earlier interglacials, where abundant fossils provide ample testimony of the rich and diverse fauna. This implies not only that the 'wildwood' (*sensu* Rackham 1976) would have been more open, at least in patches, than one might have supposed from a superficial study of most present-day forests, but also that the flora (and the invertebrate and avian fauna) would have adapted to this partly open environment over a period of perhaps millions of years. It is thus not surprising that the great majority of plant species that occur in woodlands, including most of the lichens and bryophytes, are restricted today to relatively well-illuminated situations such as the edges of glades, or to open stands of trees, or to borders of rides or trackways.

This account, and the examples given, refer mainly to Britain and to western Europe, the areas with which the author is most familiar. However, the principles involved apply more widely and brief mention is made of preliminary studies in other parts of the world.

8.2 Types of woodland structure in Europe today

Apart from wholly artificial stands of alien planted trees, largely conifers, we can recognize a number of different structural types of semi-natural woodlands. These include (1) coppice, with or without standard trees; (2) high forest; and (3) pasture–woodlands of various kinds.

8.2.1 Coppice-woodlands

Coppice-woodlands form an ancient type of human modification of the wildwood. There is clear evidence that some woodlands were being managed in this way from about 4000 BC (Coles and Orme 1976, 1977). Coppiced poles of ash (*Fraxinus excelsior*), oak (*Quercus* spp.), and hazel (*Corylus avellana*) were used in the construction of trackways of wattle hurdles across a bog in the Somerset Levels at that time. From the even sizes of the poles, they were evidently derived from managed coppice-woodland.

Coppice-woodlands, however ancient (and many of them may well have a continuous history as woodland from the time of the wildwood), are normally very poor in epiphytic lichens and bryophytes. Although there may be long continuity of a woodland habitat there is often little or no continuity of mature tree habitat. Some old sweet chestnut (*Castanea sativa*) coppices, in Kent and Sussex particularly, may however have interesting bryophyte communities on the old coppice-stools, with such species as *Dicranum flagellare*, *D. montanum*, *D. tauricum*, *D. fuscescens*, and *Herzogiella seligeri* growing on the decayed peat-like wood of the stools. Most traditional coppice-woodlands contain standard trees which provide more continuity of older bark surfaces for lichen growth than is the case of the coppiced underwood which was cut every 10–15 years. The boles of the standard trees, however, are subjected to drastic environmental changes over time in the coppicing cycle. As the coppice underwood grows up, the boles of the standard trees become increasingly shaded; few lichens can tolerate this. Then, after cutting of the underwood, not only are the boles of the standards suddenly strongly illuminated, but atmospheric humidity within the wood falls dramatically as well. So, in general, standard trees in coppice do not provide a habitat for other than the commonest and most stress tolerant epiphytic lichens and bryophytes.

Many coppice-woodlands in Britain today are totally neglected, as coppice management (except for some sweet chestnut areas) is generally no longer economic. Such woods become very dense and dark with overgrown coppice poles up to 50 or more years old, and even terricolous bryophytes tend to decline under these conditions.

However, some old coppice-woodlands have wide rides, and along these may occur large oak or ash trees. Here, the combination of adequate light, shelter from excessive desiccation, and long continuity of the individual trees can result in relatively rich epiphytic floras. Not only may there be a luxuriant cover of the more widespread epiphytic lichens and bryophytes, but also of some species more characteristic of less altered ancient woodland. Notably, the lichens *Arthonia vinosa, Pachyphiale carneola, Biatorina atropurpurea, Normandina pulchella, Thelotrema lepadinum, Usnea cornuta,* and *U. ceratina,* and bryophytes like *Orthotrichum lyellii, Leucodon sciuroides,* and *Frullania tamarisci* may occur in old woodland sites which are relatively free from atmospheric pollution. Many, indeed most, ancient coppice-woodlands, however, have very low species totals of lichen epiphytes. One extreme case occurs in the Bradfield Woods in Suffolk where, in an area of coppice-with-standards woodland of about 1 km², only 15 epiphytic lichens can be found. This is a site which is considered to have been woodland since prehistoric times. There is a remarkably rich ground flora and shrub layer containing more than 50 species which are regarded as vascular plant indicators of ancient woodland.

8.2.2 High forests

This term is employed here to describe tall woodland of standard trees grown in close canopy. While such woodland structure still occurs in some places in untouched forest today in Europe, it is nearly always the result of management by man, either by planting of trees as an even-aged commercial crop, or by manipulation of pre-existing stands. Many former coppice-woodlands have been converted to high forest by singling; selection of the best stem on a multi-stemmed coppice stool for retention and cutting out of the others. 'False high forest' formed in this way is, at least initially, even-aged and is often structurally unstable as there tends to be a weak point where the bole of the new standard tree joins the former coppice stool, so foresters generally prefer to replant formerly uneconomic coppice-woodland.

Most high forests in Europe are too dense and dark to support rich epiphytic communities. Many of the great 'Forêts Domaniales' (state forests, often former Royal hunting forests) of lowland France are of this type. These ancient, formerly irregular and gladed forests have mostly been converted from pasture-woodland by fencing out domestic animals, and natural regeneration has filled in the gaps. Very tall, straight-stemmed trees in close canopy dominate these forests. They tend however to be managed on a 'selection felling' basis whereby trees are removed when they attain a trunk diameter of 0.5–0.8 m. Hence, though there is some continuity of trees over a long period of time, there is little continuity of really old trees and very little illumination of the boles within the dense interiors of these forests.

Such forests are poor in epiphytes other than the widespread common species with good dispersal mechanisms. Mid-nineteenth-century records indicate that many French forests, from eastern Brittany across to the Ile-de-France around Paris, had rich epiphytic floras with *Lobaria* and *Sticta* species. Such lichens have largely disappeared today, though the cover of the common *Parmelia* species is often quite good still, especially along rides and in the tree crowns.

8.2.3 Pasture-woodlands

Pasture-woodlands are, by definition, woodlands that are, or have been in the recent past, grazed by animals. For fuller details see Rackham (1976, 1980) and Harding and Rose (1986). The Domesday Book (AD 1086) gives a unique picture of the distribution of woodland in many (though not all) parts of England, and to some extent its use (Darby 1952–77; Rackham 1980). In some counties a distinction was made between *silva minuta* and *silva pastilis*; Rackham interprets these as coppiced woodland and pastured woodland respectively. It is clear, therefore, that these two types of woodland management were already well established by the Norman Conquest. Some of these would have been areas primarily for hunting, such as the Saxon parks at Woodstock in Oxfordshire and at Clarendon in Wiltshire. Others would seem to have been areas for grazing domestic stock, or used for swine pannage; some were probably used for all these purposes. In Norman times and subsequently, pasture-woodlands became more widespread and highly organized. It is possible even today to recognize four main categories of pasture-woodlands, and these are described below.

Forests and chases

These were areas of land where Forest Laws applied. They were by no means wholly wooded, and indeed often contained enclosed farmed land within their legal limits. The woodland (if any) within them would have included areas of pasture-woodland, which in some cases (e.g. the New Forest, Forest of Dean) were very extensive. (In general 'forests' were under Royal control and 'chases' under baronial or ecclesiastical control, but the terms have often been rather loosely applied.)

The basic purpose of these areas was to provide places in which deer (either native red deer or the introduced fallow deer) were protected and conserved to provide fresh venison in winter, especially for the Royal Court and for the wealthier subjects. Animal husbandry at the time was so inefficient that it was not generally possible to provide adequate supplies of beef, mutton, or pork in the winter months. The romantic popular accounts of such areas in the literature have over-emphasized the concept of hunting for sport; such areas were primarily of practical importance for the production of fresh meat in winter for those who could afford it; the poorer classes had to manage largely with salted meat, or none at all in winter.

Forests and chases were usually extensive, and (apart from included areas of farmland or coppices) were unfenced. Many of them seem to have contained relics of the former wildwood as their nuclei, so providing some continuity of woodland habitat, and probably some ancient trees, from prehistoric times. Rackham (1980) records 142 medieval forests, of which about 80 were wooded, each containing, on average, some 5000 acres (2000 ha) of pasture-woodland. This is only a small part of the total area of England but highly significant in terms of the survival of epiphytic lichens and bryophytes from the past.

Parks

Medieval parks (usually, but not always, for keeping deer) grew from some 35 listed in the Domesday Book to as many as 1900 by the fourteenth century (Cantor and Hatherley 1979); that is, roughly one to every four parishes in England. Some may have been emparked or enclosed on former open land or even former farmland; but Brandon (1963) in his studies of the origin of parks in Sussex, produced evidence that many parks were formed on areas of 'waste', and probably contained relict patches or areas of the wildwood still persisting, albeit in modified form. Parks, unlike forests and chases, were enclosed by earthwork banks topped with wooden palings to keep the deer inside.

Wooded commons

Large areas of unenclosed land with common grazing rights existed in medieval England, and indeed down to the times of the Enclosure Acts of the nineteenth century. Such common grazing rights also existed in many Royal Forests, and some parks. Many of these commons carried, and indeed where they survive still carry, some woodland. Most have now disappeared, but fine examples of wooded commons still exist at Burnham Beeches in Buckinghamshire, Ebernoe Common in Sussex, and White-parish Common in Wiltshire.

Winter-grazed woodlands

These form a fourth type of wood-pasture, mainly seen on the valley sides in upland Britain and in some montane areas in continental Europe. Horner Combe on Exmoor is a fine surviving example in England, while there are still many such places in Scotland. They are little grazed in summer, but in winter the sheep (or deer) descend into them when the uplands become too bleak and inhospitable.

8.3   The implications of pasture-woodlands for the persistence of rich epiphytic floras

Entirely 'natural' woodlands (totally undisturbed by human activities) are no longer to be found in the British Isles, and in Europe are confined now to some more remote montane forests. The greatest numbers of epiphytic lichens and bryophytes are today confined to some of our remaining pasture-woodlands, whether or not they are still grazed. What do the various types of pasture-woodlands have in common that accounts for this richness, both in terms of numbers of species per unit area and the presence of numerous taxa found nowhere else?

Perhaps the most important factor is the presence, in most pasture-woodlands, of numerous mature to over-mature trees. Such areas in the medieval period were not primarily concerned with the growth of commercial timber supplies on any large scale, if at all. This situation changed from the seventeenth century onwards when vast quantities of oak timber were required for naval construction purposes. In the Royal Forests and in many parks, a certain amount of selective removal of timber occurred, as and when large trees were required by their owners or managers, but most of the trees were retained as shelter for the deer. On wooded commons, the Lords of the Manors had the right to remove timber trees, and such documentation as exists (not always a great deal) indicates that there was much more felling from time to time on the commons. In the case of the Sussex wooded common of The Mens, it seems clear that over a long period of time most parts of the common were cut over, so that there has been comparatively little continuity of ancient tree stands, though there had probably always been some wooded areas (Tittensor 1978). This correlates well with the fact that, though old wooded commons tend to have far richer epiphytic floras than coppice-woodlands or managed high forests, these floras are usually far poorer than those of the remaining least-altered medieval deer parks and Royal Forests, where there appears to have been greater continuity of old trees over time, and less disturbance of the canopy.

In nearly all pasture-woodlands, however, supplies of small wood were available as a renewable resource. These supplies were obtained by the practice of *pollarding*. Unlike coppicing, there is little direct evidence of the antiquity of this practice, but it must go back to very early times. It may have been introduced by the Normans through their adoption of Classical and Islamic traditions and techniques of emparking as a result of their conquest of Sicily, but it may be far older. The principal tree species pollarded were the two oaks (*Quercus robur* and *Q. petraea*) in Britain and France, but in southern Europe (Provence, Italy, Greece) sweet chestnut (*Castanea sativa*) was, and still is, important in this practice. Beech (*Fagus sylvatica*) was the main pollard tree at Burnham Beeches, and also in St Leonards and Worth Forests in Sussex, and in parts of the New Forest in Hampshire. Ash (*Fraxinus excelsior*) pollards are frequent on the richer soils in such places as the valleys of the New Forest (Fig. 8.1), while in Essex, at Hatfield and Epping Forests for example, hornbeam (*Carpinus betulus*) still forms an important element of the old pollard woodlands. Holly (*Ilex aquifolium*) was important locally too.

In pollarding, young trees, mostly of a diameter of about 10–20 cm, were beheaded at a height out of reach of grazing animals (Fig. 8.2), so that the regrowth could not be grazed, as would happen with coppicing nearer to ground level. The regrowth of numerous poles could then be harvested when large enough to be of use. Pollarding is thus a way of obtaining coppice type wood within pasture-woodland without the animals eating the regrowth. In some parks and forests there were enclosed coppices (within banks and pales) in the pasture-woodlands. From the point of view of epiphytes, such practices must have widened habitat diversity.

Important also to the epiphytes is that pollarding resulted in the formation of enormous ancient boles. These not only provided very long continuity of habitat for those species with slow colonizing ability, but also, because of the numerous niches created on an old pollard, led to considerable species diversity. An old pollard bole usually has deeply creviced rain tracks which retain moisture and suit lichens such as *Thelopsis rubella*; more or less horizontal knobs and platforms on the spreading roots where lichens such as *Leptogium lichenoides* may grow among mosses like *Homalothecium sericeum*; dry overhangs beneath the very swollen crown of the bole, suitable for the lichens *Lecanactis premnea* and *L. lyncea* which appear to dislike being directly wetted by water running over them; and well-illuminated gently sloping surfaces on bole and limbs suitable for *Lobaria* spp. and *Parmelia* spp. etc. The rare moss *Zygodon forsteri* is confined, in Britain, to the edges of water-holding knot-holes and rain tracks on old beech pollards at Burnham Beeches, Epping Forest, and The New Forest.

Fig. 8.1 *Fraxinus excelsior* pollard, South Ocknell Wood, The New Forest, England (February 1990).

Another feature of most active pasture-woodlands, whether of pollarded trees or not, is that they tend to have well-spaced trees with widely spreading crowns. Also, grazing limits the establishment of new trees; so pasture-woodlands tend to be well illuminated, especially as compared to ungrazed high forests managed primarily for large timber production. Most epiphytic lichens (and some rarer bryophytes) demand a combination of good illumination and some shelter especially from northerly or north-easterly winds, which tend to be very desiccating; these conditions tend to prevail in pasture-woodlands.

8.4 More recent changes in pasture-woodland management

Comparatively few ancient pasture-woodlands remain intact in Britain;

Fig. 8.2 Newly pollarded hollies (*Ilex aquifolium*), Anses Wood, The New Forest, England (March 1990).

even fewer are still grazed in the traditional way. The New Forest still has some 9000 acres (c. 3600 ha) of ancient pasture-woodlands and contains (besides extensive areas of planted or replanted, normally ungrazed, commercial forest) the largest forest area left in lowland Europe (at least north of the Alps and Pyrenees) which is still managed by grazing of ponies, cattle, and deer. It is perhaps not a coincidence that it still has the largest epiphytic lichen flora known in any comparable area in Europe, comprising 312 species. This total includes many rarities; two species (*Catinaria laureri*, *Parmelia minarum*) are known nowhere else in the British Isles, while others (such as *Pertusaria velata*, *Rinodina isidioides*, *Porina coralloidea*, *P. hibernica*, *Agonimia octospora*, *Thelopsis rubella*, *Wadeana dendrographa*, *Phyllopsora rosei*, and *Tomasellia lactea* (the last confined to *Ilex*) are frequent to common in the Forest but rare and very sparsely scattered elsewhere in Britain, and indeed rare in western Europe (Rose and James 1974).

Few other lowland Royal Forests in Britain retain much of their original character, due largely to extensive clearance and replanting, but also in some areas due to cessation of grazing, which has led to explosive regeneration of young trees. This has resulted in some cases, even where the ancient trees have been left, in the formation of dense stands of high forest which causes the disappearance of many or all of the more local and restricted species. For example, at Savernake Forest in Wiltshire, inter-

Table 8.1 A list of selected woodland sites in Britain illustrating: (a) the total number of lichen taxa in each site (most sites are approximately 1 km² in area); (b) values of the Revised Index of Ecological continuity based on 30 taxa; (c) values of the New Index of Ecological Continuity based on 70 taxa

Name and grid reference	(a)	(b)	(c)
(i) Ancient woodland areas in the New Forest; ten of the richest sites are listed. These primary pasture-woodlands are now unique in lowland western Europe.			
Mark Ash Wood (41/2407)	178	115	44
Bramshaw Wood (41/2516)	166	115	45
Great Wood (41/2515)	148	105	41
Sunny Bushes (41/2614)	103	90	31
Redshoot Wood (41/1808)	148	100	40
Stricknage Wood (41/2612)	146	100	43
Busketts Wood (41/3010)	175	105	43
Hollands Wood (41/3005)	140	100	41
Wood Crates (41/2708)	170	95	40
Stubbs and Frame Wood (41/3503)	168	110	49
(ii) Relatively intact medieval deer parks. Eighteen of the best remaining examples are cited.			
Arlington Park, Devon (21/6240)	213	130	45
Dunsland Park, Devon (21/4005)	163	115	38
Whiddon Park, Devon (20/7289)	163	100	36
Boconnoc Park, Cornwall (20/1460)	191	145	50
Trebartha Park, Cornwall (20/2577)	162	130	38
Mells Park, Somerset (31/7148)	143	80	23
Melbury Park, Dorset (31/5605)	218	110	43
Lulworth Park, Dorset (30/8582)	150	75	29
Longleat Park, Wiltshire (31/8143)	159	90	34
Parham Park, Sussex (51/0614)	190	65	26
Eridge Park, Sussex (51/5635)	185	95	30
Ashburnham Park, Sussex (51/6914)	172	80	27
Brampton Bryan Park, Hereford (32/3571)	176	65	22
Dynevor Park, Dyfed (22/6122)	137	85	30
Dolmelynllyn Park, Gwynedd (23/7224)	157	125	44
Inverary Park, Strathclyde (27/1110)	134	90	29
Drummond Park, nr. Crieff, Tayside (27/8518)	162	65	21
Cawdor Castle, nr. Nairn, Highland (28/8448)	131	60	13
(iii) Old Royal Forests other than the New Forest. Only Savernake retains relatively intact areas and is little affected by air pollution.			
Savernake Forest, Wiltshire (41/2366)	165	85	28
Windsor Forest, Berkshire (41/9373)	75	20	3
Wychwood Forest, Oxfordshire (42/3316)	128	40	9
Epping Forest, Essex (51/4–9–)	38	10	0
(iv) Mature hardwood plantations, mainly of oak. Dalegarth Wood was clear-felled and replanted c.1770 with oak; Brockishill Inclosure was replanted c.1840 but adjoins the ancient wood of Busketts (see above) so some recolonization has been possible; South Bentley Wood was replanted after felling c.1700 and adjoins the ancient Anses Wood: here lichen recolonization has been very significant. The other sites were replanted as high forest in the early nineteenth century but are remote from other ancient woods; this fact correlates well with their low index values.			
Dalegarth Wood, Cumbria (35/1700)	68	20	8
Nagshead Inclosure, Forest of Dean, Gloucestershire (32/6209)	16	0	0

Table 8.1   *(cont.)*

Name and grid reference	(a)	(b)	(c)
Cranford Cross, Devon (21/5521)	24	5	2
Brockishill Inclosure, New Forest (41/3010)	80	50	14
Pondhead Inclosure, New Forest (41/3107)	33	30	8
South Bentley Wood, New Forest (41/2312)	82	70	22

(v) Ancient wooded commons of medieval origin with old pollards. Continuity less than in (ii) but much greater than in (iv) or (vii).

Ebernoe Common, Sussex (41/9726)	111	50	15
The Mens, Sussex (51/0232)	75	50	12
Burnham Beeches, Buckinghamshire (41/9585)	94	35	6

(vi) Ancient woods, formerly pastured deer park, but now modified to coppice with standards although retaining some stands of ancient standard trees in places. Character and lichen floras intermediate between groups (ii) and (vii).

Eastdean Park Wood, Sussex (41/9011)	110	70	18
Pads Wood, Sussex (41/7816)	74	55	16

(vii) Woodlands managed as coppice (with some standards) for many centuries; standards not very old.

Bradfield Woods, Suffolk (52/9357)	15	0	0
Brenchley Wood, Kent (51/6442)	44	5	0
Nap Wood, Sussex (51/5832)	54	25	5
Hayley Wood, Cambridgeshire (52/2952)	35	5	0
Foxley Wood, Norfolk (63/0522)	15	5	1
Park Coppice, Coniston, Lancashire (34/2995)	16	0	0

(viii) Ancient pasture-woodlands in upland Britain, little grazed in summer but acting as important overwintering sites for sheep or deer. Few pollards present, except locally.

Horner Combe, Somerset (21/8944)	176	125	45
Barle Valley, Somerset (21/8–3–)	170	140	46
Camasine Woods, Loch Sunart, Highland (17/7561)	227	120	43
Walkham Valley, Devon (20/5473)	122	85	27
Low Stile Wood, Cumbria (35/2312)	219	130	51
Yew Scar, Gowbarrow, Cumbria (35/4120)	117	100	30
Coed Crafnant, Gwynedd (23/6129)	159	125	47

planting, largely with oaks, of the former open stands of ancient oak pollards has led to a loss of species diversity through shading of the old pollards. Conservation work, however, is about to be undertaken involving clearance of young oaks and scrub on the south-western sides of the old oaks to let more light on to the boles. It is hoped that the result will be a significant improvement in the lichen communities, because similar conservation work (clearing holly shading old beech trees) has already, in two years, led to marked increase in growth and cover of *Lobaria pulmonaria* in Rushpole Wood in the New Forest.

Very little is now left in lowland western Europe of pasture-woodland areas. There remain a few areas in the Forest of Fontainebleau south-east of Paris, where the old structure has been conserved in some '*Réserves Biologiques*'. These are only grazed now by a sparse deer population and the activities of wild boar, but they retain much of the epiphyte richness that must have been a feature of this old French Royal hunting forest, but which has been largely lost elsewhere in that extensive forest through planting up to high forest (Rose 1990). Some old beech forests in Brittany still retain the original pasture-woodland structure and lichen richness, though as at Fontainebleau there is now only grazing by deer and wild boar.

If we turn to the deer parks, once so numerous and a distinctive feature of Britain and Normandy, we find that very few retain their original form. The best examples left, which still have their fallow deer and retain much of their medieval structure, are at Boconnoc Park in Cornwall, Melbury Park in Dorset, and Parham and Eridge Parks in Sussex (see Table 8.1 for further data on their species totals and rare species). Dynevor Park in Dyfed and Inverary Park in Strathclyde are perhaps the best examples in Wales and Scotland respectively.

Other old deer parks still have deer, but have been structurally altered by felling and/or replanting, as at Petworth Park in Sussex, and they are now relatively poor in epiphytes. Others again have lost their deer, but have in parts retained much of their ancient structure and ancient oaks. Some of these remain outstanding, even at a European level, for their lichens. Outstanding examples are Brampton Bryan Park in Hereford, Mells Park in Somerset, Longleat Park in Wiltshire (now with exotic animals as a 'safari' park), and Whiddon Park in Devon (still grazed by cattle, though the deer have been removed). Nothing really comparable to the British medieval deer parks now remains in those extensive parts of western Europe that I have surveyed, so the international importance of this habitat in Britain for lichen epiphytes (and also for saproxylic fungi and invertebrates) is outstanding. However, there still remain a few old deer parks of similar character in northern Spain.

Many former medieval deer parks remain in a modified form. The rough ancient *Agrostis capillaris–Galium saxatile* turf has often been improved, and often only comparatively few of the oaks remain. But in many cases, such parks (whether still with deer or not, or with other types of grazing) are still of lichenological interest. This includes parks created from former farmland in later centuries. They are of interest not only because some ancient trees may remain but because extensive planting was frequently undertaken in the 'landscape gardening' era by landscape architects such as Brown and Kent. The planted trees in these parks are now quite mature, but only rarely do they have the interest of the less modified parks. Some however have groves, avenues, or isolated trees of elm (*Ulmus*), ash, and sycamore

(*Acer pseudoplatanus*). These trees have base-rich bark of high pH, and interesting communities of a different nature from those of ancient English forests. The community found on such trees in open situations (particularly where there are cattle present whose excreta reaches the bark in the form of dust) is known as the *Xanthorion parietinae* (James *et al.* 1977). It contains, besides the orange *Xanthoria parietina*, the spectacular *Anaptychia ciliaris*, several species of *Physcia sensu lato*, and *Parmelia acetabulum*. As a natural community this is characteristic of some dry, open, sunny forests in southern Europe; it may possibly have spread to Britain as a result of human activities. The *Xanthorion* sometimes occurs on trees with normally more acid bark such as oaks and limes (*Tilia*) if there are enough grazing animals present to enrich the bark with nutrients. It is not however a forest epiphyte community in the British Isles. This community is now in heavy decline, like the *Lobarion* communities of the original forest, under the influence of air pollution, herbicides, and excessive eutrophication due to the use of artificial fertilizers in spray or dust form. The death of mature elms due to Elm Disease has also led to the great reduction of this lichen community (and of bryophyte members too, such as *Leptodon smithii*) but it can still be seen in such places as Montacute Park in Somerset, and in a number of Kentish parks out of range of major pollution sources.

The *Lobarion pulmonariae* alliance (James *et al.* 1977) seems to have been the major original epiphyte community in forests at the time of the wildwood, as far as one can judge from (1) what is to be seen in remote little modified forest fragments in our upland valleys; (2) in the least altered, unpolluted montane forests of the mountain massifs of central Europe such as the Vosges, Black Forest, Pyrenees, Apennines, and Yugoslavia; and (3) from what remains in our least modified medieval deer parks and in the New Forest. This community contains a very large number of species of lichens. A few of the more conspicuous species are listed in Table 8.2. These are, or were, all widespread lichens in what remains of little modified pasture-woodland forests in western Europe. Even today, they still occur very widely in unpolluted areas, and many of them extend far eastwards and southwards into Russia, Greece, Turkey, Spain, and southern Italy. The *Lobarion* is by no means a purely oceanic community, though it has retreated westwards with forest modification. With the spread of even moderate pollution and moderate structural change the foliose species of the *Lobarion* are the first to retreat. Many of the crustose species listed in Table 8.2 may, however, be able to maintain themselves in partially disturbed woodland, forming the community known as the 'pre-Lobarion'. This community may even persist, in sparse and attenuated form, on old standards in coppice woodland (see above). For a general account of the present status of the *Lobarion* in Britain and Europe see Rose (1988). For

Table 8.2   Some major lichens of the epiphytic *Lobarion pulmonariae* community

Foliose species

Lobaria pulmonaria	P. triptophylla
L. virens (= laetevirens)	Peltigera horizontalis
L. amplissima	P. collina
L. scrobiculata	Nephroma laevigatum
Sticta limbata	N. parile
S. sylvatica	Parmelia crinita
Pannaria conoplea	P. reddenda
Parmeliella plumbea	

Crustose species

Pachyphiale carneola (= cornea)	Thelopsis rubella
Catillaria (= Biatorina) atropurpurea	Thelotrema lepadinum
Arthonia vinosa	Dimerella lutea
Catillaria sphaeroides	

Lichen nomenclature in Tables 8.2–8.4 follows Hawksworth *et al.* (1980); where more recent names have been adopted the synonyms used in Hawksworth *et al.*(1980) are cited in parentheses.

a detailed discussion of the role of historical factors in the modification of the original *Lobarion* communities in a former deer park, Nettlecombe Park in Somerset, see Rose and Wolseley (1984).

A limited number of epiphytic bryophytes may play an important role in the *Lobarion* community, but most of these are far less 'faithful' to ancient pasture-woodland than are many of the lichens (Fig. 8.3). Those species which are most faithful to old woodlands are the following: *Zygodon baumgartneri, Pterogonium gracile, Neckera crispa,* and, only as epiphytes, *Porella arboris-vitae, Frullania tamarisci, F. fragilifolia,* and *Antitrichia curti-pendula.* Other characteristic (but less faithful) bryophyte species found in *Lobarion* communities in old woodlands include *Leucodon sciuroides, Ortho-trichum lyellii, O. stramineum, Leptodon smithii* (very southern in Britain), and the common species *Homalothecium sericeum, Isothecium myosuroides, I. myurum, Neckera complanata,* and *N. pumila* (see Table 8.5).

Another, very different, lichen community occurs on ancient trees (mainly oak, sometimes beech in ancient parkland and forest pasture-woodlands). This is the *Lecanactidetum premneae* (James *et al.* 1977) which is confined to well-lit trees usually over 300 years old. It is best regarded as a 'post-climax' community of trees whose bark has become too dry and brittle to support the *Lobarion* any longer. It persists on isolated ancient trees in open non-woodland situations where medieval parkland formerly existed. The following lichens are faithful to this community: *Lecanactis premnea, L. lyncea, L. amylacea,* and *Opegrapha prosodea.* Other characteristic species include *Schismatomma decolorans, S. cretaceum,* and *Arthonia impolita.* This community is internationally important because it is only in the ancient pasture-woodlands (or relics of them) in England, Wales and

Fig. 8.3   Bole of old *Fagus sylvatica* with rich epiphytic lichen and bryophyte flora including *Lobaria pulmonaria* and *Pterogonium gracile*, by Highland Water, Wood Crates, The New Forest, England (March 1990).

Ireland that it persists in any quantity in a European context. On the continent there are just too few ancient oaks of sufficient age, except at Fontainebleau where the community is imperfectly developed, perhaps due to the very dry climate and soils.

8.5 Indicator species

A consideration of the data presented so far in this chapter has led me to develop the concept of 'ancient woodland indicator' lichen epiphytes (Rose 1974, 1976). A useful index for the identification of ancient non-polluted

Table 8.3   Lichen epiphytes used to calculate the Revised Index of Ecological Continuity (Rose 1976)

Arthonia vinosa	Pannaria conoplea
Arthopyrenia ranunculospora	Parmelia crinita
(= A. cinereopruinosa)	P. reddenda
Catillaria (Biatorina) atropurpurea	Parmeliella plumbea or P. triptophylla
C. sphaeroides	Peltigera collina
Dimerella lutea	P. horizontalis
Enterographa crassa	Porina leptalea
Haematomma elatinum	Pyrenula chlorospila or P. macrospora
Lecanactis lyncea	Rinodina isidioides
L. premnea	Schismatomma quercicola (= Lecidea
Lobaria amplissima	cinnabarina auct. pro parte)
L. pulmonaria	Stenocybe septata
L. scrobiculata	Sticta limbata
L. virens (= L. laetevirens)	S. sylvatica
Nephroma laevigatum	Thelopsis rubella
Pachyphiale carneola (= P. cornea)	Thelotrema lepadinum

pasture-woodlands is the Revised Index of Ecological Continuity (Rose 1976). This employs 30 lichens (Table 8.3) which, experience has shown, are almost or totally faithful to ancient woodlands. Because these 30 species are not all widespread throughout Britain, and because (due to various environmental changes caused by man) one would not necessarily expect all of them to occur in one single woodland area, I have taken the occurrence of 20 out of the 30 species as an excellent indication of a woodland with ecological continuity with at least early medieval times. The initial choice of 30 species was based on sites such as the New Forest and a few parks known to be very ancient from documentary evidence.

Documentary evidence of woodland antiquity is often not available, so the use of such an Index may serve as valuable evidence of great age, or otherwise, of an apparently ancient woodland, forest, or park. For reasons given earlier, this approach is of little use in most coppice-woodlands. Some of these are known to be extremely ancient, but the lichens present are of little use in such cases. Other lines of evidence (soil profiles, ancient boundary banks, ancient estate maps, and documents) must be used for dating in such cases, and should of course also be used (where available) as supplementary or confirmatory evidence even when lichenological data suggest a great age for a site. High values for lichen indices do not, of course, prove a direct connection with the lichen flora of the wildwood, but they indicate the probability that a woodland, forest, or park was formed at a time when fragments of very ancient woodland may have existed near to the site. Many lichens today have very poor dispersal powers over long distance (except in parts of the western Scottish Highlands). However, it is possible that in the pure air and undrained landscape of earlier times, lichens were able to disperse far more freely than they have been able to do in the last few hundred years, and especially since industrial pollution, drainage, and enclosure became widespread.

The Revised Index of Ecological Continuity is calculated for a given site as follows:

$$RIEC = n/20 \times 100$$

where n is the number of species present at the site out of the list of species given in Table 8.3. Some exceptional sites have an RIEC value of over 100; Boconnoc Park in Cornwall, for example, has 29 of the RIEC species present, so that the RIEC is 145. The whole concept is based upon statistical probability. The presence of a few RIEC species at a site has little significance, except perhaps to suggest that the woodland there is probably older than other nearby woods. Arguably, the greater the number of RIEC species present, the greater is the probability that a site is very ancient. If the value of the Index is over 50, it is considered likely that it is early medieval in origin, but values up to about 75 may indicate recent disturbance of an ancient site e.g. Nettlecombe Park in Somerset with an RIEC of 75 is known to be ancient but has suffered severe changes in the last half century (ploughing of much, but not all, of the Park and felling of many of the ancient trees, though many survive). Table 8.1 gives a list of examples of woodland sites, showing a range from low to high RIEC values, and total species numbers (mostly in 1 km² or less).

A high value for the RIEC could of course be reached if only part of a woodland site is ancient. Hence it is important, in listing species, to survey areas which, from the point of view of disturbance or replanting, are as homogenous as possible. Some species in the list are certainly more valuable indicators of continuity (e.g. the *Lobaria* spp.) than others. But it was considered impossible to devise a method for giving different numerical values to different species, so all have equal weightings. Any other system would be too complex.

In recent years it has become clear (1) that more sensitive indices, using larger numbers of species, are desirable and (2) that regional indices for different parts of Britain (and for different regions of Europe) would be very useful. The New Index of Ecological Continuity (NIEC, Table 8.4) covers lowland Britain north to Galloway. It also works well in western Norway and in western France from Pas-de-Calais through Brittany to the western Pyrenees and into Spanish Navarra. The NIEC involves 70 species. This is more than twice as many as used for the RIEC, but far more is now known about the taxonomy and ecology of epiphytic lichens than in 1976, so it has been possible to incorporate all these extra species into the new

Table 8.4 Lichen epiphytes used to calculate the New Index of Ecological Continuity (NIEC). Presence of the 'main' species is used to calculate the NIEC with a maximum possible total of 70. The 'bonus' species are counted only where the principal interest is conservation status (see text)

Main species

Agonimia octospora
Arthonia astroidestra Coppins (= A. stellaris auct. pro parte)
A. vinosa
Arthopyrenia antecellans
A. ranunculospora (= A. cinereopruinosa auct. pro parte)
Arthothelium ilicinum
Bacidia biatorina
B. epixanthoides
Buellia erubescens
Catillaria (= Biatorina) atropurpurea
C. sphaeroides
Cetrelia olivetorum
Chaenotheca spp. (count only one sp. but not C. ferruginea)
Cladonia caespiticia
C. parasitica
Collema furfuraceum or C. subflaccidum
Dimerella lutea
Enterographa sorediata
Haematomma elatinum
Heterodermia obscurata (W)
Lecanactis amylacea
L. lyncea
L. premnea
L. subabietina
Lecanora jamesii
L. quercicola
Lecidea sublivescens
Leptogium cyanescens
L. lichenoides
L. teretiusculum (W)
Lobaria amplissima
L. pulmonaria
L. scrobiculata
L. virens (=laetevirens)
Megalospora tuberculosa

Micarea alabastrites or M. cinerea
M. pycnidiophora
Nephroma laevigatum
N. parile
Ochrolechia inversa
Opegrapha corticola
O. prosodea
Pachyphiale carneola
Pannaria conoplea or P. rubiginosa
Parmelia crinita
P. reddenda
Parmeliella atlantica or P. plumbea
P. jamesii
P. triptophylla
Peltigera collina
P. horizontalis
Pertusaria multipuncta
P. velata
Phaeographis dendritica or P. inusta or P. lyellii*
Phyllopsora rosei
Polyblastia allobata
Rinodina isidioides
Schismatomma niveum
S. quercicola or Pertusaria pupillaris (= Lecidea cinnabarina auct. angl.)
Stenocybe septata
Sticta limbata
S. fuliginosa or S. sylvatica
Strangospora ochrophora
Thelopsis rubella
Thelotrema lepadinum
Usnea ceratina
U. florida
Wadeana dendrographa
Zamenhofia (= Porina) coralloidea
Z. (= Porina) hibernica

Bonus species

Anaptychia ciliaris (Devon only)
Arthonia arthonioides
A. anglica
A. zwackhii
A. anombrophila (= cinereopruinosa auct.)
Bacidia circumspecta
B. subincompta
Catillaria (= Catinaria) laureri
Caloplaca lucifuga
Collema fragrans

P. sampaiana
Parmelia arnoldii
P. horrescens
P. minarum
P. sinuosa
P. taylorensis
Parmeliella testacea
Pseudocyphellaria crocata
P. thouarsii agg.
Ramonia spp. (any)
Schismatomma graphidioides

Table 8.4 (*cont.*)

C. nigrescens	*Sphaerophorus globosus* (S. England
C. subnigrescens	only)
Cryptolechia carneolutea	*S. melanocarpus* (S. England only)
Gyalecta derivata	*Sticta dufourii* (*S. canariensis,*
Leptogium burgessii	blue-green algal morphotype)
L. cochleatum	*Teloschistes flavicans*
Megalaria grossa (in S. England)	*Tomasellia lactea*
Opegrapha fumosa	*Usnea articulata* (in New Forest and
O. multipuncta	Sussex where rare)
Pannaria mediterranea	*Zamenhofia* (= *Porina*) *rosei*

With main species, only one species should be counted where alternatives are cited
* only counts as a bonus species in south-east England where rare

Index to give a more sensitive parameter. Computation of the NIEC is slightly more complex than for the RIEC. Firstly, the NIEC value is derived as the total number of main species (Table 8.4) out of a maximum of 70. To the NIEC value is added the total number of 'bonus' species present (see foot of Table 8.4) to obtain a final Index figure denoted by 'T'.

Calculations of the new Index for a large number of sites reveal that it is far more effective in sorting out woodlands in terms of their conservation importance (sites with T values < 20 are of limited importance), for the emphasis is now more on grading woodlands in terms of their conservation status than in trying to assess their relative age and lack of disturbance; the RIEC has already achieved the latter. The NIEC is in fact not intended to replace the RIEC entirely but to be a complementary tool of higher sensitivity, useful for the more expert lichenologist. In addition, a series of Regional Indices are in preparation. One covers the special characteristics of the euoceanic lichen flora of the woods of western Scotland; another, south-western Ireland; a third, the highly calcifuge communities of leached bark in the acidic oak-birch woods in very high rainfall areas of upland western Britain, communities hardly seen elsewhere in Europe. A fourth index covers the rather contintental oakwoods of eastern Scotland; this is almost equally useful in the very similar old oakwoods of Denmark. Finally, Dr B. J. Coppins and myself have prepared an Index for the assessment of the old Caledonian pinewoods which takes into account too the biogeographical variation in these forests from the west to the east coast of Scotland.

Our limited knowledge about ancient woodland bryophytes is summarized in Table 8.5 which lists epiphytic, terricolous, and lignicolous taxa which are most abundant in ancient woodland in lowland Britain. With terricolous species the continuity of old bark surfaces is not an important issue, but other factors, such as continuity of a shading tree canopy, may be more important than they are for the epiphytes. Many of the species in Table 8.5 are not restricted to old forest in euoceanic parts of western Britain and alternative ancient woodland indicators will need to be found

Table 8.5   Ancient woodland bryophytes of lowland England. Based on an unpublished list prepared for the British Bryological Society by R. C. Stern with minor additions. Nomenclature follows Corley and Hill (1981).

Liverworts

Barbilophozia attenuata	*Plagiochila asplenioides*
Bazzania trilobata*	*P. porelloides*
Chiloscyphus pallescens or *C.*	**Porella arboris-vitae** +
polyanthos*	*Radula complanata*
Frullania fragilifolia +	**Saccogyna viticulosa***
F. tamarisci +	**Scapania gracilis***
Lejeunea cavifolia +	*S. undulata**
Marsupella emarginata	**Trichocolea tomentella***

Mosses

Anomodon viticulosus	**Leucobryum juniperoideum**
Brachythecium populeum	*Leucodon sciuroides* +
B. plumosum	*Leptodon smithii* +
Cirriphyllum piliferum	**Mnium stellare ***
Dicranum majus	*Neckera complanata*
D. montanum +	**N. pumila** +
Eurhynchium pumilum	*Orthotrichum lyellii* +
E. striatum	**O. stramineum** +
Fissidens exilis	**Plagiothecium latebricola** +
Herzogiella seligeri +	*P. undulatum*
Homalia trichomanoides	**Pterogonium gracile** +
Homalothecium sericeum +	*Rhizomnium punctatum*
Hookeria lucens *	**Rhytidiadelphus loreus**
Hylocomium brevirostre	*R. triquetrus*
H. splendens	*Taxiphyllum wissgrillii*
Hyocomium armoricum *	*Tetraphis pellucida*
Isothecium myosuroides +	*Thamnobryum alopecurum*
I. myurum	**Zygodon baumgartneri** +
I. striatulum	

Notes: (1) Species in **bold** print are normally only found in ancient, little-disturbed woodland. (2) Other species are commonly (or in some cases occasionally) found in ancient woodland but are not confined to it. (3) Asterisked species are confined to special habitats such as streambanks, rocks and stones in and by streams, sheltered gills. (4) Species only counted as epiphytes are marked +

The richness of a woodland can be assessed as follows:

Conservation value	Midlands and E. Anglia	Rest of lowlands
Low	<10	<15
Moderate	10–15	15–20
High	15–20	20–25
Exceptional	>20	>25

in this region. More study is required before bryophytes can be used as confidently as lichens in assessing the ecological continuity of woodlands.

8.6 Old forest species and forest management outside Europe

In North America few lists of lichens or bryophytes indicative of lack of forest disturbance have been published, but work is in progress. Professor S. Selva (Fort Kent, University of Maine) has produced an unpublished list of possible indicator lichens for 'old-growth forest' at Big Reed Pond Reserve in Maine. This includes numerous species that also occur in European forests and are regarded there as old forest indicators. Examples are *Biatorella microhaema*, *Bryoria capillaris*, *Catillaria atropurpurea*, *C. laureri*, *Cetrelia olivetorum*, *Chaenotheca* spp., *Dimerella lutea*, *Lobaria pulmonaria*, *Menegazzia terebrata*, *Nephroma parile*, *Pachyphiale fagicola*, *Pannaria rubiginosa*, *Pertusaria velata*, and *Pyrenula laevigata*. The list also includes some species which, in Europe, are definitely not regarded as faithful to old forest: *Dimerella diluta*, *Parmelia glabratula*, and *Ramalina farinacea*. Many other species occur in North American forests that are not known in Europe but which appear to be more or less faithful to ancient forest areas.

Perhaps the biggest problem, however, in much of eastern North America is to find 'old-growth' (i.e. primary) forest areas today. Most of the north-east forest was clear-felled at one time or another, and while some of the regrowth is now up to 200 years old, there is very little untouched forest. Some areas occur in north Maine, in north Vermont, and above all in the southern Appalachians (Blue Ridge Mountains, Great Smoky Mountains National Park), but so far these have not been researched in depth, or if they have, the work is not yet published. Where old growth areas persist in the eastern USA, and also in eastern Canada, the species spectra are strikingly different and richer than those in the secondary forests.

In south-east Asia, Wolseley (1991) noted that *Lobarion*-like communities were characteristic of old montane forests. The work of Kantvilas (1985, 1988) and Kantvilas *et al.* (1985) similarly reveals that many lichen species and communities are confined to untouched primeval rainforests in Tasmania. In New Zealand, Galloway (1988, personal communication) has indicated the fidelity of many species of *Pseudocyphellaria* to particular types of ancient forest, particularly *Nothofagus* forest, but so far lichen indices have not been devised for these southern temperate areas. It is suggested that such studies should be undertaken.

8.7 Conclusions

Over the last 20 years it has become clear that the epiphytic communities of the older British woodlands are of great conservation importance. This is true above all of the old pasture-woodlands of varied types, which are so scarce in a European perspective that the better British examples—above all the New Forest woodlands and the best of the few remaining deer parks—are now a unique conservation resource on an international scale. However, these woodlands require careful management. Wherever possible, grazing of some kind—if deer are impracticable, then cattle, ponies, or in special cases, sheep—should be reintroduced. There are numerous plans already in this direction. Without pasturing, the unique character of these woodlands will be lost for ever. The creation of new pollards is very desirable, including not only oak, beech, ash, and hornbeam pollards, but also, in appropriate cases, holly pollards, a feature hardly known outside the British Isles. New plantings of trees, or aided natural regeneration, is necessary in many cases where the tree stocking is down to rather small numbers of ancient trees without younger successors. Fortunately, the severe storms of recent years have done remarkably little damage to the ancient oaks of most of the British pasture woodlands, but more younger trees are needed in some areas to bridge the 'generation gaps' before it is too late.

References

Brandon, P. F. (1963). The common lands and wastes of Sussex. Unpublished Ph.D. thesis. University of London.

Cantor, L. M. and Hatherley, J. (1979). The medieval parks of England. *Geography*, **64**, 71–85.

Coles, J. M. and Orme, B. J. (1976). The Sweet Track, railway site. *Somerset Levels Papers*, **2**, 34–65.

Coles, J. M. and Orme, B. J. (1977). Neolithic hurdles from Walton Heath, Somerset. *Somerset Levels Papers*, **3**, 6–29.

Corley, M. F. V. and Hill, M. O. (1981). *Distribution of bryophytes in the British Isles. A census catalogue of their occurrence in vice-counties*. British Bryological Society, Cardiff.

Darby, H. C. (ed.) (1952–1977). *The Domesday geography of England*, (6 vols). Cambridge University Press, Cambridge.

Galloway, D. J. (1988). Plate tectonics and the distribution of cool temperate southern Hemisphere macrolichens. *Botanical Journal of the Linnean Society*, **96**, 45–55.

Godwin, H. E. (1975). *The history of the British flora: a factual basis for phytogeography*, (2nd edn). Cambridge University Press, Cambridge.

Grigson, C. (1978). The Late Glacial and Early Flandrian ungulates of England and Wales—an interim review. In *The effect of man on the landscape: the lowland zone*, Council for British Archeology, research report No. 21, (eds S. Limbrey and J. G. Evans), pp. 46–56. Council for British Archeology, London.

Harding, P. T. and Rose, F. (1986). *Pasture-woodlands in lowland Britain.* ITE (NERC), Huntingdon, UK.

Hawksworth, D. L., James, P. W., and Coppins, B. J. (1980). Checklist of British lichen-forming, lichenicolous and allied fungi. *The Lichenologist*, 12, 1–115.

James, P. W., Hawksworth, D. L., and Rose, F. (1977). Lichen communities in the British Isles: a preliminary conspectus. In *Lichen ecology*, (ed. M. R. D. Seaward), pp. 295–413. Academic Press, London.

Kantvilas, G. (1985). Studies on Tasmanian rain forest lichens. Unpublished Ph.D. thesis. University of Tasmania.

Kantvilas, G. (1988). Tasmanian rainforest lichen communities: a preliminary classification. *Phytocoenologia*, 16, 391–428.

Kantvilas, G., James, P. W., and Jarman, S. J. (1985). Macrolichens in Tasmanian rain forests. *The Lichenologist*, 17, 67–83.

Rackham, O. (1976). *Trees and woodland in the British landscape.* Dent, London.

Rackham, O. (1980). *Ancient woodland, its history, vegetation, and uses in England.* Edward Arnold, London.

Rose, F. (1974). The epiphytes of oak. In *The British oak, its history and natural history*, (eds M. G. Morris and F. H. Perring), pp. 250–73. Classey, Faringdon.

Rose, F. (1976). Lichenological indicators of age and environmental continuity in woodlands. In *Lichenology: progress and problems*, (eds D. H. Brown, D. L. Hawksworth, and R. H. Bailey), pp. 279–307. Academic Press, London.

Rose, F. (1988). Phytogeographical and ecological aspects of *Lobarion* communities in Europe. *Botanical Journal of the Linnean Society*, 96, 69–79.

Rose, F. (1990). The epiphytic (corticolous and lignicolous) lichen flora of the Forêt de Fontainebleau. *Bulletin Société Botanique de France*, 137, 197–209.

Rose, F. and James, P. W. (1974). Regional studies on the British lichen flora. I: The corticolous and lignicolous species of the New Forest, Hampshire. *The Lichenologist*, 6, 1–72.

Rose, F. and Wolseley, P. A. (1984). Nettlecombe Park—its history and its epiphytic lichens: an attempt at correlation. *Field Studies*, 6, 117–48.

Tittensor, R. M. (1978). A history of The Mens: a Sussex woodland common. *Sussex Archaeological Collections*, 116, 347–74.

Wolseley, P. A. (1991). Observations on the composition and distribution of the 'Lobarion' in forests of South East Asia. In *Systematics, conservation, and ecology of tropical lichens*, (ed. D. J. Galloway), pp. 217–43. Clarendon Press, Oxford.

9

The vanishing tropical rain forest as an environment for bryophytes and lichens

S. ROB GRADSTEIN

9.1 Introduction

Tropical forests make up about half of the world's closed forests, yet are rapidly being destroyed. By the end of the 1970s, the Food and Agriculture Organization (FAO) of the United Nations estimated that about 46 per cent of the tropical forests were already gone and the remainder were disappearing at a rate of over 2 per cent per year, either by clear-cutting or by shifting cultivation (Raven 1988). The conclusion was that all tropical rain forests would be destroyed by the middle of the next century, while the seasonal forests, occurring in regions with more pronounced fluctuations in rainfall and usually more suitable for human inhabitation, would already have disappeared long before.

Tropical forests, because of their complexity and variety of microhabitats, usually harbour a rich diversity of bryophytes and lichens. Even though they are often small and inconspicuous, especially in the lowland forest, they may play a significant role in the forest ecosystem. Thick bryophyte mats on trees capture rain water, especially in the montane forest, and help to keep humidity in the forest high. They serve as a substrate for the establishment of vascular epiphytes, especially orchids, and offer shelter to a great variety of invertebrates (insects, snails) and micro-organisms, including the nitrogen-fixing blue–green algae. Bentley and Carpenter (1984) demonstrated that nitrogen fixation on leaves by blue–green algae is proportional to the biomass of epiphyllous hepaticae covering the leaves. Apparently, the moist environment created by the tiny liverworts is favourable to the establishment and growth of the blue–green algae. The microbial fixation in turn may cause unusually high concentrations of nitrogen in the epiphytic bryophytes, exceeding those in the tissues of the host trees (Grubb and Edwards 1982). Nitrogen fixation in the forest is also enhanced by canopy-inhabiting lichens with blue–green algae as photobionts (Forman 1975).

Bryophytes and lichens are also valuable as sensitive pollution monitors and are sources of unique chemical compounds, some of which show significant antibiotic or other pharmacologically interesting activity (Rundel 1978; Asakawa 1990). Most chemical work has dealt with temperate species and very little is known about the compounds of tropical bryophytes and lichens (Spörle 1990). It would certainly be worthwhile to focus on the bryophytes and lichens of the tropics in the search for new medicinal and agro-chemical products.

While increasing attention is being paid to the taxonomy and ecology of tropical bryophytes and lichens (e.g. Richards 1984; Sipman and Harris 1989; Buck and Thiers 1989; Allen and Magill 1990; Frahm and Gradstein 1990; Galloway 1991), very little is known about the impact on them of large-scale forest destruction in the tropics. What is known is mostly based on casual observation and 'field lore' acquired by collectors through years of experience. In this chapter I shall consider the consequences of deforestation for bryophytes and lichens in the light of what is known about their habitats, species richness, dispersal, and distribution. The discussion will focus mainly on the rain forest as more information is available about the bryophytes and lichens of rain forest than for other forest types in the tropics.

9.2 Bryophytes and lichens in the rain forest

9.2.1 Habitats

Tropical rain forests are quite varied in structure and floristic composition. At lower altitudes, below 1500 m, the forest canopy is usually complex and composed of trees of different height which may often reach up to 50–60 m. Woody species are numerous and include many different trees and lianas. The canopy, moreover, usually harbours a rich variety of epiphytes. With increasing elevation the forest canopy becomes lower and less complex in structure, woody species diversity becomes reduced, and epiphytic growth becomes much more luxuriant.

Most of the bryophytes and lichens of the tropical rain forest are epiphytes. A terrestrial layer is poorly developed at lower elevation due to the presence of a thick layer of dead fallen leaves covering the forest floor, but on the acid, humic soils of the montane forest terrestrial bryophytes may be conspicuous. Terrestrial lichens are only abundant in low, rather open evergreen forests or scrub on nutrient-poor sandy soil, such as the

Amazonian caatinga or campina. Dense, spongy mats of *Cladina* and *Cladonia* on the soil surface of these forests are reminiscent of boreal pine forests.

Trees, treelets, shrubs, saplings, and woody lianas are colonized by epiphytes and some species grow on living leaves. Subtle differences in water supply, nutrients, light, and inclination of the substrate affect the ability of the bryophytes to establish themselves and therefore tree bases, trunks, ascending branches, and twigs often have different species. Based on the vertical stratification of epiphytic bryophytes and lichens in the rain forest, a distinction can be made between 'sun epiphytes', which occur mainly in the tree crowns, and 'shade epiphytes' which are largely restricted to the undergrowth of the forest (Richards 1954). Most of the lichens of the rain forest are sun epiphytes whilst bryophytes are common in the shaded understorey as well as in the tree crowns. Some species have very wide vertical amplitudes on the trees and may be considered ecological 'generalists', for example *Archilejeunea fuscescens*, *Calymperes erosum*, *Ceratolejeunea maritima*, *Cheilolejeunea rigidula*, *Chiodecton erosum*, *Crocynia biatorina*, *Eschatogonia prolifera*, *Neckeropsis undulata*, and *Phyllopsora santensis* in Guyanan rain forest (Fig. 9.1).

In the lowland rain forest Lejeuneaceae are the most important family of bryophytes and lichens in terms of species number. In the rain forest of Guyana about 30 per cent of all the species of bryophytes and macrolichens are Lejeuneaceae. Conspicuous but less rich in species are the Calymperaceae, Hookeriaceae, Hypnaceae, Orthotrichaceae, and Sematophyllaceae

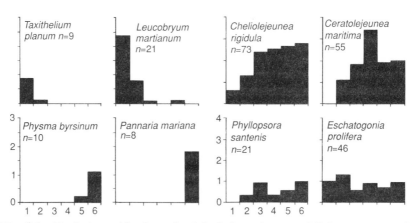

Fig. 9.1 Vertical stratification of epiphytic bryophytes and lichens on trees in lowland rain forest of Guyana, shown as abundance (vertical axis) per tree height zone (after Cornelissen and Ter Steege (1989), with permission). *Left*: shade epiphytes (*top*) and sun epiphytes (*bottom*); *Right*: generalists.

among the mosses, the Lepidoziaceae, Plagiochilaceae, and Frullaniaceae among the liverworts, and the Parmeliaceae and Coccocarpiaceae among the macrolichens. Liverworts usually dominate over mosses. Arthoniaceae, Graphidaceae, Phyllopsoraceae, Thelotremataceae, Trypetheliaceae, and various families of foliicolous lichens are important groups of crustose lichens (Sipman and Harris 1989).

The occurrence of bryophytes and lichens on the surface of living evergreen leaves is a very characteristic feature of lowland and lower montane rain forests. Above 2000 m these foliicolous taxa become rare. Bryophytes only grow on the upper surface of leaves and are therefore commonly called 'epiphyllous' but as lichens also occur on the lower surface the term 'foliicolous' is more appropriate. Many foliicolous species also grow on twigs or other impermanent substrata and are only facultatively foliicolous. The majority of the obligate foliicolous bryophytes are members of the family Lejeuneaceae; foliicolous lichens include members of many different families. Foliicolous bryophytes require shade and high humidity and are therefore largely confined to the understorey and the lower part of the canopy. Foliicolous lichens often prefer drier, more open habitats, such as forest margins and regenerating gaps where light intensity levels are higher than in surrounding areas due to opening up of the canopy. In a small area of Guyanan rain forest, Sipman (1991) recorded about 100 different species of foliicolous lichens; a slightly lower number occurred in Amazonian lowland rain forest at Araracuara, Colombia. Foliicolous bry-ophytes are fewer in number of species although it is not uncommon to collect more than 50 species in a small area of wet neotropical rain forest (Richards 1984). On a single leaf an average number of 5–25 lichen species and 3–13 species of bryophytes may be found, with a maximum of 50 for lichens and 20 for bryophytes (T. Pócs, personal communication).

Montane and subalpine rain forests—ranging from about 1500 to 3000 or 4000 m near the Equator, but lower in elevation at increasing latitudes and on islands—are structurally simpler than the lowland forest and usually have a much more luxuriant epiphytic vegetation and a very different flora. Moreover, the forest floor may be covered with dense bryophyte carpets. The well-developed terrestrial layer of the montane rain forest, as com-pared to its virtual absence in lowland rain forest, is perhaps one of the most striking differences between the two forest types. The lower temperat-ures and higher light levels in the montane forest and the availability of plentiful water due to frequent clouds and fog favour the accumulation of dead organic material on the ground and the abundant growth of the epiphytic and terrestrial cryptogams. Tree trunks and branches may be covered with a dense fur of bryophytes, up to 15 or 20 cm thick and made up of tall turf, feather-type, and pendant growth forms. In the neotropics liverworts of the genera *Plagiochila, Bazzania, Herbertus, Lepidozia, Lepicolea,* and *Trichocolea* usually prevail in the wetter montane forests whereas in drier forests mosses are more common, including *Macromitrium, Meteoridium, Mittenothamnium, Papillaria, Porotrichum, Porotrichodendron, Prionodon densus,* and *Squamidium.* In Asian montane forests robust mosses are often predominant, e.g. *Dicranoloma, Hypnodendron, Braunfelsia, Dicnemon,* and various Pterobryaceae. The crustose lichen families common in the lowlands become scarce and are replaced by such groups as Megalosporaceae, Pertusariaceae, and Pyrenulaceae. Moreover, foliose and fruticose lichens of the families Pannariaceae, Stictaceae, Lobariaceae, Usneaceae, and Parmeliaceae become prominent (e.g. Wolseley 1991).

The amount of epiphytic bryophyte mass in the montane rain forest has been measured by several authors and varies considerably depending on altitude, forest structure, and local climate. It can be as much as 44 tonnes dry weight per ha in very 'mossy' upper montane forest (Hofstede and Wolf, personal communication) as compared to about 2 tonnes in submontane rain forest (Pócs 1982). The large masses of epiphytic bryophytes and lichens are known to function as captors of large amounts of rain water. Pócs (1980) estimated that the interception of rainfall in upper montane forest of Tanzania with a very mossy canopy was about 2.5 times higher than in the lower montane forest in the same area and totalled over 50 per cent of the annual precipitation. Much lower interception values were measured by Veneklaas and van Ek (1991) in Colombian montane forests, which had similar amounts of epiphytic bryophyte mass. Interception in the lower montane forest of Colombia was 12.4 per cent and in the upper montane forest 18.3 per cent. The very different interception values measured in the upper montane forests of the two areas are probably explained by the very different canopies of the investigated forests. While the Tanzanian upper montane forest was a dwarf elfin forest with a dense, umbrella-shaped canopy and an almost continuous upper surface of epiphytic bryophytes, the canopy of the Colombian mossy forest was higher and much more open, with bryophytes in big clumps on canopy branches, forming a very discontinuous epiphyte surface. The continuous bryophyte layer of the dense canopy should have been able to capture a much higher percentage of the precipitation flux than the clumped bryophytes in the open canopy.

Many different cryptogamic communities occur in the rain forest in relation to forest type, microhabitat, substrate, height on the tree, etc. (Jovet-Ast 1949; Richards 1954; Kürschner 1990). Specialized communities are seen on leaves, palm stems, tree ferns, and bamboos (Pócs 1978, 1982), and some communities appear to be host specific. Cornelissen and Ter Steege (1989) compared the epiphytic vegetation of two tree species, *Eperua falcata* and *E. grandiflora,* in Guyanan lowland rain forest and could recognize 13 different cryptogamic communities. These included a *Leuco-bryum martianum* community of trunk bases, a *Graphina virginea* crustose lichen community of shaded, overhanging surfaces of trunks, a *Neurolejeu-nea seminervis* community of thick branches in the inner canopy, consisting of dense, species-rich bryophyte mats, and a *Pannaria mariana* community of fine twigs in the upper canopy with many lichens and small liverworts. The latter community was particularly well developed on *Eperua grandiflora*, whereas on *E. falcata* the twig community was dominated by *Diplasiolejeunea rudolphiana*. Epiphytic communities also appeared to be different at lower levels on the two host trees. Differences in bark texture and bark chemistry are probable factors responsible for the host specificity of the cryptogamic communities on the two *Eperua* spp.

A study by Frahm (1990) on the substrate ecology of epiphytic bryophytes in lowland and montane rain forests of Mt Kinabalu, Borneo, confirmed that bark texture (smooth, fissured, flaky, or striped) is an important factor in determining the biomass and floristic composition of the communities. There was no significant correlation between epiphytic vegetation and bark pH and all epiphytes were classified as acidophytic. Basiphytic epiphytic vegetations seem to be rarer in the tropics than in temperate regions. Pócs (1990) recorded the occurrence of a basiphytic community of epiphytic bryophytes and lichens in montane rain forests on the slopes of young, active volcanoes in Tanzania. The substrates on which the epiphytes grew proved rich in Na, K, and Mg. The high amounts of these nutrients were believed to be due to volcanic activity in the area, resulting in a deposition of alkaline dust and precipitation on the trees. Veneklaas (1990) measured the precipitation nutrients in a montane rain forest in the Andes at the time of a volcanic eruption and found that inputs of Ca, Mg, and SO_4 were very high for 2–3 weeks following the eruption. The conclusion from these and other studies was that repeated volcanic eruptions may significantly alter the nutrient conditions of the habitat and cause changes in the epiphytic vegetation of the forest.

9.2.2 Species richness

Our understanding of the species richness of the cryptogamic flora of the rain forest is still backward as compared with flowering plants. No complete inventory of the bryophyte and lichen flora of a piece of tropical forest has ever been published. Most studies only deal with part of the flora, leaving 'difficult' groups such as crustose lichens and Lejeuneaceae unidentified. Moreover, attention is usually restricted to the lower part of the forest whilst the higher portions of the trunks and the canopy branches are neglected due to their inaccessibility. Sampling of felled trees or fallen branches is useful to obtain information on the flora of the tree crowns, but some species may be missed due to the rapid desiccation and decay of the fallen tree.

The most complete inventory of the bryophyte and lichen flora of a piece of rain forest is the study by Montfoort and Ek (1990) on the cryptogamic epiphytes of a virgin, moist lowland forest of French Guiana. Using special mountaineering techniques for access into the canopy, as described by Ter Steege and Cornelissen (1988), 28 mature standing trees (belonging to 22 species) were sampled from the bases of the trunks up to the highest canopy twigs. In total, 363 species were identified: 66 mosses, 88 liverworts (including 71 Lejeuneaceae), and 209 lichens. These numbers are the highest thus far recorded in tropical lowland forest. The total number of species would have been even higher if lianas, shrubs, rotten logs, and other substrates in the understorey were included in the study; additional lichen species were found in the study area in these habitats by A. Aptroot (personal communication). On a single tree Montfoort and Ek found 83 species on average (19 mosses, 31 liverworts, and 33 lichens). There was a notable difference in species density between bryophytes and lichens, as five trees yielded about 75 per cent of the total number of bryophytes collected and only 30 per cent of the lichens (Fig. 9.2). Lichens were apparently much more sparsely distributed in the forest than the bryophytes. It would be unwise to generalize these results to wider forest patches or to rain forests of other regions, but a first indication of species density of bryophytes and lichens in the rain forest can now be given as a reference.

An undisturbed dry evergreen lowland rain forest in Guyana inventoried by Cornelissen and Ter Steege (1989), using the same sampling technique, yielded fewer species but microlichens and foliicolous species were not taken into account. On 11 trees, belonging to two species of *Eperua* ('walaba'), 110 species were identified: 26 mosses, 53 liverworts, and 31 macrolichens. Species density was not determined.

It is generally assumed that the bryophyte and lichen flora of tropical lowland forests is poorer in species than that of montane forests but Gradstein *et al.* (1990) have argued that this may not always be true. Inventories carried out in rich montane forests of Colombia by J. H. D. Wolf and students from the University of Amsterdam, using sampling techniques similar to those applied in the Guianas, yielded not very many more species and sometimes even less than in the Guyanan lowland forest (J. H. D. Wolf, personal communication). The main differences between the cryptogamic flora of lowland and montane rain forest floras seem to be their very different taxonomic make-up and the much more luxuriant growth in the montane forest. As argued above, most previous studies in the rain forest suffered from neglect of the rich canopy flora. Cornelissen

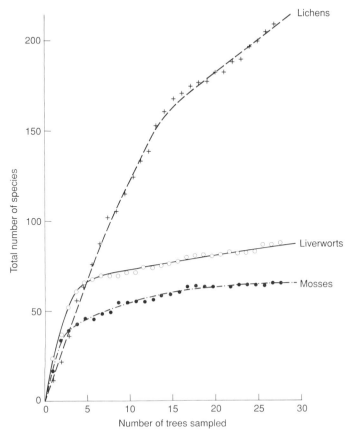

Fig. 9.2   Species density of epiphytic bryophytes and lichens in lowland rain forest of French Guiana (after Montfoort and Ek 1990, with permission).

and Gradstein (1990) found that in the lowland rain forest of Guyana about 70 per cent of the epiphytic bryophyte and macrolichen species growing on trees are found in the canopy. While some of them also occurred in the understorey, about 50 per cent of the bryophyte species and 86 per cent of the macrolichens were only gathered from the tree crowns and the upper portions of the trunks, over 10 m above ground level. In contrast, only 14 per cent of the bryophyte species and none of the lichens were exclusive to the understorey. These findings seem to be at variance with observations by Richards (1954, 1984) who found fewer species of bryophytes in the canopy than in the understorey of the rain forest. The different results may be explained by the fact that Richards also included the epiphytic flora of small trees, shrubs, and lianas in the undergrowth whereas information on the canopy flora was obtained from felled trees and fallen branches.

9.2.3 Dispersal

Rain forests are dynamic communities and the various substrates available
to the bryophytes and lichens may be frequently disturbed by natural
causes. Most trees do not live for more than 200–300 years and the life
span of leaves in the undergrowth, which are the main habitat of the
foliicolous species, is probably about 1.5–3 years, while those of the canopy
trees is usually only about a year (Richards 1988). Tree falls or breakage of
tree crowns by storms or lightning strikes are common in the forest and
cause frequent habitat disturbance and succession in the forest community.
Another disturbance factor is volcanic activity, which may cause alkalinifi-
cation of the habitats due to deposition of ashes (see above).

   Richards (1988) has pointed out that because of the impermanence of
the substrates, short-distance dispersal is important for the forest species.
Rain splash may function as a main agent of dispersal for the species in the
undergrowth, where the air is rather still, whereas wind dispersal may be
effective for species living in the outer canopy.

   Sipman and Harris (1989) noted a predominance of rather large spore
sizes among rain forest lichens and of thick spore walls, for instance in
Graphidaceae, Pyrenulaceae, Thelotremataceae, and Trypetheliaceae.
Foliicolous species, however, generally have thin-walled spores and this
would allow them to germinate rapidly, a necesssity for fast colonization of
the ephemeral leaf surfaces. In foliicolous liverworts rapid establishment is
promoted by the presence of a protonema within the spore wall, allowing
for rapid germination immediately after spore release from the capsule. All
species of Lejeuneaceae have such endosporous protonemata, and this
might well be an explanation for their successful diversification in the rain
forest. Formation of reproductive organs on protonemata or juvenile
gametophytes in some foliicolous taxa (*Ephemeropsis, Metzgeriopsis, Aphano-
lejeunea*) is another strategy to speed up the reproductive cycle. This
strategy is also seen among the terrestrial species of *Pogonatum* sect.
Racelopus, which grow on impermanent substrata of the rain forest floor
such as pebbles and small patches of bare soil (Touw 1986).

   Among the mosses of the rain forest, Richards (1984) observed that sun
epiphytes usually have larger spores than the shade epiphytes, so that the
latter might have enhanced chances for dispersal. Spore size differences
seem less relevant among the liverworts but Van Zanten and Gradstein
(1988) found that spores of some shade epiphytes were much less drought-
resistant than those of sun epiphytes and were unable to germinate after
only a few days of desiccation, even though the geographical ranges of the
investigated liverwort species were usually very wide. A striking example
was *Stictolejeunea balfourii* which grows in highly sheltered habitats in

lowland rain forests, on tree bases and roots close to the ground. The spores of this species lose their capacity of germination within a day of desiccation, yet the species has a very widespread, pantropical distribution. As the species is obviously incapable of long-range dispersal, the very wide range of *Stictolejeunea balfourii* has been explained by assuming that the species is a very ancient taxon which spread step-by-step during and after the Cretaceous via the migrating land masses. The principles of dispersal of the rain forest bryophytes and lichens certainly remain very poorly understood and need much more work.

9.2.4 Ranges

Most families and genera of lichens of the rain forest are widespread throughout the tropics. The bryophyte families are usually pantropical as well but the genera are often more restricted in distribution. There are many genera of mosses restricted to the rain forests of Asia and Australasia, whilst rain forests of the New World tropics are rich in endemic genera and subgenera of liverworts, especially of Lejeuneaceae (Gradstein and Pócs 1989; Schuster 1990; see also Chapter 4).

At the species level, assessment of the ranges is hazardous because of the very fragmentary knowledge of the taxonomy and distribution of the rain forest flora. A consideration of the subject should therefore focus on taxa for which recent taxonomic revisions are available. Sipman and Harris (1989) checked the distribution of about 250 rain forest species of lichens which had been treated in monographs and found that about 30 per cent were pantropical and the remaining species were restricted to the neotropics or the palaeotropics. For the purpose of this chapter it is the rain forest species with very limited distributions which are of particular interest, as they are among the first threatened with extinction when the forest vanishes. Examples of narrow endemics listed by Sipman and Harris are *Parmelina antillensis* (West Indies), *Relicina luteoviridis* (Sabah), *Relicina precircumnodata* and *Calenia leptocarpa* (Philippines), and 13 species of Trypetheliaceae (Amazonian lowlands forest). In addition, a number of species of the genus *Laurera* are restricted to single islands in the Malesian archipelago and three species of *Megalospora* (*M. albescens*, *M. granulosa*, and *M. weberi*) are known only from the subalpine dwarf forest of Papua New Guinea where a surprising number of lichen endemics occur (Aptroot and Sipman 1991).

Among the rain forest bryophytes, pantropical species are much fewer than in lichens even though an increasing number of widespread, transoceanic species has become known as a result of recent monographic work. The Calymperaceae and the Lejeuneaceae subfamily Ptychanthoideae, which contain many rain forest epiphytes and have recently been studied on a worldwide basis (Gradstein 1987, 1991; Reese 1987*a*, *b*), contain

Table 9.1 Rare and endangered bryophytes and lichens of the tropical rain forest. Species occurring in areas undergoing rapid deforestation are marked by an asterisk.

Name	Distribution	Life zone	No of collections	Taxonomic position
Musci				
*Bryomela tutezona Crosby & Allen	Panama	montane	1	type sp. of genus
Calymperes mitrafugax Florsch., C. platyloma Mitt., C. smithii Bartr.	Northern Amazonia and Guyanas	submontane	few	closely related spp.
*Chaetomitrium schofieldii Tan & Robinson	Mindanao	lowland	1	
Dicnemon robbinsii (Bartr.) Allen	Papua New Guinea	montane	1	
*Hageniella hattoriana Tan	Mt Kinabalu, Sabah	lowland	1	
*Hypnodendron beccarii (Hampe) Jaeg.	Borneo	lower montane	many	sect. Phoenicobryum
Hypnodendron brevipes Broth.	Papua New Guinea and Louisiade Archip.	submontane	5	ibid.
Leucomium steerei Allen	Eastern Guyana Highlands	lower montane	2	
*Merrilliobryum fabronioides Broth.	Northern Luzon	montane	few	monotypic genus
Pterobryella papuana (C. Muell.) Jaeg.	New Guinea	montane	few	monotypic genus
*Syrrhopodon isthmi Reese	Panama	submontane	1	
*Syrrhopodon sarawakensis (Dix.) Reese	Borneo and Malay Peninsula	lowland	few	monotypic genus
*Taxitheliella richardsii Dix.	Sarawak	lowland	1	monotypic genus
Hepaticae				
*Caudalejeunea grolleana Gradst.	Northern Madagascar	lowland	2	monotypic subgenus
Dactylolejeunea acanthifolia Schust.	Dominica	submontane	2	monotypic genus
*Eopleurozia simplicissima (Herz.) Schust.	Sarawak	submontane	3	isolated taxon
Haesselia roraimensis Grolle & Gradst.	Mt Roraima, Guyana	submontane	5	isolated taxon
*Luteolejeunea herzogii (Buchloh) Piippo	Pacific coast of northern South America	lowland-submontane	few	monotypic genus
Nowellia wrightii Grolle	Eastern Cuba	lower montane	few	monotypic family
Phycolepidozia exigua Schust.	Dominica	submontane	1	monotypic family
*Plagiochila wolframii Inoue	Northern Peru	subalpine	1	good endemic sp.
*Perssoniella vitreocincta Herz.	New Caledonia	montane (Araucaria forest)	few	monotypic family

Schistochila undulatifolia Piippo	Papua New Guinea	submontane	few	monotypic section
*Spruceanthus theobromae (Spruce) Gradst.	Ecuador (Prov. Los Rios)	lowland	3	the only neotropical sp. of Asiatic genus
*Symbiezidium madagascariensis Steph.	Madagascar and Seychelles	lowland	2	
*Thysananthus evansii Fulford	Belize, Costa Rica	coastal lowland	3	monotypic subgenus
Thysananthus mollis Steph.	Papua New Guinea	montane	10	
Lichens (by Aptroot, Utrecht)				
*Cladonia modesta Ahti & Krog	Kenya	lower montane	2	
*Coccocarpia fulva (Sm.) Arvidsson	New Caledonia	lowland	2	
Compsocladium archboldianum Lamb	New Guinea	montane	several	monotypic genus
Exiliseptum ocellatum (Müll. Arg) Harris	Amazonian Brazil	lowland	1	monotypic genus
Graphis isidiifera Hale	Dominica	submontane	1	isolated taxon
Menegazzia minuta James & Kantvilas	Northern Tasmania	lowland	few	
*Myelorrhiza jenjiana Verdon & Eliz	Queensland	lowland	several	genus with 2 spp.
Polystroma fernandezii Clem.	Guyanas	lowland	many	monotypic genus
*Sphaerophorus madagascareus Nyl. ex Crombie	Madagascar	lower montane	few	isolated taxon
*Thelotrema magnificum (Berk. & Broome) Hale	Sri Lanka	lowland	many	

together about 325 species of which about 5 per cent are more or less pantropical and a handful of others occur disjunctively in the neotropics and Africa. Within the limits of each continent, the species of these families are usually widespread and only a few qualify as rain forest endemics, e.g. *Syrrhopodon isthmi* Reese, known only from the Cerro Jefe near Panama City, and *Spruceanthus theobromae* (Spruce) Gradst., endemic to coastal Ecuador (Prov. Los Rios). Further examples are given in Table 9.1.

9.3 Impact of rain forest destruction on bryophytes and lichens

Almost everywhere in the tropics forests are being logged or cleared (Fig. 9.3) and vast areas of rain forest have been converted into plantations and farmlands. Remnants of the forest appear in the landscape as regenerating secondary communities, consisting of a few, fast growing tree species. Where selective logging is taking place, depleted forests are being created consisting of a mixture of primary forest elements and weedy secondary elements which have colonized the gaps left by the removal of the timber trees (Richards 1952). Forest destruction is alarming in areas with rapid population growth such as the Philippines, large parts of Indonesia, Sri

Fig. 9.3 Road construction through virgin submontane rain forest, Serranía del Pilon Lajas, Dept. Beni, Bolivia, *c.* 700 m altitude (photograph by the author).

Lanka, West Africa, parts of Central America, and the Atlantic coast of Brazil. The rain forest remnants of Madagascar and New Caledonia, with their high levels of endemism, are also on the verge of disappearance (Myers 1980, 1983).

The cryptogamic flora of the disturbed environment is very different from that of the primary forest, even though the precise differences are poorly documented. Under the higher light intensities in the open habitats, bryophytes and lichens may grow almost randomly, and sometimes profusely, on trunks, poles, and branches near the ground, on rocks and boulders, road banks, etc. The impression may be gained of that of a rich cryptogamic flora. Indeed, Lambley (1991) observed an increased diversity in the altered environments in Papua New Guinea: 'Secondary habitats created by man increase the diversity of the lichen flora. Rubber plantations, which are usually sited below 500 m, provide an interesting contrast to the rain forest they replace. The well-spaced uniform stands provide a light, but still humid environment, which is favourable for many lichens including *Parmotrema cristiferum*, *Usnea* spp., *Physma* spp., *Relicina* spp., *Leptogium* spp., *Coccocarpia* spp., and *Pannaria*. In highland valleys where man has cleared most of the forest between 1400–2500 m, lichen habitats are often scarce and confined to isolated trees and fences. However, hedges of *Cordyline* ... support a surprisingly rich lichen community in which *Pseudocyphellaria argyracea*, *P. crocata*, *Lobaria isidiosa*, *Collema* spp., and *Pannaria* feature' (p. 80). Epiphytic bryophytes may grow in abundance in *Citrus* and coffee plantations to the extent that they become a pest and are periodically removed by the growers to protect their crop (Pócs 1982). Petit and Symoens (1974) recorded luxuriant terrestrial bryophyte carpets in plantations of *Cupressus lusetanica* and *Acacia mearnsii* at mid-montane elevations in Burundi. Twenty-eight species were recorded, most of them mosses, and different terrestrial communities could be recognized in plantations of different ages. Very dense plantations and *Eucalyptus* groves in the same area were usually devoid of terrestrial bryophytes. Apparently, shading and litter produced by the trees were important factors affecting the establishment of the terrestrial bryophytes. It would have been interesting to compare the bryophyte vegetation of these plantations with that of natural forests in the area.

Most authors contend that the secondary forests and plantations have an impoverished species richness compared with the primary forest. For instance, secondary rain forests generally lack the rich Thelotremataceae flora of the virgin forests (Sipman and Harris 1989). Species of *Relicina* are disappearing in Australia due to rain forest destruction as they need mature rain forest trees for successful establishment (Elix 1991). In Hawaii, Smith (1991) noted a marked uniformity of the lichen flora of planted Myrtaceaedominated forests as compared with the natural rain forest. The lack of old stands and the monospecific character of the plantations had led to a strong depletion and alteration of the flora, with a few species becoming locally dominant, e.g. *Coccocarpia erythroxyli*, *Parmotrema cristiferum*, and *Ramalina exiguella* on *Schinus terebinthifolius*. An interesting observation was that some local Hawaian endemics had successfully invaded the newly created habitats: the rare endemic *Ramalinopsis manii* had become associated with planted *Aleurites moluccana* introduced from Polynesia and *Heterodermia obesa* was locally common in plantations of *Psidium guajava*. There was no evidence, however, that alien lichens had invaded the area and displaced native species, a common feature in the vascular flora of Hawaii.

The above observations on lichens are paralleled by the bryophytes. Pócs (1982 and personal communication.) noted that epiphytic bryophyte communities of cultivated trees (cocoa, coffee, *Citrus*, mango, rubber, etc.) may resemble those of the natural forest trees, but are usually much poorer in species. A comparison between the epiphytic bryophyte floras of primary rain forest and planted forests on the slopes of Mt Kilimanjaro, Tanzania, showed that 90 per cent of the native forest species did not occur in the plantations and that only 5 per cent of the epiphytic biomass had remained. The depletion of the epiphytic vegetation may be less when the plantation forests are near to the virgin forest, e.g. in clearings. In the Guianas, Florschütz-de Waard and Bekker (1987) noted a marked similarity between the epiphytic floras of the rain forest canopy and of scrubby vegetation in the same general area. Sun epiphytes and generalists were the main taxa surviving on the shrubs and in the plantations and some of these, e.g. species of Cryphaeaceae, *Erpodium*, *Fabronia*, *Frullania*, *Groutiella*, *Lejeunea*, *Leucodontopsis geniculata*, and *Lopholejeunea*, may even be more common on the shrubs and cultivated trees than in the forest. On wet slopes of the northern Andes, cut road banks are invaded by canopy liverworts of the montane forest such as *Dicranolejeunea axillaris*, *Frullania brasiliensis*, *Frullania convoluta*, *Frullanoides densifolia*, *Herbertus acanthelius*, *Jamesoniella rubricaulis*, *Omphalanthus filiformis*, and *Taxilejeunea pterigonia*. These species behave as weedy pioneers in the newly created, man-made habitats.

Hyvönen *et al.* (1987) checked the occurrence in disturbed and undisturbed habitats of 43 rain forest mosses on the Huon Peninsula of Papua New Guinea and found that 29 species grew in both environments, whereas 14 had not been recorded outside the virgin forest. Most of the species restricted to the virgin forest seemed to be shade epiphytes. A few sun epiphytes appear unable to re-establish in secondary growth or plantations, such as *Holomitrium arboreum* in Guyana (Richards 1984) and *Frullania nodulosa*, *Mastigolejeunea turgida*, *Pycnolejeunea contigua*, and *Thysananthus spathulistipus* in West Africa (A. J. Harrington, personal communication). It should be realized, however, that species growing in one area only in primary forest may occur outside the forest in other areas. In northern South America, *Frullania nodulosa* is not restricted to the forest canopy but is also common in savannah bush. *Neurolejeunea seminervis, Schiffneriolejeunea amazonica,* and *Acanthocoleus aberrans* in the Guianas have only been found in the canopy of the forest, but elsewhere in the neotropics they also occur in scrub.

It would thus appear that shade epiphytes are more adversely affected by forest destruction than sun epiphytes, which may be able to re-establish in secondary growth and plantations after forest clearing. Some of these may also grow on soil and exposed rocks. Hyvönen *et al.* (1987) and Norris (1990) suggested that species richness of bryophytes may increase, due to the invasion of weedy species, in areas with shifting cultivation and small-scale damage of the forest. Species richness would become much reduced, however, if traditional slash-and-burn agriculture were replaced by modern large-scale farming methods.

Changes due to forest destruction might be particularly severe at the community level but have scarcely been documented. Norris (1990) recorded the disappearance of a mist-induced *Braunfelsia* bryophyte community of the mossy montane forest of New Guinea. This community forms large balls on the upper canopy branches of virgin mossy forests on foggy ridges, and is composed of tall-turf forming mosses of the genera *Braunfelsia, Brotherobryum, Dicnemon, Cladopodanthus,* and *Schistomitrium.* The colonies soon decay after tree fall and are lacking in disturbed forest in similar environments, presumably due to excessive dehydration of the opened canopy. Structurally the *Braunfelsia* community resembles bryophyte communities of mossy forests elsewhere in the tropics—a detailed description of a similar tall turf community in the Andes was given by van Leerdam *et al.* (1990)—but what makes the *Braunfelsia* community unique is its flora, which is largely made up of endemic taxa. Endangered mist-dependent cryptogamic communities also include the rich bryophyte and lichen vegetations of isolated, foggy hill tops in the otherwise very dry *Brachystegia*-wooded landscape of tropical East Africa. These vegetation types are increasingly destroyed by man through bushfires and have almost been eliminated in some parts of Tanzania, where they covered very large areas as recently as the early 1970s (T. Pócs, personal communication).

Foliicolous communities comprise another category of threatened cryptogamic vegetation. Notwithstanding their remarkable ability for rapid colonization, most foliicolous species appear to be sensitive to disturbance of the forest because of their preference for sheltered habitats. Sipman (1991) observed that opening of the canopy by logging caused foliicolous lichens in the undergrowth to become discoloured and moribund. Some foliicolous species re-occur in secondary growth and plantations but the species richness in these habitats appears to be much lower than in the primary forest, even though foliicolous growth may be luxuriant in *Citrus* plantations (Pócs 1982).

As pointed out earlier, large-scale disturbance of the forest may also occur due to natural catastrophes, for instance by volcanic eruptions. Probably the best known example is the recovery of the rain forest of the volcanic island of Krakatoa after its destruction by the great eruption of 1883 (see Richards (1952), p. 269–82). Hurricanes, heavy rains, and floods can cause devastation of the forest especially in coastal areas and on islands where such events may occur frequently. The forests of Dominica, the wettest and most densely wooded island in the West Indies, are affected by hurricanes about once every 15 years and were largely destroyed in 1979 by Hurricane David (Evans 1986). Torrential rains in unusually warm 'El Niño' years can cause large-scale destruction of habitats on Pacific islands. Weber and Beck (1985) gave a detailed account of the effects of the extraordinary El Niño of 1983 on the cryptogamic vegetation of the Galapagos Islands. Forest floors were swept clean of bryophytes, and lichens on rocks and bark became waterlogged and rotted or fell away from the substrates. The authors estimated that it would take a century or more for the cryptogamic vegetation to recover.

While much attention has been paid to the process of regeneration of rain forest after clearance (e.g. Richards 1952; Lieth and Werger 1989), virtually nothing is known about the rate of recovery of the cryptogamic vegetation. Fast-growing, weedy species might return soon but the slow-growing species, restricted in their occurrence to the mature rain forest trees, would probably take longer to reappear. In 1988 I observed in Guadeloupe that a few common liverworts of the montane rain forest, namely *Dicranolejeunea axillaris*, *Omphalanthus filiformis*, and *Taxilejeunea pterigonia*, were extremely abundant on shrubs in an area of young, regenerating woodland which had been completely destroyed by volcanic eruption twelve years before (Beaudoin 1985). They were the same species which commonly invade road banks in the Andes (see above) and should be considered weedy pioneers. On the neighbouring island of Dominica small twig epiphytes such as *Diplasiolejeunea rudolphiana*, *Ceratolejeunea* spp., and *Prionolejeunea* sp. were very common in young, regenerating brushwood dominated by large-leaved monocotyledons (*Heliconia*: Zingi-beraceae), in an area where montane forest had been devastated by Hurricane David nine years earlier.

How cryptogamic species fit into the secondary succession of the rain forest remains to be determined. The few available data from non-tropical regions suggest that lichens may recover more slowly than bryophytes. Rose (1976) calculated that it may take 500 years for the natural lichen vegetation to re-establish in temperate forest in Britain. Bryophyte recovery in temperate forest has been reported to be 80–100 years in regrowth of *Sequoia* forest of northern California (Norris 1987). Chapman and King (1983) demonstrated that in a regenerating subtropical rain forest of Australia some bryophyte species had returned 25 years after cutting. The more scattered occurrence of lichen species in primary tropical rain forest as compared with bryophytes (Montfoort and Ek 1990) might reflect their slower recovery. Generalization of this point, however, is dangerous as there are many different kinds of rain forest and bryophyte and lichen richness varies considerably. A careful comparison of rain forest stands of different ages would have to be carried out to obtain more insight into the process of recovery of the cryptogamic flora after clearing of the forest. Based on their general structure and physiognomy, rain forests of up to 60–80 years in age are usually considered secondary, but forests older than 80 years may become indistinguishable from primary forests (Brown and Lugo 1990).

9.4 Conservation of rain forest bryophytes and lichens

Protection of their habitats is essential to save the bryophyte and lichen communities of the rain forests from extinction. Even though some forest species can survive in the disturbed landscape, especially the ecological generalists and the sun epiphytes, others will vanish and rich cryptogamic communities become lost. In many tropical countries virgin forest areas are set aside as parks or reserves and when well managed these protected areas could serve for the survival of the rain forest communities. The preliminary data on species density of bryophytes and lichens in a tropical rain forest suggest that small reserves could be adequate to conserve cryptogams, especially the bryophytes. Forest reserves should be large enough, however, to ensure rejuvenation of the host trees. Furthermore, forest reserves should encompass all the different life zones as the tropical cryptogamic flora shows very distinct altitudinal diversification (see e.g. Van Reenen and Gradstein 1983; Enroth 1990; Krog 1991). Whether the rain forest species will survive in the reserves remains uncertain, however. According to Pócs (personal communication) some species are disappearing from preserved forest fragments in East Africa due to local climatic change and desiccation caused by large-scale forest destruction in the region.

In areas where human population growth is rapid and the needs for natural resources pressing, maintenance of forest reserves and conservation of species may be very difficult to achieve. In the Philippines, Tan and Rojo (1989) recorded that about 80 per cent of the virgin rain forest had disappeared and that forest reserves are also threatened. The well-known

Mt Makiling forest reserve near Manila was recently allowed to be partially logged and bulldozed for a geothermal energy exploration study. In Panama the richest bryophyte areas include the premontane and montane rain forests of Fortuna in the north of the country and forests around Santa Fé at the eastern end of the Cordillera de Talamanca. Fortuna is the site of an endemic *Hookeriopsis* and two endemic *Plagiochila* species whilst Santa Fé is the type site of *Bryomela tutezona* (Table 9.1). Although efforts are being made to prevent deforestation in the area and Fortuna has been declared a national forest reserve, road building and settlements are causing increasing threats to the natural vegetation. Most parks lack sufficient personnel and appropriate equipment to control the forest areas (N. Salazar, personal communication).

Arvidsson (1991) has pointed out that botanical gardens in the tropics could function as refugia for some forest lichens, at least temporarily. There should be pieces of mixed forest in these gardens, with old growth trees and a variety of habitats and substrates, to allow the cryptogamic vegetation to thrive. Moreover, these habitats should be shielded from the deleterious effects of air pollution.

The effective protection of the rare, endangered species of the rain forest is a challenge to conservation. To pinpoint the endangered bryophytes and lichens is a very tedious affair because the forests have been so poorly collected and the taxonomy and distribution of most of their species is very incompletely known. Species rare in one area may be common elsewhere or may survive in secondary habitats after disturbance or destruction of the forest (Hyvönen *et al.* 1987). *Buxbaumia javanica* is a very rare and endangered moss of the Philippines according to Tan *et al.* (1986), but in New Guinea the species is not uncommon (D. H. Norris, personal communication).

The concept of the rare species includes narrowly endemic species and those which are widely distributed but are nowhere common. Examples of the latter category from the neotropics are the liverworts *Blepharolejeunea saccata* (Steph.) van Slag. & Kruijt, a species widespread in lower montane rain forest yet known from only a handful of collections, and *Verdoornianthus griffinii* Gradst, recorded three times in Amazonian lowland rain forest, at distances thousands of kilometres apart. Many foliicolous lichens seem to fit this category, as they are often pantropical yet known from only one or two collections in each continent (e.g. Serusiaux 1984). As the rarity of these species may be due to undercollecting they will not be considered further.

Narrowly endemic species are relatively rare in bryophytes and lichens and tend to occur in regions with considerable habitat diversity. The rain forests of New Caledonia and New Guinea are known to be particularly rich in endemics and a list of 37 bryophytes, from a limited number of families, endemic to primary rain forest of New Guinea was provided by Piippo *et al.* (1987). Their figure is to be used with caution, however, because many of the listed endemic species have not yet been monographed and could fall into synonymy after revision. The species listed in Table 9.1 are a random selection of what seem to be 'good' endemic species—treated in monographs—which are threatened with extinction if the rain forest vanishes. Some occur in areas undergoing rapid deforestation (Myers 1983) and should therefore be considered 'endangered' according to the International Union for Conservation of Nature and Natural Resources (IUCN) Red Data Book Categories (species in Table 9.1 marked by an asterisk). Others occur in areas which are not threatened with immediate deforestation but should be considered at risk; according to the IUCN categories these may be classified as 'rare'. It goes without saying that the list of species is preliminary and needs updating by further exploration and taxonomic work.

9.5 Summary and conclusions

The preceding sections may have made it sufficiently clear that the massive forest destruction in the tropics has a deleterious effect on the bryophytes and lichens. Studies in various tropical regions indicate that the flora of secondary forests and plantations is usually much impoverished compared with the virgin forest. Shade epiphytes of the forest undergrowth seem more seriously affected than canopy epiphytes, although some canopy species may also become rare or disappear after disturbance. Depletion of the rich habitat diversity of the primary forest as well as the predominance in disturbed habitats of fast-growing tree species, seem important reasons why species of the primary forest are unable to re-establish in secondary growth. Changes at the community level should be particularly severe but have scarcely been documented.

Even though no species of bryophytes and lichens have as yet been reported as extinct, a number of endemic species should be considered endangered and threatened with extinction when the forest vanishes. In order for these and other species to be preserved, habitat samples should be set aside in all life zones throughout the tropics. These forest reserves would probably not need to be very large, at least for bryophytes, as preliminary data on species density in the rain forest suggest, but should be large enough to ensure protection and rejuvenation of the forest and the host trees. The safeguarded areas may not only serve as refugia for the endangered species, but should also be laboratories for the study of the natural richness in bryophytes and lichens of the rain forest.

There must be much more research on the impact of tropical forest decline on bryophytes and lichens to help replace speculation with data. We need many more inventories to find out which species are locally common and which are rare. We also need more comparisons between the floras of undisturbed forest and those of secondary growth or plantations, such as done by Hyvönen *et al.* (1987) in New Guinea and Pócs (1982) in Africa, to determine the amount of floristic change. The flora of the forest canopy should not be neglected in these studies. Understanding of the changes and the processes of regeneration could be obtained by long-term observations using plots in forests of different ages. There should also be much more work on the taxonomy and habitat preferences of the rain forest taxa to find out which species are truly endemic and to pinpoint those which are unable to re-establish in disturbed habitats. These are base-line studies which have been promoted for many years—they were main goals of the UNESCO Man and the Biosphere Program (see UNESCO MAB report nr. 16, 1974)—yet they have been inadequate as far as the bryophytes and lichens are concerned.

One of the greatest problems is that there are too few knowledgeable bryologists and lichenologists living or stationed in the tropics. They are usually working in isolation with insufficient library and herbarium facilities and are often overburdened with other duties. Yet they should be in a much better position to do in-depth floristic and ecological studies than researchers from temperate areas. If their work was better coordinated, progress could possibly be more rapid. Organizations such as the Sociedad Latinoamerica de Briología have an important role to play to coordinate and stimulate cryptogamic work in the tropics. Because of the speed at which forests are vanishing in the tropics, more intensive study of the bryophytes and lichens of the tropical rain forest is urgently needed.

Acknowledgements

I am much indebted to my colleages Mr A. Aptroot, Dr C. Delgadillo, Dr D. J. Galloway, Dr A. J. Harrington, Dr T. Pócs, Dr N. Salazar, Dr H. J. M. Sipman, and Dr B. J. Tan who have supported me in various ways with information, references and other materials useful to the preparation of the manuscript. I am also grateful to Mr D. Montfoort for drawing Fig. 9.2, and to Mr A. Aptroot for the list of endangered lichens.

References

Allen, B. and Magill, R. E. (eds) (1990). Proceedings of the Tropical Bryology Conference, 31 July–3 August 1989, Missouri Botanical Garden, St Louis. *Tropical Bryology*, **2**, 1–284.

Aptroot, A. and Sipman, H. J. M. (1991). New lichens and lichen records from New Guinea. *Willdenowia*, **20**, 221–56.

Arvidsson, L. (1991). On the importance of botanical gardens for lichens in the Asian tropics. In *Tropical lichens: their systematics, conservation, and ecology*, (ed. D. J. Galloway), pp. 193–9. Clarendon Press, Oxford.

Asakawa, Y. (1990). Terpenoids and aromatic compounds with pharmacological activity from bryophytes. In *Bryophytes, their chemistry and chemical taxonomy*, (eds H. D. Zinsmeister and R. Mues), pp. 369–410. Clarendon Press, Oxford.

Beaudoin, R. (1985). Analyse de la distribution des bryophytes épiphylles sur la sourfrière de Guadeloupe. *Comptes Rendues de la Societé de Biogéographie*, **61**, 47–57.

Bentley, B. L. and Carpenter, E. J. (1984). Direct transfer of newly-fixed nitrogen from free-living epiphyllous microorganisms to their host plant. *Oecologia*, **63**, 52–6.

Brown, S. and Lugo, A. E. (1990). Tropical secondary forests. *Journal of Tropical Ecology*, **6**, 1–32.

Buck, W. R. and Thiers, B. M. (1989). Review of bryological studies in the tropics. In *Floristic inventory of tropical countries*, (eds D. G. Campbell and H. D. Hammond), pp. 485–93. New York Botanical Garden, New York.

Chapman, W. S. and King, G. C. (1983). Floristic composition and structure of a rainforest area 25 years after logging. *Australian Journal of Ecology*, **8**, 415–23.

Cornelissen, J. H. C. and Gradstein, S. R. (1990). On the occurrence of bryophytes and macrolichens in different lowland rain forest types at Mabura Hill, Guyana. *Tropical Bryology*, **3**, 29–35.

Cornelissen, J. H. C. and Ter Steege, H. (1989). Distribution and ecology of epiphytic bryophytes and lichens in dry evergreen forest of Guyana. *Journal of Tropical Ecology*, **5**, 131–50.

Elix, J. A. (1991). The lichen genus *Relicina* in Australia. In *Tropical lichens: their systematics, conservation, and ecology*, (ed. D. J. Galloway), pp. 17–34. Clarendon Press, Oxford.

Enroth, J. (1990). Altitudinal Zonation of Bryophytes on the Huon Peninsula, Papua New Guinea. *Tropical Bryology*, **2**, 61–90.

Evans, P. G. H. (1986). Dominica Multiple Land Use Project. *Ambio*, **15**, 82–9.

Florschütz-de Waard, J. and Bekker, J. M. (1987). A comparative study of the bryophyte flora of different forest types in West Suriname. *Cryptogamie, Bryologique et Lichénologique*, **8**, 31–45.

Forman, R. T. T. (1975). Canopy lichens with blue–green algae: a nitrogen source in a Colombian rainforest. *Ecology*, **56**, 1176–84.

Frahm, J.-P. (1990). The ecology of epiphytic bryophytes on Mt Kinabalu, Sabah (Malaysia). *Nova Hedwigia*, **51**, 121–32.

Frahm, J.-P. and Gradstein, S. R. (1990). The ecology of tropical bryophytes: a bibliography. *Tropical Bryology*, 3, 75–8.

Galloway, D. J. (ed.) (1991). *Tropical lichens: their systematics, conservation, and ecology*. Clarendon Press, Oxford.

Gradstein, S. R. (1987). The Ptychanthoideae of Latin America: an overview. *The Bryologist*, 90, 337–43.

Gradstein, S. R. (1991). Diversity and distribution of Asian Lejeuneaceae subfamily Ptychanthoideae. *Tropical Bryology*, 4, 1–14.

Gradstein, S. R. and Pócs, T. (1989). Bryophytes. In *Tropical rain forest ecosystems*, (eds H. Lieth and M. J. A. Werger), pp. 311–25. Elsevier, Amsterdam.

Gradstein, S. R., Montfoort, D., and Cornelissen, J. H. C. (1990). Species richness and phytogeography of the bryophyte flora of the Guianas, with special reference to the lowland rain forest. *Tropical Bryology*, 2, 117–25.

Grubb, P. J. and Edwards, P. J. (1982). Studies of mineral cycling in a montane rain forest in New Guinea. III: The distribution of mineral elements in the above-ground material. *Journal of Ecology*, 70, 623–48.

Hyvönen, J., Koponen, T., and Norris, D. H. (1987). Human influence on the mossflora of tropical rainforest in Papua New Guinea. *Symposia Biologica Hungarica*, 35, 621–9.

Jovet-Ast, S. (1949). Les groupements de muscinées epiphytes aux Antilles françaises. *Revue Bryologique et Lichénologique*, 18, 125–46.

Krog, H. (1991). Lichenological observations in low montane rainforest of eastern Tanzania. In *Tropical lichens: their systematics, conservation, and ecology*, (ed. D. J. Galloway), pp. 85–94. Clarendon Press, Oxford.

Kürschner, H. (1990). Die epiphytische Moosgesellschaften am Mt Kinabalu (Nord-Borneo, Sabah, Malaysia). *Nova Hedwigia*, 51, 1–75.

Lambley, P. W. (1991). Lichens of Papua New Guinea. In *Tropical lichens: their systematics, conservation, and ecology*, (ed. D. J. Galloway), pp. 69–84. Clarendon Press, Oxford.

Lieth, H. and Werger, M. J. A. (eds) (1989). *Tropical rain forest ecosystems, biogeographical and ecological studies*. Elsevier, Amsterdam.

Montfoort, D. and Ek, R. C. (1990). *Vertical distribution and ecology of epiphytic bryophytes and lichens in a lowland rain forest in French Guiana*. Institute of Systematic Botany, Utrecht.

Myers, N. (1980). *Conversion of tropical moist forests*. National Academy of Sciences, Washington.

Myers, N. (1983). Conversion rates in tropical moist forests. In *Tropical rain forest ecosystems*, (ed. F. B. Golley), pp. 289–300. Elsevier, Amsterdam.

Norris, D. H. (1987). Long-term results of cutting on the bryophytes of the *Sequoia sempervirens* forest in Northern California. *Symposia Biologica Hungarica*, 35, 467–73.

Norris, D. H. (1990). Bryophytes in perennially moist forests of Papua New Guinea: ecological orientation and predictions of disturbance effects. *Botanical Journal of the Linnean Society*, 104, 281–91.

Petit, E. and Symons, F. (1974). Les bryophytes des bois artificiels de *Cupressus* et d'*Acacia* au Burundi. Analyse factorielle de la végétation bryophytique. *Bulletin du Jardin Botanique National de Belgique*, **44**, 219–47.

Piippo, S., Koponen, T., and Norris, D. H. (1987). Endemism of the bryophyte flora in New Guinea. *Symposia Biologica Hungarica*, **35**, 361–72.

Pócs, T. (1978). Epiphyllous communities and their distribution in East Africa. *Bryophytorum Bibliotheca*, **13**, 681–713.

Pócs, T. (1980). The epiphytic biomass and its effect on the water balance of two rain forest types in the Uluguru Mountains (Tanzania, East Africa). *Acta Botanica Hungarica*, **26**, 143–67.

Pócs, T. (1982). Tropical forest bryophytes. In *Bryophyte ecology*, (ed. A. J. E. Smith), pp. 59–104. Chapman and Hall, London.

Pócs, T. (1990). The exploration of the East African bryoflora. *Tropical Bryology*, **2**, 177–91.

Raven, P. H. (1988). Tropical floristics tomorrow. *Taxon*, **37**, 549–60.

Reese, W. D. (1987*a*). *Calymperes* (Musci: Calymperaceae): World ranges, implications for patterns of historical dispersal and speciation, and comments on phylogeny. *Brittonia*, **39**, 225–37.

Reese, W. D. (1987*b*). World ranges, implications for patterns of historical dispersal and speciation, and comments on phylogeny of *Syrrhopodon* (Calymperaceae). *Memoirs of the New York Botanical Garden*, **45**, 426–45.

Richards, P. W. (1952). *The tropical rain forest*. Cambridge University Press, Cambridge.

Richards, P. W. (1954). Notes on the bryophyte communities of Lowland Tropical Rain Forest, with special reference to Moraballi Creek, British Guyana. *Vegetatio*, **5–6**, 319–28.

Richards, P. W. (1984). The ecology of tropical forest bryophytes. In *New manual of bryology*, (ed. R. M. Schuster), Vol. 2, pp. 1233–70. Hattori Botanical Laboratory, Nichinan.

Richards, P. W. (1988). Tropical forest bryophytes: synusiae and strategies. *Journal of the Hattori Botanical Laboratory*, **64**, 1–4.

Rose, F. (1976). Lichenological indicators of age and environmental continuity in woodlands. In *Lichenology: progress and problems*, (eds D. H. Brown, D. L. Hawksworth, and R. H. Bailey), pp. 279–307. Academic Press, London.

Rundel, P. W. (1978). The ecological role of secondary lichen substances. *Biochemical systematics and ecology*, **6**, 157–170.

Schuster, R. M. (1990). Origins of neotropical Leafy Hepaticae. *Tropical Bryology*, **2**, 239–64.

Serusiaux, E. (1984). New species or interesting records of foliicolous lichens. *Mycotaxon*, **20**, 283–306.

Sipman, H. J. M. (1991). Notes on the lichen flora of the Guianas, a neotropical lowland area. In *Tropical lichens: their systematics, conservation, and ecology*, (ed. D. J. Galloway), pp. 135–50. Clarendon Press, Oxford.

Sipman, H. J. M. and Harris, R. C. (1989). Lichens. In *Tropical rain forest ecosystems*, (eds H. Lieth and M. J. A. Werger), pp. 303–9. Elsevier, Amsterdam.

Smith, C. W. (1991). Lichen conservation in Hawaii. In *Tropical lichens: their systematics, conservation, and ecology*, (ed. D. J. Galloway), pp. 35–45. Clarendon Press, Oxford.

Spörle, J. (1990). Phytochemische Untersuchungen an ausgewählten panamaischen Lebermoosen. Dissertation, Universität des Saarlandes, Saarbrücken.

Tan, B. C. and Rojo, J. P. (1989). The Philippines. In *Floristic inventory of tropical countries*, (eds D. G. Campbell and H. D. Hammond), pp. 44–62. New York Botanical Garden, New York.

Tan, B. C., Fernando, E. S., and Rojo, J. P. (1986). An updated list of endangered Philippine plants. *Yushania*, **3**, 1–5.

Ter Steege, H. and Cornelissen, J. H. C. (1988). Collecting and studying bryophytes in the canopy of standing rain forest trees. In *Methods in Bryology*, (ed. J. M. Glime), pp. 285–90. Hattori Botanical Laboratory, Nichinan.

Touw, A. (1986). A revision of *Pogonatum* sect. *Racelopus*, sect. nov., including *Racelopus* Dozy & Molk., *Pseudoracelopus* Broth., and *Racelopodopsis* Thér. *Journal of the Hattori Botanical Laboratory*, **60**, 1–33.

UNESCO-MAB (1974). International working group on Project I: Ecological effects of increasing human activities on tropical and sub-tropical ecosystems. *MAB Report Series*, **16**, 1–96.

Van Leerdam, A., Zagt, R. J., and Veneklaas, E. J. (1990). The distribution of epiphyte growth-forms in the canopy of a Colombian cloud-forest. *Vegetatio*, **87**, 59–71.

Van Reenen, G. B. A. and Gradstein, S. R. (1983). A transect analysis of the bryophyte vegetation along an altitudinal gradient on the Sierra Nevada de Santa Marta, Colombia. *Acta Botanica Neerlandica*, **32**, 163–75.

Van Zanten, B. O. and Gradstein, S. R. (1988). Experimental dispersal geography of neotropical liverworts. *Beihefte zur Nova Hedwigia*, **89**, 41–94.

Veneklaas, E. (1990). Nutrient fluxes in bulk precipitation and throughfall in two montane tropical rain forests, Colombia. *Journal of Ecology*, **78**, 974–92.

Veneklaas, E. and Van Ek, R. (1991). Rainfall interception in two tropical montane rain forests, Colombia. *Hydrological Processes*, **4**, 311–26.

Weber, W. A. and Beck, H. T. (1985). Effect on cryptogamic vegetation (lichens, mosses, and liverworts). In *El Niño en las Islas Galápagos: El Evento de 1982–1983*, pp. 342–61. Fundacíon Charles Darwin, Quito.

Wolseley, P. A. (1991). Observations on the composition and distribution of the 'Lobarion' in forests of South East Asia. In *Tropical lichens: their systematics, conservation, and ecology*, (ed. D. J. Galloway), pp. 217–43. Clarendon Press, Oxford.

10

Impact of agriculture on bryophytes and lichens

Dennis H. Brown

10.1 Introduction

Two major agricultural chemical practices that can have an impact on bryophytes and lichens are discussed in this chapter. The use of pesticides (e.g. herbicides, fungicides, and insecticides) has been important in improving the productivity of modern farming. Sometimes such chemicals are deliberately employed to control mosses and liverworts, especially in amenity grass such as lawns and sports turf or around potted plants. On other occasions spray drift or longer distance dispersal may be an inadvertent result of otherwise acceptable application procedures. Nutrient fertilizer usage, including animal waste, is the second major influence and may also be the result of controlled or relatively accidental agricultural practices.

Unfortunately, although problems associated with excessive additions of the above chemicals are generally acknowledged as existing in the environment, there is remarkably limited published literature. Often the results of investigations are presented as reports to the body providing sponsorship for a study or remain as unpublished academic theses. It is very probable that the present literature survey has failed to locate valuable studies; potential authors and editors should be encouraged to publicize relatively preliminary studies that may add to a clearer picture of the problems involved. Very few substantiated generalizations can yet be made concerning the impact of agricultural practices on the cryptogamic flora or about sensitivities of species to particular chemicals. This chapter is a preliminary attempt to bring the existing literature together.

10.2 Agrochemicals

10.2.1 Analyses of field samples

Bryophytes and lichens have a substantial reputation as effective biomonitors of metal emissions, whereby plants either grown *in situ* or transplanted to contaminated sites are analysed for their metal content. Some studies have shown that they may be equally capable of trapping and possibly accumulating organic molecules, and general reviews have been provided by Thomas (1986*a*) and Herrmann (1990). Although not strictly agrochemicals, polycyclic aromatic hydrocarbons (PAHs) are frequently included in analyses of organic pollutants and include fluoranthene and benzopyrene. They are non-polar molecules mostly derived from the combustion of fossil fuels and are frequently associated with particulate matter (Thomas 1981; Thomas *et al.* 1984). Polychlorinated biphenyls (PCBs) have many non-agricultural uses but are associated with pesticide use and are more volatile than PAHs. A number of pesticides are chlorinated hydrocarbons such as hexachlorobenzene (HCB) and hexachlorohexane (HCH). It should be noted that γ-HCH (Lindane) can be converted to α-HCH. As the former is more water soluble, the ratio α/γ-HCH is higher in the atmosphere than in mosses where the residues correlate with rain- or meltwater concentrations rather than atmospheric gas phase content (Thomas 1986*b*).

Although Carlberg *et al.* (1983) recovered two- to five-fold greater concentrations of chlorinated organic compounds in the epiphytic lichen *Hypogymnia physodes* than the terricolous moss *Hylocomium splendens*, their data are insufficient to distinguish between the significance of plant groups or differences in the exposure of the habitat to movement of air masses. The latter is probably of greater significance, as no clear differences between lichen and moss chlorinated hydrocarbon concentrations were detected in Antarctica (Bacci *et al.* 1986). The detection of chlorinated hydrocarbons in lichens and mosses from Antarctica confirms their value as monitors of long-distance transport (Bacci *et al.* 1986). Antarctic values were comparable to Swedish (Thomas *et al.* 1984) and Finnish (Herrmann and Hübner 1984) samples but in more industrial regions the values may be 50- to 100-fold higher. Levels of chlorinated hydrocarbons were not significantly higher in lichens than in the leaves or gymnosperms and angiosperms in Tuscany (Italy) where differences were considered to be due to mean exposure times of the samples, the lipid content of the sample being of minor importance (Gaggi *et al.* 1985). Thomas *et al.* (1984) found that *Hypnum cupressiforme, Racomitrium lanuginosum,* and to a lesser extent *Cladonia rangiferina,* contained amongst the highest levels of PAHs (probably particulate) compared to various angiosperms and gymnosperms. There were fewer differences in chlorinated hydrocarbons between the various plant groups but *Betula* bark was especially enriched. Thomas *et al.* (1984) also reported that two-year-old *Pinus* needles had slightly higher polychlorinated hydrocarbon concentrations, but concluded that enrichment factors were unreliable due to high volatilization rates.

Thomas (1983) reported a correspondence between trace pollutant concentrations in mosses and those recovered with suspended particle sampling devices. A concentration factor (plant/air) of $1–3 \times 10^5$ was reported for PCBs (Villeneuve *et al.* 1988) in lichens from southern France and Antarctica. Thomas (1981, 1984) and Simon (1985), respectively, reported enrichment factors between aerial and aquatic sources and mosses. PAHs or DDT showed relatively high enrichments against total precipitation whereas HCH isomers showed low enrichment, unless compared to values found associated with suspended particles (Thomas 1981, 1984). HCH enrichment was low from river water, while PAH enrichment was high and possibly involves adsorption processes (Simon 1985). Thomas (1984) gives a detailed discussion of multiple regression models of moss uptake related to precipitation amount and pollutant concentrations.

There is considerable variation in the amounts of organic pollutants recovered in different species of mosses and lichens (Bacci *et al.* 1986). The greater recovery of fluoranthene from *Cetraria delisei* and *Stereocaulon alpinum* compared to mosses analysed from Spitsbergen was attributed to their larger surface area retaining particulate matter with associated PAH (Thomas 1986*b*).

Villeneuve *et al.* (1988) found no relationship between plant concentration of PCBs and altitude, but insecticides of the DDT series and toxaphene showed increasing concentrations with altitude. This was attributed to different exposures to agricultural sources, rather than occult precipitation, which has been postulated to account for comparable increases of radionuclides and metals with altitude. Moreover, Carlberg *et al.* (1983) noted that differences between recovery patterns for organic and metal pollutants may reflect the poor washout ratios for many of the former, implying that rainfall scavenges aerial organic pollutants inefficiently. This highlights the necessity of knowing the form of the pollutant investigated and the mechanism of uptake before one can satisfactorily predict the relationship between plant and environmental levels. The low water solubility and high lipo-affinity of many agrochemicals suggests that many of these chemicals reach the troposphere as vapours that are only weakly absorbed by particulate matter. Where comparisons are made between specific chemicals, differences may be attributed to variations in supply, uptake mechanism, and retention (Carlberg *et al.* 1983). Simon (1985) showed that breakdown of 3,4-benzopyrene occurred on the surface of the aquatic moss *Fontinalis* but was unable to ascribe this to metabolic or photochemical processes.

Monitoring surveys have shown a decline in organic pollutants from south to north in Norway (Carlberg *et al.* 1983). PAHs in Iceland (Thomas and Schunke 1984) and Spitsbergen (Thomas 1986*b*) probably derive from local fossil fuel sources, whereas chlorinated hydrocarbons reflect long distance transport. This is due to the mainly pastoral nature of their agriculture and the greater proportion of α-HCH compared to γ-HCH, which is the reverse of the central European situation (Thomas and Schunke 1984). Herrmann and Hübner (1984) showed high concentrations of PAHs and PCBs in the centre of a Finnish town, with the former showing a more rapid decrease away from the centre. This was attributed to PAHs being adsorbed on particles while PCBs remained in the gas phase. Terrestrial distribution patterns partly reflect the extent of agricultural activities in Germany (Thomas and Herrmann 1980), and water analyses have demonstrated local emission sources in German rivers (Simon 1985). While Garty *et al.* (1982) detected PCBs in transplanted samples of the lichen *Ramalina duriaei* after one year, they concluded that traffic was a more important source than agriculture. Thomas (1986*a*) considered that the high exchange rates of HCHs between air and plant makes long-term accumulation by mosses unlikely.

Uptake of artificially supplied herbicides showed that higher concentrations of glyphosate than triclopyr could be recovered from the lichen *Cladonia rangiferina* growing in field plots. Both herbicides were recovered in higher concentrations than those found in ericaceous plants from the same plots (Siltanen *et al.* 1981). Loss of glyphosate residues with time was observed. Larson *et al.* (1985) also observed uptake and loss of chlorinated hydrocarbons, with greater losses of α-HCH compared to γ-HCH from dry plants and enhanced loss when *Pleurozium schreberi* was treated with artificial rainwater. Eronen *et al.* (1979) reported losses of MCPA from *Pleurozium* in field samples.

10.2.2 Physiological effects

There have been relatively few studies of the physiological effects of agrochemicals on mosses and lichens and most of these have concentrated on a limited range of herbicides. As part of a study on the effects of various chemicals on the lawn moss *Rhytidiadelphus squarrosus*, Mabb (1989) investigated uptake of a number of radioactively labelled herbicides. She showed that positively charged ^{14}C-paraquat, like Fe (which is often used to kill lawn mosses as ferrous sulphate), was rapidly bound to the anionic exchange sites of the cell wall, with a smaller proportion recovered from an apparently intracellular site. Pre-incubation with Fe reduced the quantity of paraquat bound to the cell walls. Transfer from the wall to the cell interior appeared possible and leaves bound more paraquat than did stem tissue. Virtually no uptake of the anionic 2,4-D or dichlorophen was detected. Such behaviour is to be expected from the high cation, but low anion, exchange capacity of bryophyte cell walls (Clymo 1963; Brown 1984). Damage by these herbicides probably results from the uptake of the undissociated molecule at a suitably neutral pH.

Photosynthesis has been shown to be inhibited by applications of dichlorophen to the moss *R. squarrosus* and liverwort *Marchantia polymorpha* (Brown *et al.* 1986). Dark CO_2 release was stimulated at low levels but inhibited at higher concentrations and this, linked with the release of intracellular K and Mg ions, suggests membranes as a primary site of action. Mabb (1989), working with the same moss, concluded from an analysis of photosynthesis–irradiance curves that dichlorophen and paraquat both inhibited photosynthetic electron transport, although the former was shown rapidly to reduce chlorophyll *a* concentrations. Paraquat, like 2,4-D, did not immediately damage photosynthetic reactions.

Low concentrations of dichlorophen stimulated respiration in the lichens *Cladonia portentosa* and *Peltigera horizontalis* and photosynthesis in the latter, but inhibited both processes at higher concentrations (Howes 1986). Nitrogen fixation in cyanobacterial lichens has also been shown to be sensitive to herbicides in laboratory experiments. Low concentrations of atrazine and ioxynil, but not diuron, stimulated nitrogen fixation in *Peltigera aphthosa*, but not in *Stereocaulon paschale*. High concentrations, however, were inhibitory to both nitrogen fixation and photosynthesis (Kallio and Wilkinson 1977). Increases in 'respiration' were observed, but Brown *et al.* (1986) suggested that this might be due to uncoupled electron transport resulting in excessive but energetically valueless decarboxylation caused by membrane damage. However, Hällbom and Bergman (1979) failed to detect any depression of nitrogen fixation in *Peltigera praetextata* treated in the field with a number of herbicides shown to damage the isolated cyanobacterium in culture. They considered that this might reflect mycobiont protection of the photobiont of the lichen and/or problems associated with field application and uncontrollable interactions with climatic variables.

Stjernquist (1981) also reported photosynthetic stimulation on the addition of half-normal glyphosate and one-eighth-normal 2,4-D dosages to *Dicranum polysetum* and *Pleurozium schreberi*, respectively. There was much variation in moss sensitivity to a variety of herbicides. *Pleurozium* was most sensitive to 2,4-D, *Hylocomium splendens* to MCPA and *Sphagnum squarrosum* to triclopyr. The acrocarpous moss *Pohlia nutans* was only slightly affected by all of the herbicides tested; it is also amongst the most tolerant mosses to heavy metal pollution (Folkeson and Andersson-Bringmark 1988).

Herbicide damage was progressive with *Polytrichum commune*, which may reflect either slower penetration through the more pronounced cuticle or uptake from the soil via the more developed conducting system (Stjernquist 1981). Field experiments showed that recovery, often by regrowth, was possible after initial damage, but the relative abundance of the different species could be altered in ways apparently related to photosynthetic sensitivity determined in the laboratory, diurnal photosynthetic production and energy reserve supplies and climatic variables. The complex pattern reported by Stjernquist (1981) deserves much fuller study. It should be noted that Rudolph and Samland (1985) found that the synthesis of sphagnum acid (found exclusively in *Sphagnum* spp.) was inhibited by low concentrations of glyphosate (0.5 mM), which also caused the accumulation of shikimate, as a result of blocking the shikimic acid pathway to this cinnamic acid derivative.

10.2.3 Herbicide effects on semi-natural populations

A number of studies have been made on the eradication of nuisance mosses and liverworts from lawns, orchards and flower pots in nurseries and greenhouses. Mosses and liverworts may reduce the vigour of container-grown vascular plants, which are watered frequently. Waterlogging may be an initial stimulus to moss growth in lawns, and herbicide application is often a more convenient control measure than soil improvement. Low nutrient supply may also be a contributory factor in reducing grass vigour. In the past, lawn mosses were controlled by mercury salts (Blandy 1954; Jackson 1961; Woolhouse 1972; Brown and Whitehead 1986), but the use of such dangerous chemicals is no longer acceptable.

Dichlorophen (Jackson 1961; Woolhouse 1972; van Himme and Stryckers 1981; Brown *et al.* 1986) is an effective and relatively selective herbicide for eliminating bryophytes and is present in many commercial formulations. Several other herbicides have been proposed for moss control. In field trials, endothal limited moss development over a number of years, apparently by desiccation at all stages of the life cycle, whereas bensulide was almost without effect (Brauen *et al.* 1986). In pot cultures the fungicides thiram and captan were reasonably effective, and the algicide chinonamid was fast-acting but with poor residual activity (van Himme and Stryckers 1981). The same authors found the herbicide chloroxuron effective against mosses but not liverworts, whereas the reverse was observed with binapacryl. Although not specific for bryophytes, Wilson and Hughes (1985) observed that oryzalin eliminated *Marchantia polymorpha* whereas mogeton controlled both this liverwort and *Funaria hygrometrica* without damage to many woody and herbaceous ornamental genera and cultivars in field and nursery pot cultures. Mogeton was most effective in combination with diphenamid or oryzalin, but all were more efficient than oxadiazon, used as a standard.

Ronoprawiro (1975) found glyphosate an effective moss control agent under tea bushes, but combinations of 2,4-D, paraquat, diquat, diuron, MSMA, and captafol proved relatively ineffective, although the latter was of some value in controlling mosses growing on the trunks and branches of tea bushes. Pihakaski and Pihakaski (1980) reported that glyphosate affected the chloroplast structure and photosynthetic activity of *Pellia epiphylla* in laboratory experiments. Atkinson *et al.* (1980) and van Himme and Stryckers (1981) showed that simazine was relatively ineffective in moss control. Atkinson *et al.* (1980) advocated its use as a soil stabilizer in orchards where it had been observed to control angiosperm weeds but permitted the growth of a complete cover of mosses, including *Polytrichum* species not previously seen in the area (Bond 1976). Skender (1983) suggested that abundant moss growth, following pre-treatment with atrazine or dalapon, was sufficient to depress the return of weed species. Balcerkiewick and Rusińska (1987) also reported considerable growth of bryophyte weed species, resulting in a flora resembling that initially developing after fire damage, when the herbicides Gramoxone, Reglone, atrazine or simazine were used in a variety of habitats. The success of bryophyte growth was attributed to reduced competition from angiosperms. These authors reported substantially weaker effects, compared to angiosperm damage, when Gramoxone was applied directly to mosses and Ougham (1983) found that atrazine had little effect on the growth of *Marchantia* thalli. Gilbert (1977) reported a thick covering of the lichens *Cladonia furcata* and *Baeomyces rufus* on paths treated with simazine and three *Peltigera* species at another site treated with Prefix.

10.2.4 Herbicide effects on natural communities

The published data on the effects of deliberate pesticide applications appears to be confined to the preliminary studies of Gilbert (1977). He treated sandstone rocks and walls, limestone walls, concrete posts, acid heathland, and oak trees with a single application of 2,4-D, paraquat, or MCPA. Only MCPA at five times the manufacturer's recommended concentration caused clear damage to lichens on sandstone rocks or walls and oak trees although some damage was possibly also seen in the acid heath community with all herbicides supplied at five times the recommended application rate. Stjernquist (1981) discussed damage to field populations of mosses following 2,4-D applications and their recovery after one year. Browning was seen with all mosses studied, but regrowth from the apex occurred with *Pleurozium* and *Dicranum polysetum*. Glyphosate damage developed more slowly, but was probably more extensive. Both herbicides were used at the manufacturers' recommended dosage.

Mabb (1989) observed visible damage to *R. squarrosus* in herbicide-treated turf using dichlorophen, paraquat, 2,4-D, chloroxuron with Fe salts, Gloquat C, thiram, oxadiazon, acifluorfen, and oxyfluorfen. The detrimental effect of paraquat and 2,4-D on higher plants in the turf resulted in an increase in moss cover, while chloroxuron with an iron salt, dichlorophen, oxadiazon, and thiram all caused a long-term reduction in moss cover when applied at greater than the manufacturers' recommended rate.

K. R. Payne (personal communication) treated sarsen stones with dilutions of commercial herbicide formulations and followed changes to the percentage frequency of lichens. The formulations used were Roundup (glyphosate), Alistell (2,4-D + MCPA + linuron), Broadshot (2,4-D + dicamba + triclopyr), and asulam. Changes in percentage cover were only observed at 'weed wipe' concentrations and not with diluted solutions; Roundup was the least and Broadshot the most damaging. Visual damage was reported to thalli of the foliose *Parmelia glabratula* ssp. *fuliginosa* and *P. conspersa*, which tended to become algal covered, and crumbling from the rock was reported for the crustose species *Lecanora orosthea* (apparently the most sensitive crustose species), *L. rupicola*, *Buellia saxorum*, *Aspicilia caesiocinerea*, and *Rinodina atrocinerea*. Perkins and Marrs (1990) used a single application of similar herbicide formulations on heathland lichens (asulam, triclopyr, and fosamine) and pasture-woodland trees (MCPA, glyphosate, triclopyr, and asulam with and without the oil-based carrier Dessipron) at up to five times (trees) or ten times (heathland) the recommended dosages. While *Calluna*, grasses and broad-leaved plants were affected by the herbicides there was no visual damage to heathland lichens (mainly *Cladonia* and *Peltigera* species) after one year. However, the lichens on trees (*Evernia prunastri*, *Ramalina farinacea*, *Parmelia caperata*, and *P. perlata*) became discoloured and shrivelled within 15 days with 5 × MCPA. This was most obvious when lichens were sprayed to run-off at these concentrations. Damage was progressive with some thallus loss, but also some signs of recovery, after seven months. Five-fold glyphosate added until run-off caused some damage within one month, which increased with time. Triclopyr initially caused little damage under any conditions, but some browning of *R. farinacea* was detected after seven months with the five-fold concentrated to run-off treatment. Similarly, asulam with Dessipron had a limited effect.

Alstrup (1991) reported the results of a substantial investigation on herbicide and fungicide damage to lichens growing on north- or south-facing siliceous walls in Denmark. The herbicide mixtures tested were

Herbatox Combi 3 (2,4-D + dichloroprop + MCPA—all as dimethamino salts), Dantril (ioxynil + bromoxynil + dichloroprop + MCPA) and the fungicides Tilt turbo (propioconazol + tridemorph) and Rival (fenpropimorph + procloraz) at up to 10 times the recommended dose. Monitoring involved both visual assessment and measurement of the largest diameter of thalli apparently free of competitive interactions with other lichens. The results showed that lichen sensitivity to the pesticides differed between species and chemicals. Often the herbicides were less damaging than the fungicides. Many lichens appeared to be more sensitive to Dantril than Herbatox. Tilt turbo applications damaged most lichens but many species tolerated Rival. Some species were sensitive to low dosages of both classes of pesticides (e.g. *Candelariella vitellina*), others more sensitive to the fungicides (e.g. *Acarospora fuscata*, *Parmelia glabratula*, *Physcia caesia*), or herbicides (e.g. *Lecidea soredizoides*), and some very little affected by either (e.g. *Lepraria incana*, *Psilolechia lucida*). Many species were found to be sensitive to only one chemical or showed modest sensitivity to all pesticides tested. Amongst the least damaged were sorediate species, which are probably water-repellent; some of these showed enhanced growth after spraying. A number of lichenicolous fungi were found to be present after pesticide treatments, perhaps suggestive of thalli weakened by these chemicals.

It appears that herbicides, used at normal concentrations and in volumes comparable to those of spray drift, usually have very limited inhibitory effects on lichens. However, damage to bryophytes, by those herbicides to which they are sensitive, may occur at the manufacturers' recommended dose. More work needs to be undertaken to substantiate this, and to link it with analyses of residues recovered in lichens and bryophytes. Uptake, retention, and possible metabolism of residues (Siltanen *et al.* 1981) needs further investigation, particularly in comparing lichens exposed to different air-flow patterns (terricolous or epiphytic). Sensitivity to different chemicals may be related to chemical exposure or to actual plant species.

10.3 Fertilizers

There have been many references to 'nitrophilous' cryptogamic species or communities, the implication being that growth is dependent upon nitrogenous substances. Similar assumptions have been made about 'ornithocoprophilous' communities of bird-perching places. It appears that these terms have been developed because of the association of particular lichens and bryophytes with potentially N-enriched sites. Masse (1966) reported higher N contents of 'ornithocoprophilous' lichens compared to species not affected by bird excrement. However, such sites are also frequently enriched with other elements, for example Ca or PO_4^{3-}, or increased pH, and the form of N may be critical. Most reports emphasize that 'nitrophilic' communities are associated with NH_4^+ rather than NO_3^- enrichment ('ammoniophily'). Most research assumes a direct chemical or physical effect on the cryptogamic species involved, damage due to 'eu-' or 'hypertrophication' being a reflection of the inability of particular species to cope with the deviations beyond their normal environmental tolerance limits. James (1973), however, suggested that enhanced algal growth on the surface of lichens may be a cause of damage, perhaps reducing the light intensity reaching the lichen photobiont. The osmotic effects of enhanced chemical depositions in extreme cases of fertilizer additions should also be considered (Brown *et al.* 1987). Although most studies have tended to emphasize a single chemical or physical condition, comprehending the complex of conditions involved may be more relevant than attempting to identify a single dominant controlling factor. Barkman (1958), James (1973), and James *et al.* (1977) should be consulted for detailed discussions of the older literature.

10.3.1 Uptake of fertilizer constituents

Nitrogen compounds may be present as NH_3 (or more usually NH_4^+ ions, especially on particles), which directly results in an increased pH. These neutral to alkaline conditions may then stimulate its conversion to nitrates by nitrifying bacteria; NH_3 is retained as such under initially acidic conditions. The result of 'acid rain' or atmospheric conversions of oxides of N, emitted by natural or man-made processes, results in the formation of HNO_3 or nitrates. Although oxides of N may derive from the burning of fossil fuels, Turner (1991) pointed to a correlation with N-fertilizer usage and emphasized that this may be quantitatively a more important source of tropospheric combined N. Assimilation of NH_4^+ ions by a plant frequently results in acidification of its immediate cellular environment, whereas the reverse is true for NO_3^- assimilation, due to the biochemical imbalance of hydrogen ion requirements during formation of organic N compounds by the two different pathways.

In order to understand cryptogamic reactions to elevated levels of N-compounds it is appropriate to consider their N-utilizing capabilities. Both inorganic forms of N are assimilated by lichens (Crittenden 1983, 1989) and bryophytes (Rieley *et al.* 1979; Weber and van Cleve 1981) from dilute solutions in the field. Many field and laboratory studies have shown that NH_3 is usually more readily absorbed than is NO_3^- by both lichens (Smith 1960*a*; Lang *et al.* 1976; Hällbom and Bergman 1983; Crittenden 1983, 1989) and bryophytes (Schwoerbel and Tillmans 1974; Simola 1975; Miyazaki and Satake 1985; Schuurkes *et al.* 1986). This difference is in part a consequence of NO_3^- assimilation requiring the presence of the inducible enzymes NO_3^- and NO_2^- reductase. While Schuurkes *et al.* (1986) failed to detect NO_3^- utilization by *Sphagnum flexuosum*. Deising (1987) showed NO_3^- reductase to be an inducible enzyme in a range of other *Sphagnum* species. Its activity was not significantly controlled by NH_4^+ ions, but its induction was paralleled by increases in reductant-generating systems. Rudolph *et al.* (1987) showed that growth was supported by NO_3^-, but high levels of NH_3 inhibited growth and depressed NO_3^- reductase and net photosynthesis. Although Salomon (1914) and Smith (1960*a*) found that cyanobacterial lichens were especially poor at taking up NO_3^-, Hällbom and Bergman (1983) showed considerable NO_3^- uptake by both *Peltigera praetextata* and *P. aphthosa*. In laboratory studies the latter authors showed that both inorganic ions depressed N_2 fixation in *P. praetextata*, but only NH_4^+ ions were effective with *P. aphthosa*. In bi-partite cyanobacterial lichens (e.g. *P. praetextata*) the NH_3 released by the N_2-fixing cyanobacterium is incorporated into amino acids via glutamate dehydrogenase in the fungal cells (the cyanobacterium having minimal glutamine synthetase activity), whereas in tri-partite species (e.g. *P. aphthosa*) it is also assimilated by glutamine synthetase and glutamate synthase in the chlorophycean alga (Rai 1988). Meade (1984) reported the occurrence of all the above enzymes of NH_3 assimilation in a range of bryophytes, but was unable to detect glutamate dehydrogenase in any *Sphagnum* species.

Bryophytes (Simola 1975, 1979; Schofield and Ahmadjian 1972), lichens (Smith 1960*b*) and isolated lichen symbionts (Schofield and Ahmadjian 1972) are capable of utilizing a range of organic N compounds. Asparagine and NH_3 have been shown to be more actively absorbed by the algal layer of dissected lichens (Smith 1960*c*). A number of laboratory studies have shown the induction of urease activity on the addition of urea to chlorophycean lichens while others indicate that this enzyme appears to be constitutive in cyanophycean species (Vicente and Legaz 1988). It is probable that the bulk of the thallus activity is associated with the green algal cells. In an experiment with either a paste or suspension of bird droppings, Armstrong (1984) found that the growth of *Parmelia conspersa* was enhanced by both, and the paste increased *Xanthoria parietina* growth, had no effect on *Physconia grisea*, but damaged *Parmelia glabratula* ssp. *fuliginosa*. A suspension of uric acid had no effect on either of the *Parmelia* species or *Xanthoria*, suggesting it was not the N-compound that was responsible for the changes. Armstrong suggested that other elements besides N in the bird droppings, for example Ca, could be important, particularly because *Xanthoria* is known to respond positively to additions of alkaline dust (Gilbert 1976). Masse (1969) reported the occurrence of uricase in lichens.

The kinetics of PO_4^{3-} uptake by the lichen *Hypogymnia physodes* was studied by Farrar (1976) who showed that intracellular uptake was an active, carrier-mediated process. High concentrations of PO_4^{3-}, as sometimes used in laboratory studies, damaged a number of physiological processes in *Evernia prunastri* (Brown *et al.* 1987). Other studies of the kinetics of anion uptake by lichens that employed AsO_4^{3-} as a representative (Richardson *et al.* 1984; Nieboer *et al.* 1984) may not be fully reliable due to the toxic nature of this chemical at the higher concentrations used, as shown with *Hylocomium splendens* (Wells and Richardson 1985; Wells *et al.* 1987). The same reservations must apply to ion competition experiments using potentially toxic SeO_4^{2-} and SO_4^{2-}. Sulphite effects on *Sphagnum* are discussed in more detail in Chapter 12.

It is unfortunate that we still have a limited understanding of the uptake, retention, redistribution, and loss of nutrients by bryophytes and lichens under normal field conditions (Brown and Bates 1990). Farrar (1976) estimated that the P status of *Hypogymnia physodes* could be sustained by 1 h of rainfall per week containing normal rainfall PO_4^{3-} concentrations. Equivalent calculations have been made which suggest that mosses may (e.g. Weetman and Timmer (1967) using *Hylocomium*) or may not (Damman (1978) using *Sphagnum*; Tamm (1953) using *Hylocomium*) be able to obtain all of their nutrients from rainfall. None of these calculations actually establish the reality of the speculations. Rieley *et al.* (1979) observed little uptake of P from rainfall passing through mosses loosely arranged in funnels in a woodland. Chapin *et al.* (1987) found that with the feather moss *Hylocomium splendens* only half of the applied radioactive PO_4^{3-} sprayed on natural mats was retained by the plant. The use of the non-specific fungicide Captan resulted in a greater retention of radioactivity in the moss. From these data, and root pruning experiments, they deduced that loss of P from the moss occurred through mycorrhizal fungi, although their conclusions have been questioned (Brown and Bates 1990).

10.3.2 Responses to uncontrolled fertilizer emissions

Søchting (1991) transplanted *Hypogymnia physodes* into the vicinity of a pig farm and demonstrated an enhanced N content that increased towards the source of N, presumed to be NH_3 in dry deposition. Søchting and Johnsen (1990) showed patterns of N content in *Cladina* spp. from throughout Europe, reaching 10.3 mg g^{-1} in The Netherlands, that reflected the anticipated inputs of N from atmospheric pollution. In many regions this input of N was considered to derive from oxides of N introduced in rainfall, which poses a threat to many, normally N-poor, heathlands. A particular discolouration and die-back phenomenon has been noted in podetial mats of species such as *C. portentosa*, which were found to be enriched in N compared to undamaged specimens (Søchting and Johnsen 1987). The same authors have shown a reduction in lichen cover in permanent quadrats, especially those in regions of intensive farming.

In The Netherlands, recent changes in the epiphytic flora have been correlated with NH_3 emissions (van Dobben 1987; de Bakker and van Dobben 1988; de Bakker 1989), following earlier acidification damage due to SO_2 (van Dobben 1983). (Ammonia emissions were estimated by multiplying cattle numbers by a specific emission factor (van der Voet and Udo de Haes 1985).) As the floristic change has been from species preferring acid bark to those mostly associated with neutral bark, a pH rather than N basis for the change was postulated. pH was strongly correlated with NH_3 emissions whereas the bark NH_4^+ concentrations were not; bark NH_4^+ was better correlated with tree circumference. Species diversity increased, despite the loss of acidophytic species such as *Lecanora conizaeoides* and *Hypogymnia physodes*, due to the appearance of 'nitrophytic' species such as *Candelariella vitellina*, *Lecanora dispersa* and *L. hagenii*, *Phaeophyscia orbicularis*, *Physcia adscendens*, *P. caesia*, and *Xanthoria polycarpa*. In addition a number of apparently pH-indifferent species, for example, *Buellia punctata*, *Lecidella elaeochroma*, and *Parmelia sulcata*, showed increased abundance.

Many of the species involved are representative of the phytosociological epiphytic lichen alliance the *Xanthorion* and especially the *Buellietum punctiformis* association, which is common in areas enriched with inorganic fertilizers and SO_2 (James 1973; James *et al.* 1977). In extreme conditions the algal-dominated *Pleurococcetum vulgaris* association develops. *Diploicia canescens* and *Buellia punctata* appear to be the most fertilizer-resistant lichens, with *Phaeophyscia orbicularis*, *Physcia adscendens*, *Physconia grisea*, *Ramalina farinacea*, *Evernia prunastri*, and *Xanthoria parietina* being some-what less resistant. With low levels of fertilizers *X. polycarpa*, *X. candelaria*, *Physconia pulverulacea*, and *R. fastigiata* occur, whereas *Anaptychia ciliaris* and *Physcia aipolia* may be present when eutrophication is by animal products. *Evernia* often persists near tree bases when most foliose species have been lost (F. Rose, personal communication). Certain communities within the *Xanthorion* are encountered in the absence of fertilizer inputs, but are then associated with dust-impregnated bark or on the bark of trees such as *Populus tremula*, which may be rich in Ca and have a relatively high pH. Changes in the distribution of individual species, for example *Physcia stellaris* (de Bakker 1987) and *Xanthoria elegans* (Søchting 1989), have also been related to increased NH_3 emissions. Rosentreter (1990) attributed local enrichments of orange lichens on desert shrubs to either agricultural fertilizers, agricultural dust, or dust from dirt roads. However, he also concluded that greater cover of orange lichens on sagebush was indicative of lower site productivity, and was a quicker estimate of site characteristics.

James *et al.* (1977) considered that under nutrient-enriched conditions (particularly from birds) the *Parmelion conspersae* alliance may be the acid rock counterpart of the *Xanthorion parietinae*. It comprises the *Candelariel-letum corallizae*, *Lecanoretum sordidae* and *Parmelietum glomelliferae* associations. The occurrence of species of *Candelariella*, *Physcia*, and *Xanthoria* in the first of these associations makes for obvious comparisons with the *Physcietum caesiae*, a calcareous rock association of the *Xanthorion*. James *et al.* (1977) also noted the influence of contamination of lakes and streams by fertilizer run-off, but gave no formal phytosociological status to the communities that develop in these algal-dominated conditions.

10.3.3 Responses to controlled fertilizer additions

Physiological observations

Field experiments with fertilizer additions have been used to study changes in the physiology of selected component cryptogams or the floristic balance of species. These experiments frequently involve the addition of considerable quantities of N-fertilizers or agricultural mixtures on a single occasion; some involve repeated additions over a number of growing seasons.

Hällbom and Bergman (1979) showed that N_2 fixation was inhibited in *Peltigera praetextata* treated with NH_4NO_3, and that the symbiosis began to disintegrate. Carstairs and Oechel (1978) found no changes in nutrient concentrations, net photosynthesis and dark respiration with various additions and combinations of KNO_3, KH_2PO_4, and $CaSO_4$ to *Cladina alpestris*. However, the total chlorophyll content of the tips, but not the whole thallus, decreased with an increase in the chlorophyll *a:b* ratio. Kauppi (1980) found that NH_4Cl had little effect on the chlorophyll content of *C. stellaris*, whereas it was sometimes slightly enhanced with KNO_3. The latter observation was correlated with increased photosynthesis and the demonstration of more and larger algal cells in the thallus. Fertilization appeared to cause breakdown of the symbiosis. The greatest effect was observed with the use of meltwater from snow, which contained, in addition to NH_4^+ and NO_3^-, Ca and K with a pH of 11.3.

Tomlinson (1988) found in laboratory experiments that both NH_4Cl and KNO_3 solutions were harmful to photosynthesis in *Parmelia sulcata*. She also noted that damage increased with time but, although the NH_4^+ salt was slightly more damaging than the NO_3^-, KCl was also somewhat inhibitory and suggested a 'salt effect' may be involved. Without directly testing the hypothesis, Kauppi (1980) suggested that the damaging effect of NH_4Cl on *C. stellaris* might be due to the Cl^- ion. Field observations over a 17-week period failed to show any visible damage, although respiration rates were reduced. Stjernquist (unpublished data) fumigated *Ramalina fastigiata*, *P. praetextata*, *Parmelia acetabulum*, and *Leucodon sciuroides* with NH_3 gas for 24 h with up to 1000 μg NH_3 m^{-3}. Photosynthesis in *Peltigera* decreased with increasing NH_3 concentration, whereas it increased in *Leucodon* and no clear picture emerged with the other species. All species tested showed some decline in dark respiration with increasing NH_3 concentrations. Vagts and Kinder (1990) performed a number of field experiments and monitored the effect of organic and inorganic fertilizers on the growth of *Cladonia furcata* (increase in biomass or length). They observed (a) an increase in growth with organic fertilization using rabbit faeces, (b) no direct effect of the inorganic NPK formulation Nitrophoska® although the enhanced growth of grasses may have caused indirect damage, and (c) the addition of the fertilizer $CaCN_2$ (calcium cyanamide or 'nitrolime') was extremely damaging and resulted in high efflux of potassium and ultimately death. They also showed that liming resulted in a decline in the P content of *C. furcata* and a loss of regenerative ability, although no reduction in linear growth was detected.

Stjernquist (1981) found that both *Dicranum polysetum* and *Pleurozium schreberi* showed increased assimilation with daily additions of NPK fertilizer in the field to give, respectively, annual additions of 15, 10, and 2 g m^{-2} compared to additions of 0.31 g m^{-2} N and 0.05 g m^{-2} K from rainfall. Chlorophyll contents were increased by fertilization and, while photosynthesis paralleled the changes in chlorophyll in *Pleurozium*, there was also an increase in photosynthetic efficiency in *Dicranum*. Fertilizer treatment resulted in a greater shoot density per unit area of ground and a less branched form to the shoots of *Pleurozium*. Such changes may alter the light penetration and productivity of the two mosses, and it was reported that under a closed canopy (where *Pleurozium* already had an elevated chlorophyll content) fertilization favoured *Dicranum* growth.

Tamm (1953) considered that as the highest productivity of the moss *Hylocomium splendens* occurred just outside the limits of tree crowns, this was a reflection of the greater throughfall experienced at this location. Unfortunately direct fertilization experiments were not performed to authenticate this hypothesis. As most fertilization experiments involve the addition of a solution, care must be taken to ensure that an adequate 'irrigation control' is included. Differences in water regime (Busby *et al.* 1978) may partly account for Tamm's observations (see Brown 1982; Brown and Bates 1990 for a further discussion). When Skre and Oechel (1979) supplied fertilizers at two to five times the normal input from rain they failed to detect any increase in growth and detected some negative responses. Bates (1987) failed to detect any change in the growth rate of *Pseudoscleropodium purum* treated with a complete nutrient solution in the

field. It has been shown that the distortion of the elemental composition of
P. purum by nutrient additions is a relatively temporary phenomenon, with
the plant reverting to a chemical pattern in equilibrium with the natural
rainfall conditions (Bates 1989). Natural fluctuations in the concentration
of physiological elements has been shown with lichens (Puckett 1985) and
bryophytes (Markert and Weckert 1989), which may reflect seasonal
differences in growth rate and/or nutrient supply.

Community changes

Many experiments involving the artificial enrichment of various communities with fertilizers have only incidentally considered changes in the
cryptogamic flora. O'Toole and Synnott (1971) and Synnott (1987) are
amongst the few studies where higher plant competition has been deliberately eliminated by weeding. Most other experimental systems make it
impossible to establish whether changes observed are due to direct effects
on the cryptogams or indirect ones through changes in the biomass,
shading, leachates, etc. of the other vegetational layers. Only a selection of
reports will be mentioned and further examples are contained in the
reference lists of these publications.

   Studies in relatively open grassland have shown that enhanced grass
growth was correlated with decreased moss biomass and was in proportion
to the amount of added nutrients (Michiewicz 1976; Lambert *et al.* 1986).
The former author showed *Brachythecium albicans* and *Eurhynchium swartzii*
were the least affected while *Pohlia nutans* was eliminated. The study by
O'Toole and Synnott (1971) and Synnott (1987) dealt with fertilized
exposed peat cuttings and showed that initial colonization by *Marchantia
polymorpha* (wetter sites) and *Funaria hygrometrica* (drier sites), followed by
other weedy moss species, was stimulated by P and Ca but not by N or K.
This indicated some degree of species specificity and also the influence of
water availability. Mannerkoski (1970) observed the same species following
additions of Oulu saltpetre, rock phosphate, and muriate of potash to
cleared peat surfaces.

   Forest fertilization studies may involve single or multiple additions of
nutrients. Van Dobben and Dirkse (1989) briefly reported that regular
three-yearly additions resulted in increases in herbs and *Brachythecium
oedipodium* and decreases in dwarf shrubs and *Cladonia* species; urea was
equivalent to a combination of acid and N. Nitrogen-free acid-treated plots
showed crustose lichens growing on the bare soil. *B. oedipodium* is a litter-
growing species that appeared in regularly fertilized plots (irrigated five
days per week) with *Pohlia nutans* and *Plagiothecium* species (Kellner and
Marshagen 1991). Lichens vanished and *Pleurozium schreberi* decreased
substantially with such treatment, while *Dicranum polysetum* was hardly

affected, in agreement with the latter's greater efficiency at low light intensities and the increase in rapidly vegetatively spreading herbs (*Epilobium angustifolium*, *Rubus idaeus*, etc.). The same changes in bryophyte composition were observed by Nohrstedt (1988) following urea treatments, especially when applied at high doses under dry conditions. Nohrstedt *et al.* (1988) reported some decline in N_2-fixing lichens in fertilized forests. No decrease in cover of *Stereocaulon paschale* was seen although slight discoloration of cephalodia and reduced N_2 fixation was reported. *Nephroma arcticum* cover decreased by 30 per cent after urea treatment but only 9 per cent after NH_4NO_3, while *Peltigera aphthosa* was 60 per cent decreased by both treatments.

The particular balance of environmental conditions appears important. Thus with young pine stands Persson (1981) found that *Pohlia nutans* increased with NH_4NO_3 fertilization while other forest mosses, which normally dominate the bottom vegetation layers, declined. Fertilizer-induced losses of lichens from areas where they dominated the bottom layer have been regularly reported, especially with dry fertilizer additions (Persson 1981; Eriksson 1984; Faltynowicks 1986; Hofmann 1987). However, an increase in lichen cover was reported by Gerhardt and Kellner (1986) following single additions of urea and NH_4NO_3. Higher and repeated treatments generally resulted in the loss of forest mosses with urea additions on a variety of soils, and also of the lichen *Cladina rangiferina* with urea and NH_4NO_3 on a sandy soil (Gerhardt and Kellner 1986). Malkonen *et al.* (1980) found an immediate decrease in moss cover after adding N, P, or K and an increase in grasses and herbs, especially after adding N, in a *Myrtillus*-type Norway spruce stand. Eriksson (1984) reported that urea caused a greater loss of *Cladonia rangiferina* and *Cetraria islandica* from a heath, and increase of *Vaccinium*, than did NH_4NO_3.

Comparable changes occur in drained and fertilized peatland forests, mires, and bogs; drainage alone often benefits forest bryophytes (Vasander 1987). Treatment with readily soluble PK and NPK + micronutrients resulted in loss of forest mosses but greater damage occurred to *Sphagnum* species (Jäppinen and Hotanen 1990). The production/biomass ratio of the living biomass of *Polytrichum* was unchanged while its cover decreased when treated with PK, NPK + micronutrients or NPK (urea), possibly due to uneven fertilizer distribution. Nitrogen and P contents of *Sphagnum* and forest mosses increased after fertilizer treatment, including N increases with PK-only treatment, possibly due to mobilization from the underlying peat (Jäppinen 1987). *Sphagnum* species often decrease with fertilizer treatments including wood-ash, which raises the pH (Päivänen and Seppälä 1968). Differences initially appear between hummocks and hollows (Vasander *et al.* 1988).

10.4 Concluding remarks

The effects of agricultural chemicals on lichens and bryophytes deserve further study because these organisms may prove to be sensitive, selective, and convenient biomonitors of such environmental pollutants. Pesticide and fertilizer residues can be recovered from both lichens and bryophytes, which may, however, show rapid dynamic changes. Although changes in the floristic composition of treated habitats have been observed, more carefully planned studies are needed to separate the effects of direct from indirect damage. The loosely applied expression 'fertilizer effects' needs careful definition in terms of both floristic and chemical parameters, including the distinction between pH and specific chemical effects.

Acknowledgements

I am very grateful to the many workers who have drawn my attention to the scattered literature and who gave me access to, and permission to quote from, reports or unpublished studies. In particular I wish to thank V. Alstrup, J.-P. Hottanen, S. Howes, O. Kellner, M. Kinder, H. Ougham, K. R. Payne, U. Søchting, H. Tomlinson, I. Vagts, and H. van Dobben. I also thank my wife for valuable discussions and her critical comments on earlier drafts of this manuscript.

References

Alstrup, V. (1991). Pasticidanvendelsens indvirken pa lavfloraen pa sten. Unpublished report, Institute for Ecological Botany, Copenhagen University, Copenhagen.

Armstrong, R. A. (1984). The influence of bird droppings and uric acid on the radial growth of five species of saxicolous lichens. *Environmental and Experimental Botany*, **24**, 95–9.

Atkinson, D., Abernathy, W., and Crisp, C. (1980). The effect of several herbicides on moss establishment in orchards. *Proceedings of the 1980 British Crop Protection Conference—Weeds*, 297–302.

Bacci, E., Calamari, D., Gaggi, C., Fanelli, R., Focardi, S., and Morosini, M. (1986). Chlorinated hydrocarbons in lichen and moss samples from the Antarctic Peninsula. *Chemosphere*, **15**, 747–54.

Bakker, A. J. de (1987). *Physcia stellaris* (L.) Ach. in Nederland. *Gorteria*, **13**, 210–16.

Bakker, A. J. de (1989). Effects of ammonia emission on epiphytic lichen vegetation. *Acta Botanica Neerlandica*, **38**, 337–42.

Bakker, A. J. de and Dobben, H. F. van (1988). Effecten van ammoniakemissie op epifytische korstmossen een correlatief ondezoek in de Peel. *Rijksinstituut voor Natuurbeheer Rapport* **88**/35, 1–48.

Balcerkiewick, S. and Rusińska, A. (1987). Expansion of bryophytes on areas treated with herbicides. *Symposia Biologica Hungarica*, **35**, 285–93.

Barkman, J. J. (1958). *Phytosociology and ecology of cryptogamic epiphytes*. Van Gorcum, Assen.

Bates, J. W. (1987). Nutrient retention by *Pseudoscleropodium purum* and its relation to growth. *Journal of Bryology*, **14**, 565–80.

Bates, J. W. (1989). Retention of added K, Ca and P by *Pseudoscleropodium purum* growing under an oak canopy. *Journal of Bryology*, **15**, 589–605.

Blandy, R. V. (1954). The control of mosses in lawns and sports turf. *Journal of the Science of Food and Agriculture*, **5**, 397–400.

Bond, T. E. T. (1976). *Polytrichum* spp. in an apple orchard on herbicide-treated soil. *Proceedings of the Bristol Naturalists' Society*, **35**, 69–72.

Brauen, S. E., Goss, R. L. and Nus, J. L. (1986). Control of acrocarpous moss with endothal. *Journal of the Sports Turf Research Institute*, **62**, 138–40.

Brown, D. H. (1982). Mineral nutrition. In *Bryophyte ecology*, (ed. A. J. E. Smith), pp. 383–444. Chapman and Hall, London.

Brown, D. H. (1984). Uptake of mineral elements and their use in pollution monitoring. In *The experimental biology of bryophytes*, (eds A. F. Dyer and J. G. Duckett), pp. 229–55. Academic Press, London.

Brown, D. H. and Bates, J. W. (1990). Bryophytes and mineral cycling. *Botanical Journal of the Linnean Society*, **104**, 129–47.

Brown, D. H., Beckett, R. P. and Legaz, M. E. (1987). The effect of phosphate buffer on the physiology of the lichen *Evernia prunastri*. *Annals of Botany*, **60**, 553–62.

Brown, D. H., Ougham, H., and Beckett, R. P. (1986). The effect of the herbicide dichlorophen on the physiology and growth of two bryophytes. *Annals of Botany*, **57**, 201–9.

Brown, D. H. and Whitehead, A. (1986). The effect of mercury on the physiology of *Rhytidiadelphus squarrosus*. *Journal of Bryology*, **14**, 367–74.

Busby, J. R., Bliss, L. C., and Hamilton, C. D. (1978). Microclimate control of growth rates and habitats of the boreal forest mosses, *Tomenthypnum nitens* and *Hylocomium splendens*. *Ecological Monographs*, **48**, 95–110.

Carlberg, G. E., Ofstad, E. B., Drangsholt, H., and Steinnes, E. (1983). Atmospheric deposition of organic micropollutants in Norway studied by means of moss and lichen analysis. *Chemosphere*, **12**, 341–56.

Carstairs, A. G. and Oechel, W. C. (1978). Effects of several microclimatic factors and nutrients on net carbon dioxide exchange in *Cladonia alpestris* (L.) Rabh. in the subarctic. *Arctic and Alpine Research*, **10**, 81–94.

Chapin, F. S., Oechel, W. C., Cleve, K. van, and Lawrence, W. (1987). The role of mosses in the phosphorus cycling of an Alaskan black spruce forest. *Oecologia*, **74**, 310–15.

Clymo, R. S. (1963). Ion exchange in *Sphagnum* and its relation to bog ecology. *Annals of Botany*, **27**, 309–24.

Crittenden, P. D. (1983). The role of lichens in the nitrogen economy of subarctic woodlands: nitrogen loss from the nitrogen-fixing lichen *Stereocaulon paschale* during rainfall. In *Nitrogen as an ecological factor*, (eds J. A. Lee, S. McNeill, and I. H. Rorison), pp. 43–68. Blackwell, Oxford.

Crittenden, P. D. (1989). Nitrogen relations of mat-forming lichens. In *Nitrogen, phosphorus and sulphur utilization by fungi*, (eds L. Boddy, R. Marchant, and D. J. Read), pp. 243–68, Cambridge University Press, Cambridge.

Damman, A. W. H. (1978). Distribution and movement of elements in ombro-trophic peat bogs. *Oikos*, **30**, 480–95.

Deising, H. (1987). *In vivo* studies on the regulation of nitrate reductase in *Sphagnum* species. *Symposia Biologica Hungarica*, **35**, 59–69.

Dobben, H. F. van (1983). Changes in the epiphytic lichen flora and vegetation in the surroundings of 's-Hertogenbosch (The Netherlands) since 1900. *Nova Hedwiga*, **37**, 691–719.

Dobben, H. F. van (1987). Effecten van ammoniak op epifytische korstmossen. In *Acute en chronische effecten van NH₃ (en NH₄⁺) op levende organismen*, (eds A. W. Boxman and J. F. M. Geelen), pp. 52–61. Nijmegen.

Dobben, H. F. van and Dirkse, G. M. (1989). Effects of experimental fertilization on forest undergrowth in Northern Sweden. *Acta Botanica Neerlandica*, **38**, 359–60.

Eriksson, O. (1984). Effekter av skogsgödsling pa renbetet och renbetningen. *Skogsfakta*, Supplement **5**, 80–7.

Eronen, L., Julkunen, R., and Saarelainen, A. (1979). MCPA residues in developing forest ecosystem after aerial spraying. *Bulletin of Environmental Contamination and Ecotoxicology*, **21**, 791–8.

Faltynowicz, W. (1986). The dynamics and role of lichens in a managed *Cladonia*-pine forest (*Cladonia-Pinetum*). *Monographiae Botanicae*, **69**, 1–97.

Farrar, J. F. (1976). The uptake and metabolism of phosphate by the lichen *Hypogymnia physodes*. *New Phytologist*, **77**, 127–34.

Folkeson, L. and Andersson-Bringmark, E. (1988). Impoverishment of vegetation in a coniferous forest polluted by copper and zine. *Canadian Journal of Botany*, **66**, 417–28.

Gaggi, C., Bacci, E., Calamari, D., and Fanelli, R. (1985). Chlorinated hydrocar-bons in plant foliage: an indication of the tropospheric contamination level. *Chemosphere*, **14**, 1673–86.

Garty, J., Perry, A. S., and Mozel, J. (1982). Accumulation of polychlorinated biphenyls (PCBs) in the transplanted lichen *Ramalina duriaei* in air quality biomonitoring experiments. *Nordic Journal of Botany*, **2**, 583–6.

Gerhardt, K. and Kellner, O. (1986). Effects of nitrogen fertilizers on the field- and bottomlayer species in some Swedish coniferous forests. *Meddelanden fran Växtbiologiska institutionen, Uppsala*, **1**, 1–47.

Gilbert, O. L. (1976). An alkaline dust effect on epiphytic lichens. *Lichenologist*, **8**, 173–8.

Gilbert, O. L. (1977). Lichen conservation in Britain. In *Lichen ecology*, (ed. M. R. D. Seaward), pp. 415–36. Academic Press, London.

Hällbom, L. and Bergman, B. (1979). Influence of certain herbicides and a forest fertilizer on the nitrogen fixation by the lichen *Peltigera praetextata*. *Oecologia*, **40**, 19–27.

Hällbom, L. and Bergman, B. (1983). Effects of inorganic nitrogen on C_2H_2 reduction and CO_2 exchange in the *Peltigera praetextata-Nostoc* and *Peltigera aphthosa-Coccomyxa-Nostoc* symbioses. *Planta*, **157**, 441–5.

Herrmann, R. (1990). Biomonitoring of organic and inorganic trace pollutants by means of mosses. In *Bryophytes. Their chemistry and chemical taxonomy*, (ed. H. D. Zinsmeister and R. Mues), pp. 319–35. Clarendon Press, Oxford.

Herrmann, R. and Hübner, D. (1984). Concentration of micropollutants (PAH, chlorinated hydrocarbons and trace metals) in the moss *Hypnum cupressiforme* in and around a small industrial town in Southern Finland. *Annales Botanici Fennici*, **21**, 337–42.

Himme, M. van and Stryckers, J. (1981). Mossenbestrijding in container- en potkulturen van siergewassen. *Medelingen van de Faculteit Landbouwwetenschappen, Rijksuniversiteit Gent*, **46**, 199–212.

Hofmann, G. (1987). Vegetationsänderungen in Kiefernbeständen durch mineral-düngung. *Hercynia*, **24**, 271–8.

Howes, S. (1986). The effect of the herbicide dichlorophen on the physiology of the lichens: *Peltigera horizontalis* and *Cladonia portentosa*. Unpublished B.Sc. thesis. University of Bristol.

Jackson, N. (1961). Moss eradication trials. *Journal of the Sports Turf Research Institute*, **37**, 264–75.

James, P. W. (1973). The effects of air pollutants other than hydrogen fluoride and sulphur dioxide on lichens. In *Air pollution and lichens*, (eds B. W. Ferry, M. S. Baddeley, and D. L. Hawksworth), pp. 143–75. Athlone Press of the University of London, London.

James, P. W., Hawksworth, D. L., and Rose, F. (1977). Lichen communities in the British Isles: a preliminary conspectus. In *Lichen ecology*, (ed. M. R. D. Seaward), pp. 295–413. Academic Press, London.

Jäppinen, J. -P. (1987). Ojituksen ja lannoituksen vaikutukset sammalten typpi- ja fosforipitoisuuksiin kahdela suomuuttumalla. *Suo*, **38**, 13–22.

Jäppinen, J. -P. and Hotanen, J. -P. (1990). Effect of fertilizer on the abundance of bryophytes in two drained peatland forests in Eastern Finland. *Annales Botanici Fennici*, **27**, 93–108.

Kallio, S. and Wilkinson, R. E. (1977). The effects of some herbicides on nitrogenase activity and carbon fixation in two subarctic lichen. *Botanical Gazette*, **138**, 468–73.

Kauppi, M. (1980). The influence of nitrogen-rich components on lichens. *Acta Universitatis Oluensis A, Scientiae Rerum Naturalium*, **101**, *Biologia*, **9**, 1–25.

Kellner, O. and Marshagen, M. (1991). Effects of irrigation and fertilization on the ground vegetation in a 130-year-old stand of Scots pine. *Canadian Journal of Forest Research*, **21**, 733–8.

Lambert, M. G., Clark, D. A., Grant, D. A., Costall, D. A., and Gray, Y. S. (1986). Influence of fertiliser and grazing management on North Island moist hill country. 4: Pasture species abundance. *New Zealand Journal of Agricultural Research*, **29**, 23–31.

Lang, G. E., Reiners, W. A., and Heier, R. K. (1976). Potential alteration of precipitation chemistry by epiphytic lichens. *Oecologia*, **25**, 229–41.

Larson, R. B., Lokke, H., and Rasmussen, L. (1985). Accumulations of chlorinated hydrocarbons in moss from artificial rainwater. *Oikos*, **44**, 423–9.

Mabb, L. P. (1989). Uptake and effects of herbicides on the lawn moss *Rhytidia-delphus squarrosus*. Unpublished Ph.D thesis. University of London.

Mälkönen, E., Kellomäki, S., and Holm, J. (1980). Typpi-, fosfori- ja kalilannoi-tuksen vaikutus kuusikon pintakasvillisuuteen. *Communicationes ex Instituto quaestionum forestalium Finlandiae*, **98**, 1–35.

Mannerkoski, H. (1970). Lannoituksen vaikutuksesta kylvölaikkujen kasvillisuu-teen. *Suo*, **21**, 80–6.

Markert, B. and Weckert, V. (1989). Fluctuations of element concentrations during the growing season of *Polytrichum formosum* (Hedw.). *Water, Air, and Soil Pollution*, **43**, 177–89.

Masse, L. (1966). Étude comparée des teneurs en azote total des lichens et de leur substrat: les espèces 'ornithocoprophiles'. *Compte Rendus de l'Academie des Sciences, Paris, Serie D*, **262**, 1721–4.

Masse, L. (1969). Quelques aspects de l'uricolyse enzymatique chez les lichens. *Compte Rendus de l'Academie des Sciences, Paris, Serie D*, **268**, 2896–8.

Meade, R. (1984). Ammonia-assimilating enzymes in bryophytes. *Physiologia Plantarum*, **60**, 305–8.

Mickiewicz, J. (1976). Influence of mineral fertilization on the biomass of moss. *Polish Ecological Studies*, **2**, 57–62.

Miyazaki, T. and Satake, K. (1985). *In situ* measurements of uptake of inorganic carbon and nitrogen by the aquatic liverworts *Jungermannia vulcanicola* Steph. and *Scapania undulata* (L.) Dum. in an acid stream, Kashiranashigawa, Japan. *Hydrobiologia*, **124**, 29–34.

Nieboer, E., Padovan, D., Lavoie, P., and Richardson, D. H. S. (1984). Anion accumulation by lichens. II: Competition and toxicity studies involving arsenate, phosphate, sulphate and sulphite. *New Phytologist*, **96**, 83–93.

Nohrstedt, H.-O. (1988). Skador pa bottenskiktet och vegetations-förändringar efter en skogsgödsling med urea. *Institutet for Skogsförbättring*, **3**, 1–13.

Nohrstedt, H.-O., Wedin, M., and Gerhardt, K. (1988). Effekter av skogsgödsling pa kvävefixerande lavar. *Institutet for Skogsförbättring*, **4**, 1–29.

O'Toole, M. A. and Synnott, D. M. (1971). The bryophyte succession on blanket peat following calcium carbonate, nitrogen, phosphorus and potassium fertilizers. *Journal of Ecology*, **59**, 121–6.

Ougham, H. (1983). Field and laboratory experiments on Panacide toxicity in bryophytes. Unpublished B.Sc. thesis. University of Bristol.

Päivänen, J. and Sepplälä, K. (1968). Hajalannoituksen vaikutus lyhytkortisen nevan pintakasvillisuuteen. *Suo*, **19**, 51–6.

Perkins, D. F. and Marrs, R. H. (1990). Effects of herbicide drift on lichens in heathland (Norfolk) and pasture woodland (Gwynedd). In *Pesticide drift and impact*, (ed. B. N. K. Davis), pp. 106–15. Unpublished Report. Institute of Terrestrial Ecology, Huntingdon.

Persson, H. (1981). The effect of fertilization and irrigation on the vegetation dynamics of a pine-heath ecosystem. *Vegetatio*, 46, 181–92.

Pihakaski, S. and Pihakaski, K. (1980). Effects of glyphosate on ultrastructure and photosynthesis of *Pellia epiphylla*. *Annals of Botany*, 46, 133–41.

Puckett, K. J. (1985). Temporal variation in lichen element levels. In *Lichen physiology and cell biology*, (ed. D. H. Brown), pp. 211–25. Plenum Press, London.

Rai, A. N. (1988). Nitrogen metabolism. In *Handbook of lichenology*, (ed. M. Galun), pp. 201–37. CRC Press, Boca Raton, Florida.

Richardson, D. H. S., Nieboer, E., Lavoie, P., and Padovan, D. (1984). Anion accumulation by lichens. I: The characteristics and kinetics of arsenate uptake by *Umbilicaria muhlenbergii*. *New Phytologist*, 96, 71–82.

Rieley, J. O., Richards, P. W., and Bebbington, A. D. L. (1979). The ecological role of bryophytes in a North Wales woodland. *Journal of Ecology*, 67, 497–527.

Ronoprawiro, S. (1975). Control of mosses in tea. *Asian Pacific Weed Science Society Conference*, 1975, 365–9.

Rosentreter, R. (1990). Indicator value of lichen cover on desert shrubs. In *Proceedings—Symposium on cheatgrass invasion, shrub die-off, and other aspects of shrub biology and management*, (eds E. D. McArthur, E. M. Romney, S. D. Smith, and P. T. Tueller), pp. 282–9. US Department of Agriculture, Forest Services Information Research Station, Ogden, Utah.

Rudolph, H., Deising, H., and Voigt, J. U. (1987). The tolerance of raised bog *Sphagnum* species in respect to inorganic nitrogen. *Symposia Biologica Hungarica*, 35, 71–80.

Rudolph, H. and Samland, J. (1985). Occurrence and metabolism of sphagnum acid in the cell walls of bryophytes. *Phytochemistry*, 24, 745–9.

Salomon, H. (1914) Uber das vorkommen und die aufnahme einiger wichtiger nährsalze bei den flechten. *Jahrbuch der Wissenchaftlicher Botanik*, 54, 309–54.

Schofield, E. and Ahmadjian, V. (1972). Field observations and laboratory studies of some Antarctic cold desert cryptogams. In *Antarctic terrestrial biology*, (ed. G. A. Llano), pp. 97–141. American Geophysical Union, Washington.

Schuurkes, J. A. A. R., Kok, C. J., and Den Hartog, C. (1986). Ammonium and nitrate uptake by aquatic plants from poorly buffered and acidified waters. *Aquatic Botany*, 24, 131–46.

Schwoerbel, J. and Tillmanns, G. C. (1974). Stickstoffaufnahme aus dem wasser and nitratreduktase-aktivität bei submersen wasserpflanzen: *Fontinalis antipyretica* L. *Archiv für Hydrobiologie, supplement 47*, 2, 282–94.

Siltanen, H., Rosenberg, C., Raatikainen, M., and Raatikainen, T. (1981). Triclopyr, Glyphosate and phenoxyherbicide residues in cowberries, bilberries and lichen. *Bulletin of Environmental Contamination and Toxicology*, 27, 731–7.

Simola, L. K. (1975). The effect of several protein amino acids and some inorganic nitrogen sources on the growth of *Sphagnum nemoreum*. *Physiologia Plantarum*, 35, 194–9.

Simola, L. K. (1979). Dipeptide utilization by *Sphagnum fimbriatum*. *Journal of the Hattori Botanical Laboratory*, 46, 49–54.

Simon, H. (1985). Untersuchungen über das anreicherungsverhalten des wasser-mooses *Fontinalis antipyretica* für umweltrelevante spurensubstanzen (chloro-kohlenwasserstoffe, PCA, spurenmetalle)—bewertung seiner eignung für die gewässergüteüberwachung mit hilfe statistischer modelle. Unpublished Diplo-marbeit. University of Bayreuth.

Skender, A. (1983). Uticaj herbicida i mahovina u suzbijanju cvjetnica na regula-cijskim objektima—obaloutvrdama. *Fragmenta herbological Jugoslavica*, **12**, 105–10.

Skre, O. and Oechel, W. C. (1979). Moss production in a black forest spruce *Picea mariana* forest with permafrost near Fairbanks, Alaska, as compared with two permafrost-free stands. *Holarctic Ecology*, **2**, 249–54.

Smith, D. C. (1960*a*). Studies in the physiology of lichens. 1: The effects of starvation and of ammonia absorption upon the nitrogen content of *Peltigera polydactyla*. *Annals of Botany*, **24**, 52–62.

Smith, D. C. (1960*b*). Studies in the physiology of lichens. 2: Absorption and utilization of some simple organic nitrogen compounds by *Peltigera polydactyla*. *Annals of Botany*, **24**, 172–85.

Smith, D. C. (1960*c*). Studies in the physiology of lichens. 2: Experiments with dissected discs of *Peltigera polydactyla*. *Annals of Botany*, **24**, 188–99.

Søchting, U. (1989). *Xanthoria elegans* spreading in Denmark. *Graphis Scripta*, **2**, 167.

Søchting, U. (1991). Laver som kvaelstofmonitorer i danske skove. Unpublished report. Institue for Sporenplanter, University of Copenhagen.

Søchting, U. and Johnsen, I. (1990). Overvagning af de danske likenheder. *URT 14. argang*, **1990**, 4–9.

Sørchting, U. and Johnsen, I. (1987). Injured reindeer lichens in Danish lichen heaths. *Graphis Scripta*, **1**, 103–6.

Stjernquist, I. (1981). Photosynthesis, growth and competitive ability of some coniferous forest mosses and the influence of herbicides and heavy metals (Cu, Zn). Unpublished thesis. University of Lund.

Synnott, D. M. (1987). The effects of drainage, shelter and fertilisers on bryophyte colonisation and succession on blanket peat in Western Ireland. *Glasra*, **10**, 83–9.

Tamm, C. O. (1953). Growth, yield and nutrition in carpets of a forest moss (*Hylocomium splendens*). *Meddelanden Från Statens Skogsforskningsinstitut*, **43**, 1–140.

Thomas, W. (1981). Entwicklung eines immissionemeißstsems für PCA, chloro-kohlenwasserstoffe und spurenmetalle mittels eiphytischer moose—angewand fur den Raum Bayern. *Bayreuther Geowissenschafiliche Arbeiten*, **3**, 144 S.

Thomas, W. (1983). Uber die verwendung von pflanzen zur analyse räumlicher spurensubstanz-immissionsmuster. *Staub-Reinhalt. Luft*, **43**, 141–8.

Thomas, W. (1984). Statistical models for the accumulation of PAH, chlorinated hydrocarbons and trace metals in epiphytic *Hypnum cupressiforme*. *Water, Air, and Soil Pollution*, **22**, 351–71.

Thomas, W. (1986*a*). Representativity of mosses as biomonitoring organisms for the accumulation of environmental chemicals in plants and soils. *Ecotoxicology and Environmental Safety*, **11**, 339–46.

Thomas, W. (1986b). Accumulation of airborne trace pollutants by arctic plants and soil. *Water Science and Technology*, **18**, 47–57.

Thomas, W. and Herrmann, R. (1980). Nachweis von chlopestiziden, PCB, PCA und schwermetallen mittels epiphytischer moose als biofilter enlang eines profils durch Mitteleuropa. *Staub-Reinhalt. Luft*, **30**, 440–4.

Thomas, W. and Schunke, E. (1984). Polyaromatic hydrocarbons, chlorinated hydrocarbons, and trace metals in moss samples from Iceland. *Lindbergia*, **10**, 27–32.

Thomas, W., Rühling, A., and Simon, H. (1984). Accumulation of airborne pollutants (PAH, chlorinated hydrocarbons, heavy metals) in various plant species and humus. *Environmental Pollution (Series A)*, **36**, 295–310.

Tomlinson, H. (1988). The effects of nitrogen fertilisers on the lichen *Parmelia sulcata*. Unpublished B.Sc thesis. University of Bristol.

Turner, R. E. (1991). Fertiliser and climate change. *Nature*, **349**, 469–70.

Vagts, I. and Kinder, M. (1990). Experimentell-okologische untersuchungen zur wirkung verschiedener standortfaktoren auf *Cladonia furcata* auf einem santrockenrasen in Bremen. Unpublished Diplomarbeit. University of Bremen.

Vasander, H. (1987). Diversity of understorey biomass in virgin and in drained and fertilized southern boreal mires in eastern Fennoscandia. *Annales Botanici Fennici*, **19**, 137–53.

Vasander, H., Lindholm, T., and Kaipaiinen, H. (1988). Vegetation patterns on a drained and fertilized raised bog in southern Finland. *Proceedings of the 8th International Peat Congress*, **1**, 177–84.

Vicente, C. and Legaz, M. E. (1988). Lichen enzymology. In *Handbook of lichenology*, (ed. M. Galun), pp. 239–81. CRC Press, Boca Raton, Florida.

Villeneuve, J. -P., Fogelqvist, E., and Cattini, C. (1988). Lichens as bioindicators for atmospheric pollution by chlorinated hydrocarbons. *Chemosphere*, **17**, 399–403.

Voet, E. van der and Udo de Haes, H. A. (1985). *Effekten van intensieve veehouderijbedrijven, CML Mededeling*, **18**, Centrum voor Milieukunde, Leiden.

Weber, M. G. and Cleve, K. van (1981). Nitrogen dynamics in the forest floor of interior Alaska black spruce ecosystems. *Canadian Journal of Forest Research*, **11**, 743–51.

Weetman, G. F. and Timmer, V. (1967). Feather moss growth and nutrient content under upland black spruce. *Pulp and Paper Research Institute of Canada, Technical Report*, **503**, 1–38.

Wells, J. M. and Richardson, D. H. S. (1985). Anion accumulation by *Hylocomium splendens*: uptake and competition studies involving aresenate, selenate, selenite, phosphate, sulphate and sulphite. *New Phytologist*, **101**, 571–83.

Wells, J. M. and Richardson, D. H. S. (1985). Anion accumulation by *Hylocomium splendens*: uptake and competition studies involving arsenate, selenate, selenite, phosphate, sulphate and sulphite. *New Phytologist*, **101**, 571–83.

Wilson, D. and Hughes, A. (1985). Evaluation of oxyzalin and mogeton for weed control in field and container grown hardy nursery stock. *1985 British Crop Protection Conference—Weeds*, **1985**, 1095–102.

Woolhouse, A. R. (1972). Moss eradication trials. *Journal of the Sports Turf Research Institute*, **48**, 102–97.

11

Ecophysiological effects of acid rain on bryophytes and lichens

Andrew M. Farmer, Jeffrey W. Bates, and J. Nigel B. Bell

11.1 Introduction

The term 'acid rain' encompasses a range of different pollutants. In its widest sense it includes wet acidic deposition, gaseous acidic and oxidant pollutants, and even ammonia and its derivatives. All of these pollutants can cause detrimental effects to lichens and bryophytes. We shall mostly discuss wet-deposited pollutants, which occur in rain, snow, and mist, the latter being termed 'occult' deposition. The effects of SO_2 on bryophytes and lichens are well reviewed elsewhere (see Hawksworth and Rose 1976; Richardson and Nieboer 1983; Richardson 1988; Fields 1988; Winner 1988) and so only a brief overview is given here. Some attention will also be given to photochemical oxidant pollutants and nitrogen oxides (NO_x), which have been the subject of recent studies with cryptogams, but have not been reviewed elsewhere. The effects of NH_x on bryophytes and lichens, which may be of increasing importance in Western Europe are considered in Chapters 6 and 10.

The various forms of atmospheric pollutant differ in importance in urban and rural regions. Wet acidic deposition has elicited most concern in regions remote from areas of pollutant production. Long-distance transport of the pollutants, as occurs when tall chimney stacks are employed, allows the complete oxidation of NO_x and SO_2 in the atmosphere, so that the deposited anions are the relatively non-toxic nitrate and sulphate, together with H^+. Such pollution is of most importance in north-western and central Europe and parts of north-eastern North America. Where transport is over a shorter distance, the SO_2 may be merely dissolved, so that bisulphite is deposited. Bisulphite is known to be highly toxic to many lichen species from studies where it is used as an analogue for SO_2 (Richardson and

Nieboer 1983). It has also been found to damage wetland bryophyte communities (Lee and Studholme, this volume). NO_x is a predominantly urban pollutant, while O_3 and NH_x reach their highest concentrations in rural areas (Woodin 1989; UKRGAR 1990). Pollutant concentrations are rapidly changing in many parts of the world. In western Europe and North America urban SO_2 levels are declining, though this is much less marked in rural areas (Laxen and Thompson 1987). In the developing world, increasing industrialization is causing an increase in sulphur emissions, however. The developed world is subject to increasing levels of NO_x and O_3 as numbers of motor vehicles increase, and increasing NH_x as agricultural production is intensified. It is these pollutants, together with the low pH conditions that they can cause, which are an increasing threat to cryptogam communities. Due to the many different pollutants it is often difficult to distinguish which are the most important in a given situation. Of particular concern, for example, is the need to distinguish between the effects of increasing levels of acidic wet deposition and nitrogen deposition and the residual effects of declining SO_2.

11.2 Sulphur dioxide

It is universally accepted that SO_2 has been the single most important pollutant causing elimination of bryophytes and lichens from urban and industrialized regions (Hawksworth and Rose 1970, 1976; Winner 1988). However, irrefutable proof of a direct damaging effect of SO_2 on cryptogams in the field has rarely been obtained. There are a number of points that have arisen from this work that are applicable to other pollutant studies. At the physiological level, lichens, in particular, have been found to show diverse responses to SO_2. Most metabolic activities are affected by the pollutant in sensitive species at sufficient exposure levels. SO_2 also interacts with other environmental factors. This is most clearly seen with pH. Low pH increases the toxicity of SO_2 action (see Chapter 6). A full understanding has been hindered by the lack of long-term fumigations of cryptogams with realistic SO_2 concentrations in controlled conditions as has been carried out for vascular plants (e.g. Rafarel and Ashenden 1991). Ronen (1986) fumigated *Ramalina duriaei* in open-top chambers and found a reduction in photosynthesis with 17.5 p.p.b. SO_2 given twice a week for 20 weeks. The only 'long-term' field fumigation has been that of Moser *et al.* (1980), but the lowest concentration of SO_2 used was 60 p.p.b., which is still very high compared to ambient levels today. There is still an obvious need for long-term fumigation experiments with SO_2 and it would be helpful for these to include studies of the same species on several substrates with different buffering capacities to differentiate between direct and indirect effects. Such studies would aid in the understanding of SO_2 pollution scales (e.g. Hawksworth and Rose 1970), which have proved very difficult to verify (Richardson 1988). It would also be possible to distinguish between the effects of long-term low-level pollutant concentrations and those of occasional peak levels, the relative importances of which are still unclear.

11.3 Photochemical oxidants

The effects of the photochemical oxidants ozone (O_3), peroxyacetylnitrate (PAN) and hydrogen peroxide (H_2O_2) have been studied for lichens, but only a limited amount of work has been done on the effects of O_3 on bryophytes. The importance of oxidant pollutants in the field was investigated by Sigal and Nash (1983) who undertook an extensive survey of epiphytic lichens in southern California, an area characterized by elevated levels of O_3 and PAN, but low levels of other atmospheric pollutants. A comparison with historical data showed that several macrolichens had been lost in the most polluted areas. These included *Bryoria abbreviata, B. fremontii, B. oregana, Alectoria sarmentosa, Calicium viride, Cetraria canadensis, Platismatia glauca*, and *Evernia prunastri*. Transplantation of *Hypogymnia enteromorpha* into areas subject to photochemical smog resulted in the degradation of the thalli, while transplants in clean areas remained healthy.

McCune (1988) undertook an analysis of lichen communities around Indianapolis using the 'index of atmospheric purity' (IAP) technique. Although SO_2 levels were low, the IAP scores correlated well with the SO_2 contours for the region, but no relationship was found with the moderately high O_3 concentrations.

Experimental studies have been undertaken to confirm the observed field sensitivities of the Californian epiphytes to oxidant pollutants. Nash and Sigal (1979) fumigated *Parmelia sulcata* and *Hypogymnia enteromorpha* with O_3 for short periods and monitored their physiology. They found photosynthesis of *P. sulcata* was affected by 12 h fumigation with 500 p.p.b. O_3, but *H. enteromorpha* was more resistant. The response only developed 3 h after fumigation. The concentrations used in this study were relatively high, but may approximate maximum levels for total oxidant pollutants in southern California. In a separate study, lower levels of O_3 (200 p.p.b. for 12 h) were found to reduce photosynthesis in *Parmelia caperata* (Ross and Nash 1983).

Other studies of the effects of O_3 on lichens have produced less clear results. Rosentreter and Ahmadjian (1977) claimed to have found an effect of O_3 on chlorophyll levels in *Cladonia arbuscula* and the *Trebouxia* phycobiont of *Cladina stellaris*. However, Brown (1980) questioned the validity of the chlorophyll extraction technique used. Brown and Smirnoff

(1978) had previously recorded no effect of 2000 p.p.b. O_3 on photosynthesis in *Cladonia rangiformis*. Nash and Sigal (1979), however, suggest that as Brown and Smirnoff only looked for an immediate effect, they would have missed the delayed photosynthetic reduction apparent in the Californian study. Sigal and Johnston (1986) also found no significant effect of O_3 on nitrogen fixation or photosynthesis in *Lobaria pulmonaria*. C. A. Jones (unpublished data) also fumigated *L. pulmonaria*, *L. amplissima*, and *Parmelia laevigata* with 80 p.p.b. O_3, 6 h per day for 6 d, but was unable to detect increased electrolyte leakage or a reduction in photosynthesis.

Fewer studies have been undertaken with PAN. Sigal and Taylor (1979) fumigated *Collema nigrescens*, *H. enteromorpha* and *P. sulcata* with 50 and 100 p.p.b. PAN for 4 h per day for 8 d. *Peltigera rufescens* was also exposed to 200 p.p.b. PAN for 1 h. Photosynthesis in *P. sulcata* was more sensitive to PAN than in *H. enteromorpha* and *C. nigrescens*, the same sensitivity order as obtained with O_3. After a 5 d fumigation the algal layer of *P. sulcata* was visibly damaged. Eversman and Sigal (1984) discovered that PAN caused an increase in starch accumulation and a decrease in the pyrenoid area of the photobiont of *P. sulcata*, although no degradation of the thylakoids was seen. Ultrastructural changes have also been found with O_3 fumigation (Eversman and Sigal 1987).

There has been only one study of the effects of H_2O_2 on cryptogams (Farmer *et al.*, unpublished data). No immediate effect of 500 μM aqueous solutions of H_2O_2 was detected on electrolyte leakage by *Evernia prunastri*, *Platismatia glauca*, *Hypogymnia physodes*, and *Usnea florida*, or on photosynthesis following two weeks of repeated mistings with 150 μM H_2O_2 solutions in controlled conditions. Sigal and Nash (1983) had earlier concluded that the first two species were affected by other oxidant pollutants in the field. The concentrations used in this study are above those normally experienced, so it is unlikely that H_2O_2 damage of these species would occur in the field.

There has been little work on the effects of oxidant pollutants on bryophytes. Comeau and Leblanc (1971) found that O_3 facilitated the regeneration of protonema from detached leaves of *Funaria hygometrica*, but the ecological significance of this is obscure. Stanosz *et al.* (1987) fumigated soils from mixed woodlands with 0–320 p.p.b. O_3 for 6 hours per day, 4 days per week for 10 weeks. At the end of this period the cover of *Ditrichum pusillum*, *D. lineare*, and *Pohlia nutans* was negatively correlated with O_3 concentration, due to a suppression of both shoot numbers and height of *D. pusillum*.

11.4 Nitrogen oxides (NO$_x$)

The effects of NO$_x$ on lichens and bryophytes have received little attention, though NO$_x$ has been found to be generally phytotoxic to higher plants (Wellburn 1990). Nash (1976) was the first to perform a fumigation experiment with NO$_x$ on four lichen species, although Schmid and Kreeb (1975) had previously noted an effect of HNO$_3$ fumes on the phosphatase activity of *Hypogymnia physodes*. Nash fumigated four species for 6 h with a range of concentrations of NO$_2$ and only demonstrated an effect on chlorophyll concentration with 4000 p.p.b. NO$_2$ or greater. As 4000 p.p.b. is very much greater than ambient pollutant levels Nash considered that NO$_2$ was unlikely to be an important lichen phytotoxin. It is possible, however, that chlorophyll levels are an insensitive indicator of damage. An indication of the potential toxicity of dissolved NO$_x$ to lichens is given by work with free-living algae. Marti (1983) found that 0.5 mM nitrite inhibited photosynthesis of cultured lichen phycobionts and Wodzinski *et al.* (1977) had previously noted that 1.0 mM nitrite greatly reduced photosynthesis in cyanobacteria, but not in green algae. A demonstration of the effects of NO$_x$ in the field is difficult as high NO$_x$ levels often occur in combination with high levels of other pollutants such as SO$_2$. However, von Arb *et al.* (1990) studied a range of physiological parameters in *Parmelia sulcata* from sites in northern Switzerland and used a multiple regression analysis to separate the effects of NO$_x$, SO$_2$ and O$_3$. Correlations of damage with increasing NO$_x$ and SO$_2$ were both strong, but those with O$_3$ demonstrated only a weak relationship.

Fumigations of bryophytes with NO$_x$ have been rare. In an early study Fairfax and Lepp (1976) gave *Hypnum cupressiforme* and *Dicranum scoparium* very high levels of NO$_2$ (800 p.p.b.) for 3 h. Differences in cation leaching following this treatment were found for material collected from different sites in north Wales, but in general *D. scoparium* was found to be more sensitive. Fumigation with NO$_2$ was also noted to increase sensitivity of these species to subsequent immersion in low pH (4.0) solutions.

There has been only one major study of the effects of long-term fumigations of NO$_x$ at realistic concentrations on bryophytes (and ferns). Ashenden *et al.* (1989) undertook fumigations with realistic NO$_2$ concentrations (60 p.p.b.) employing field-grown bryophyte colonies transferred to 'solardomes'. They had difficulties maintaining some bryophyte species in culture, but observed both positive and negative effects of NO$_2$ on *Polytrichum formosum* and *Isothecium myosuroides*. They grew *P. formosum* from October to May and found that plants given NO$_2$ initially grew better, but by early spring there was a 36 per cent reduction in the numbers of new shoots and a 46 per cent reduction in the numbers of old shoots showing regrowth. *I. myosuroides* was fumigated for 12 weeks from January to April and also showed a significant reduction in new shoot production when fumigated with NO_2. Continuing this work, Morgan *et al.* (1992) fumigated *Homalothecium* (= *Tomenthypnum*) *nitens*, *Hylocomium splendens*, *Ctenidium molluscum*, and *Pleurozium schreberi* with 40 p.p.b. NO_x, which comprised varying proportions of NO and NO_2. Short-term (24 h) fumigation with NO_2 alone or predominating in the mixture was found to induce nitrate reductase activity, but after three weeks fumigation this induction no longer occurred. When NO was most abundant it prevented induction of this enzyme. NO_x fumigation also produced a small increase in electrolyte leakage, but no significant effects on photosynthesis.

In concluding this section on gaseous pollutants it is important to stress the need for long-term fumigation studies employing realistic pollutant concentrations. This is the only way to demonstrate conclusively whether the measured levels in the field are able to cause damage to the cryptogamic flora. It is also important to look beyond the findings of SO_2 fumigation studies. Thus while NO_2 reaches high levels in winter (PORG 1990) when lichens and bryophytes are more likely to be wet and physiologically active, O_3 reaches its peaks in the summer. O_3, unlike SO_2 and NO_2 is, however, not readily soluble in water and so may readily damage cell membranes even in a dry state. Thus fumigation studies must be done with regard to ecological conditions.

11.5 Wet acidic deposition

Two major aspects of wet acidic deposition are of potential importance to bryophytes and lichens: the effects of H^+ deposition and the fertilizing effect of nitrate. Dissolved pollutants such as hydrogen peroxide may occasionally be important also.

Evidence for the deleterious effects of acidic precipitation on lichens and bryophytes comes from several sources. Field observations, comparing sites with different pollutant loadings or the same sites over time have indicated possible reasons for floristic changes. Intensive studies of the environment at contrasting sites may also further indicate causal agents of change. Finally, field and laboratory experiments have been used to investigate hypotheses arising from the field studies. The effects of acid deposition on *Sphagnum*-dominated communities are considered in Chapter 12 and those on aquatic species in Chapter 13 and Farmer (1990), so neither will be dealt with here.

11.5.1 Community changes related to increased acid deposition

Compared with the effects of SO_2 pollution, which led to wholesale decline of lichen and bryophyte communities in many areas of the developed world, field observations of decline which can been unequivocally attributed to wet acidic deposition have been relatively few.

Before describing specific examples of cryptogamic decline attributed to acidification it is worth considering briefly the ecological importance of natural variations in acidity and the likely consequences of an increase in acidity. Natural variations of substrate and water acidity are a major factor governing the composition of bryophyte and lichen communities (e.g. Sjors 1950; Barkman 1958; James *et al*. 1977; Bates 1978; Büscher *et al*. 1990). Consequently, increasing acidification brought about by pollutant deposition can be expected to favour some species and weaken others (Hawksworth 1990). Presumably the taxa at most risk from acidification *per se* are those requiring a non-acidic environment whereas acidophilous or acidicline taxa (*sensu* Bücher *et al*. 1990) might be expected to possess natural resistance to low pH and associated ionic conditions. However, calcicole species, which are normally limited to substrates or waters dominated by carbonate/carbonic acid buffer systems, are unlikely to be affected because of the high buffering capacity of the latter. It can be argued, therefore, that neutrocline taxa (*sensu* Büscher *et al*. 1990), that is those with poor resistance to high concentrations of H^+ and Al^{3+} but occurring in environments with less robust buffering properties (e.g. carbonic acid/silicate systems), are likely to be most susceptible. Roughly, this category includes species which field botanists label as 'base requiring' or 'basiphile', but which actually thrive on mildly acidic to neutral substrates (pH >5). For soil-inhabiting woodland bryophytes Büscher *et al*. (1990) deduced that the major factor preventing these species from colonizing more acidic substrates is sensitivity to Al^{3+} ions which become increasingly available below pH 5. Amongst bryophytes, with the notable exception of *Sphagnum*, taxa which naturally grow on acid rocks (Bates 1982) or soils (Büscher *et al*. 1990) possess significantly lower cell wall cation exchange capacity (CEC) than neutrocline or calcicole species. Comparable data for calcicole and calcifuge lichens are lacking. Büscher *et al*. (1990) showed experimentally that elevated tissue CEC in bryophytes is disadvantageous on acidic soils because it results in disproportionate accumulation of Al^{3+} in comparison to Ca^{2+}, which is essential for cell function although possibly toxic in excess (Kinzel 1983; Rorison and Robertson 1984). It is probably significant that in the acid peatland systems dominated by *Sphagnum*, which also has high tissue CEC, Al^{3+} is much less abundant than in acid mineral soils

(Sparling 1967). Similarly, Al^{3+} toxicity is unlikely to play an important role in the substrate relationships of epiphytes which do not normally come into contact with clay minerals.

The most comprehensive field observations of the effects of wet acidic deposition on cryptogam communities in the field have concerned the epiphytic *Lobarion pulmonariae* community. Gilbert (1986) noted that *Lobaria pulmonaria* and *Sticta limbata* were declining in abundance on oak and ash trees in Northumberland, England. These trees were remote from SO_2 pollution, but subjected to precipitation of low pH and the loss of species was found to have occurred on trees with a low bark pH. Day (1985) analysed historical floristic data for Cumbria, England, and concluded that the *Lobarion* community had shown considerable decline since the 19th century. Three cyanobacterial lichens, *Lobaria scrobiculata*, *Pannaria rubiginosa*, and *Parmeliella atlantica*, had become extinct in Cumbria and a further 11 species had declined. Some of this decline, in areas of high rainfall with high levels of H^+ loading, was very recent. A series of permanent quadrats has been established to monitor the status of the *Lobarion* in Great Britain (Looney and James 1990; Wolseley and James 1991; Looney 1991). Loss of members of the *Lobarion* community, low relative growth rate of *Lobaria* species, and declining bark pH of the phorophytes have been found in a number of regions over the period 1986–90 and much of this may be attributable to wet acidic deposition.

Similar changes have been observed in Sweden. In southern Sweden *Lobaria scrobiculata* was refound at only 14 of 280 sites where it was recorded before 1950 (Fig. 11.1; Hallingbäck 1986). *L. amplissima* had also declined sharply, being relocated in only 16 of 57 old localities (Hallingbäck and Thor 1988). *L. pulmonaria* had not shown such a great decline, although it now has a more restricted distribution in the provinces of Skåne (Hallingbäck and Olsen 1987) and Gäsene (Hallingbäck and Martinsson 1987). In each study there had been a contraction in substrate preference away from tree species with bark that is readily acidified (e.g. *Quercus*) to those that have better buffered bark (e.g. *Ulmus*). While SO_2 levels are low, rainfall in much of this area is of a low pH and it was considered that the trees with initially nutrient-poor bark had become further acidified, causing the loss of species. Hallingbäck (1989) extended this study to the epiphytic bryophyte *Neckera pumila*. This species has also declined in a number of localities in southern Sweden. The bark of trees in areas where it has been lost is more acidic than either sites in the south on which it still grows or in other less polluted parts of the country.

Arup *et al.* (1989) reported that the saxicolous and terricolous lichen flora of southern Sweden was also showing evidence of decline due to acidification. The flora recorded in 1939 was compared with that in

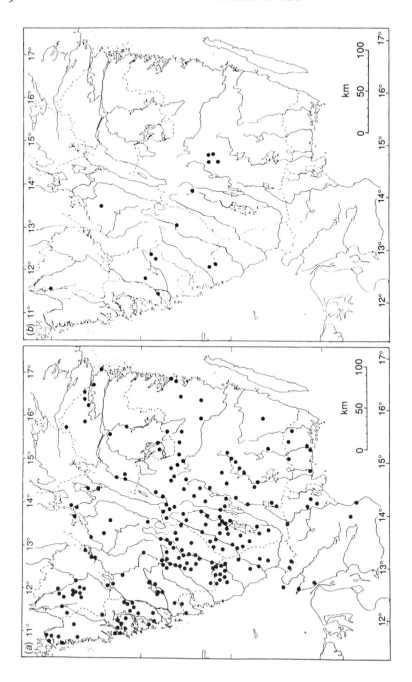

Fig. 11.1 The distribution of *Lobaria scrobiculata* in southern Sweden: (*a*) shows records before 1950 and (*b*) after 1950 (reproduced from Hallingbäck 1986).

1986/8, and recent pH data were compared with those from 1949. Twenty species were not refound and there was extensive evidence of acidification and a strong shift in the pH preference of the recent flora towards more acid conditions.

Although it has been firmly established that soil pH has decreased significantly in some areas receiving acid deposition (e.g. Falkengren-Grerup 1986), evidence for changes to terricolous communities is scanty. Two bryophytes which appear to have suffered a decline in polluted parts of south-east England are *Hylocomium splendens* and *Rhytidiadelphus trique-trus*. In Berkshire (J. W. Bates unpublished data) these mosses are now mostly confined to calcareous soils. They have declined in abundance since the last county flora survey (Jones 1953, 1955) and formerly both occurred widely on non-calcareous soils, as they still do in unpolluted areas. It is difficult to unravel the influences of changes in land-use, gaseous SO_2 pollution, and rainfall acidity in causing these declines. However, there is some evidence that SO_2 alone is not responsible. The calcifuge moss *Pleurozium schreberi* remains as a common species on acidic soils in the part of Berkshire most affected by SO_2 pollution, but in a comparison of three feather moss species undertaken in Alberta it proved more SO_2 sensitive than *H. splendens* (Winner and Bewley 1983; Winner 1988). Similarly *R. triquetrus* was the least sensitive of four common bryophytes subjected to SO_2 fumigations ranging from 200–1500 p.p.b. (Turk and Wirth 1975). In Berkshire, the three more sensitive species (*Hypnum cupressiforme*, *Grimmia pulvinata* and *Plagiomnium undulatum*) remain abundant over the areas where *R. triquetrus* and *H. splendens* have declined.

Many cryptogams are resistant to or require low pH and it may be expected that as acid-sensitive species decline, acid-tolerant taxa would increase. Day (1985) noted that *Parmelia laevigata* was abundant on the acidified bark of trees in Cumbria. Gilbert (1986) suggested that species tolerant to acid rain need not be the same as those tolerant of SO_2. Thus *Usnea* ssp. are intolerant of SO_2, but grow on bark of low pH, while *Parmelia sulcata* seemed to be tolerant of SO_2, but not of acid rain (M. R. D. Seaward, cited by Gilbert 1986). Seaward (1989) noted that *Parmeliopsis ambigua* had expanded in the UK from an original restriction to debarked conifer wood to widespread growth on the acidified bark of deciduous trees in polluted areas. Seaward (1989) also suggested that sections of the UK oligotrophic lichen flora may have expanded, including the crustose lichens *Buellia griseovirens*, *B. pulverea*, *Gyalideopsis anastomosans*, *Micarea nit-schkeana*, *M. peliocarpa*, *M. prasina*, *Mycoblastus sterilis*, *Rinodina efflorescens*, and *Scoliciosporum chlorococcum*.

Several epiphytic bryophytes have also expanded their range in Britain, perhaps as a result of bark acidification. *Dicranoweisia cirrata* is now an abundant species over much of lowland England. Comparing the paucity of early records of this species with its conspicuous abundance today, Jones (1991) believes that *D. cirrata* was absent in the Oxford area a century ago. *Dicranum montanum* and, particularly, *D. tauricum* have evidently increased considerably in some of the counties of south-east England (Gardiner 1981; J. W. Bates, unpublished observations) in more recent decades. Other expansions of epiphytic bryophytes which may be attributable to acidification are discussed in Chapter 5.

Acidification can also be expected to have caused expansion in the abundances of acidophilous terricolous species, although there is little firm evidence. Storm (1990) noted that the acidophilous bryophytes *Pleurozium schreberi*, *Hylocomium splendens* and *Polytrichum formosum* are appearing in the understorey of stands in the Black Forest of Germany where the trees have shown forest die-back, thought to be due, in part at least, to air pollution.

In The Netherlands, van Dobben *et al.* (1983) recorded a decline in the lichen *Cladina portentosa* and the moss *Dicranum scoparium* under forest canopies since 1960. These had been replaced by grasses. Studies of herbarium specimens also showed that *D. scoparium* had experienced a reduction in fruiting after 1958. Although acidification may be important in this instance, the confounding effects of SO_2 and, possibly, NH_3 cannot be eliminated. Klein and Bliss (1984) also noted a decline in the bryophyte ground cover on Camel's Hump Mountain in Vermont, a region remote from SO_2, but receiving high levels of acid deposition. Unfortunately, individual species were not distinguished in this study. Visible injury has also been noted on mosses in southern Norway (Flatbert *et al.* 1990), a region receiving high levels of wet acidic deposition, but low in gaseous pollutants. The two most damaged species were *Dicranum majus* and *D. polysetum*.

11.5.2 Fertilizing effects of acid rain

There is increasing evidence that some of the floristic changes attributed to wet acidic deposition are caused by a fertilizer effect of the nitrogen present in the nitrate constituent. Farmer *et al.* (1991*b*) found the epiphytic moss *Isothecium myosuroides* on *Quercus petraea* in Borrowdale, Cumbria to have a higher tissue N content (1.3 per cent N) than from the same phorophyte species at Loch Sunart in western Scotland (0.7 per cent N). The former site receives 0.6–0.7 g N m⁻² of wet deposited nitrate, while the latter receives only 0.2–0.3 g N m⁻² (Williams *et al.* 1989). Søchting (1990) surveyed the tissue N content of reindeer lichens in north-west Europe. In unpolluted areas, such as western Ireland and Scotland, tissue levels were 0.26–0.49 per cent N, while in areas receiving high levels of wet deposited acidity, such as southern Norway and Sweden, values were 0.70–0.73 per cent N and visible injury could be found for individuals in Denmark (Søchting 1987).

Thompson and Baddeley (1991) studied the decline of *Racomitrium lanuginosum* heaths in Great Britain. There is evidence of decline in southern parts of Scotland and in England and Wales, but not in northern Scotland. Tissue N levels were higher in regions of high N deposition and also increased with altitude. Thompson and Baddeley suggested that this was due to increased N input through occult deposition at higher elevations. Acidic deposition and intensity of grazing were probably both contributing to the decline of *Racomitrium* heath. It was suggested that increased N supply allows enhanced growth of grass species, which then eventually dominate the community. Thompson and Baddeley (1991) suggested that 23 upland communities (in the UK National Vegetation Classification) may be affected by acid deposition. Many of these are dominated or co-dominated by bryophytes and lichens and eight of the communities are considered to be internationally important.

Pitcairn *et al.* (1991) analysed 40 years' data from permanent vegetation plots at Moor House in Cumbria, northern England. This upland site currently receives high levels of N deposition in bulk precipitation and cloudwater (approximately 30 kg N ha^{-1} y^{-1}). Numbers of lichen and bryophyte species had declined in both grazed and ungrazed plots. Greatest changes were found for species in base-rich sites. Analysis of present day and 30-year-old herbarium specimens of *Sphagnum* spp. revealed a 62 per cent increase in tissue N. A significant increase in tissue N was also found at the Black Wood of Rannoch, central Highlands of Scotland (1950s–1989) and north-eastern Scotland (1974–1989). However, no change was found for a range of bryophyte species in Beinn Eighe, north-west Scotland (1950s–1989), an area with low levels of acidic deposition. Pitcairn *et al.* (1991) concluded that the combination of evidence of increasing N levels in vegetation and decline in cryptogamic cover in permanent plots in sites receiving high levels of acid deposition is strong circumstantial evidence for the damaging influence of acid deposition.

Evidence for floristic change with increasing acidity of precipitation is lacking in many instances, but it may be possible to predict the response of particular communities and species from a knowledge of current habitat preferences. There have, unfortunately, been relatively few studies of lichens and bryophytes which link community variation to variations of environmental factors. This is least true for epiphytes which have attracted studies encompassing measurements of bark pH and chemistry. Gauslaa (1985) studied oak epiphytes in Norway. Bark pH was important in defining the type of community present, with the *Lobarion* restricted to bark above about pH 5. On trees with more acid bark the *Parmelion caperatae* and *Pseudevernion* communities dominated. All three communities were found on different trees within a single wood. In theory it would be possible to predict changes in the relative abundances of the different communities given a further decline in bark pH caused by acidic precipitation.

It has been noted that trees at higher elevation in a woodland are more likely to support the acidophilous *Parmelion laevigatae* community than the *Lobarion* (Rose 1974; Wolseley and O'Dare 1990). Occult deposition contains higher levels of dissolved ions (including H^+) than bulk precipitation and such deposition increases with increasing elevation. However, the role of occult deposition in affecting epiphyte communities has not yet been critically examined, although it has been implicated in changes in terricolous cryptogamic floras (Pitcairn *et al.* 1991).

11.5.3 Experimental studies

The field studies described above have indicated that a number of species may be declining due to acid rain. However, only one of the threatened species, *Lobaria pulmonaria*, has been experimentally studied in published work. Most experimental investigations have employed species for which there is, as yet, no evidence of decline in the field. There has been much emphasis on terricolous boreal species, although there is little evidence that they are in decline. This emphasis is, however, understandable due to the enormous ecological importance of these communities and it may be wise to estimate the potential threat, even if it is never realized. More work should be focused on neutrocline rather than acidophile species, for reasons described above.

Field experiments

Several workers have investigated the potential impact of wet deposited pollutants by applying them to pristine vegetation. Most field experiments have involved boreal and fen species growing in areas unaffected by SO_2 pollution, particularly the 'feather moss' *Pleurozium schreberi* and the 'reindeer lichens' of the genus *Cladonia* subgenus *Cladina*. These studies have involved the spraying of vegetation plots with acidic solutions for periods of up to five years. This contrasts sharply with previous studies of the effects of pollutants on bryophytes and lichens, which were mostly laboratory based, of short duration, and which often employed unrealistically high exposure rates.

The contrast between sulphate and nitrate in acidic deposition was studied by Lechowicz (1987) who treated plots of *Cladina stellaris* for three years with solutions of either 2:1 or 6:1 sulphuric:nitric acid (v:v) at a range of pHs. After two years no effect on growth was found, although some discoloration occurred when the pH of the solution was 3.5. After three years there was significant growth depression in the 6:1 treatment at pH 4.5 or less, but not in the 2:1 plots. Lechowicz (1987) suggested that the fertilizing effect of the extra N in the 2:1 plots was counteracting the detrimental effects of the acid. Although growth depression was only seen after three years, the results are complicated by the fact that growth was measured on thalli then six years old, so that slight reduction in growth early in the experiment may have been missed. Hutchinson *et al.* (1986) treated similar plots bimonthly with artificial acid rain over five years and found reduced dry weight and podetial height of *C. stellaris* and *C. rangiferina* receiving pH 2.5 solutions. Photosynthesis and chlorophyll levels were also reduced, but no effects were observed at pH 3.0 or above.

Scott *et al.* (1989) carried out further studies on the relative importance of nitric and sulphuric acids. Reindeer lichens (*Cladina rangiferina* and *C. stellaris*) were sprayed with artificial rain of pH 4.2, 3.5, or 2.8 with each acid alone or with 2:1 or 1:2 combinations. Nitric acid alone at pH 2.8 caused dry weight increases of 62 per cent over controls in *C. rangiferina*, with an increase in tissue N levels. Sulphuric acid alone caused a significant reduction in dry weight. Growth of *C. stellaris* was stimulated by nitric acid at pH 4.2, but not at lower pHs. This study shows that the nature of the acid rain, and not just its pH, is important in determining the lichen response.

Similar results have also been forthcoming in field experiments with bryophytes. Hutchinson *et al.* (1986) and Hutchinson and Scott (1988) studied *Pleurozium schreberi* in the same plots as the reindeer lichens. This moss proved more sensitive than the lichens, with photosynthesis reduced by 44 per cent in the pH 3.0 treatment after five years. The shoots were also discoloured at pH 3.5. Raeymaekers (1987) also treated plots of *P. schreberi*, but only for two years. Effects were found only at the end of the second year in the pH 3.0 treatment, with a 49 per cent reduction in biomass, a 64 per cent reduction in tissue Ca and 40 per cent reduction in tissue Mg. Tissue K was not affected. The treatment also caused an increase in total tissue N, due to the nitrate supplied in the solutions. Two years' treatment at pH 3.0 also reduced tissue chlorophyll levels and reproductive capacity (Raeymaekers and Glime 1986). Bates (1989) concluded that cation exchange is important for the uptake of Mg in *Pseudoscleropodium purum* and that the displacement of Mg by increasing H^+ in rainfall may lead to nutrient deficiency. Supporting evidence was obtained by Bates and Farmer (1990) who found that displacement of exchangeable Mg following artificial addition of Ca eventually led to a reduction in intracellular Mg and growth in *Pleurozium schreberi*.

Experimental applications of acidic pollutants to wetland species have rarely been undertaken, but Rochefort and Vitt (1988) sprayed plots containing the fen bryophytes *Tomenthypnum nitens* and *Scorpidium scorpioides* with pH 3.5 1:1 sulphuric:nitric acid. *S. scorpioides* was unaffected, but *T. nitens* showed an increase in growth and chlorophyll *b* content. The two species occupy microtopographically different positions in the fen, with *T. nitens* on hummocks, isolated from much of the groundwater. Rochefort and Vitt (1988) considered that *in situ* this moss was more likely to be nutrient deficient and thus able to benefit most from the increased nitrogen application.

The field application of low pH solutions to epiphytes was carried out by Farmer *et al*. (unpublished data). They found (Table 11.1) that after 9 months of simulated acid rain applications (pH 3.5; 2:1 sulphuric:nitric acid) to oak trees in western Scotland that there was no marked change in the cellular levels or location of major cations in the epiphytes *Lobaria*

Table 11.1   Mean tissue nutrient levels (μmol g^{-1} dry wt $\pm$ SEM, $n = 4$) in two epiphytes at Loch Sunart, western Scotland, after bimonthly spraying with pH 3.5 acid rain for 9 months. Data from Farmer *et al*. (unpublished results)

Element/cellular location	*Lobaria pulmonaria*		*Isothecium myosuroides*	
	Control	pH 3.5	Control	pH 3.5
Potassium				
Intercellular	22.5	5.2	18.0	11.5
	±11.8	±2.0	±5.3	±1.9
Exchangeable	37.7	3.4	25.4	23.5
	±17.6	±1.5	±3.7	±7.0
Intracellular	164.5	262.1	118.8	145.4
	±41.1	±38.2	±3.4	±24.0
Calcium				
Intercellular	0.7	2.3	0.0	0.7
	±0.3	±1.6	±0.0	±0.2
Exchangeable	57.5	44.9	169.1	213.8
	±11.9	±6.3	±20.2	±12.1
Intracellular	23.9	11.0	20.2	18.1
	±11.3	±5.5	±3.4	±2.2
Magnesium				
Intercellular	2.3	1.4	1.0	0.8
	±1.4	±0.5	±0.4	±0.1
Exchangeable	43.1	43.5	91.6	131.2
	±1.2	±7.9	±7.9	±11.5
Intracellular	23.7	21.4	19.2	26.6
	±3.8	±2.1	±1.5	±3.2
Nitrogen				
Total	1408.6	1262.1	419.3	536.4
	±67.9	±41.8	±39.3	±41.4

pulmonaria and *Isothecium myosuroides*. There was also no change in tissue N levels in *L. pulmonaria*, but a possible small increase for *I. myosuroides*. It is likely that the temporary effect such applications have is rapidly overcome by the large volume of stemflow that occurs in this high rainfall region of the UK.

The foregoing studies illustrate that boreal and fen species possess considerable resistance to acidic deposition. Effects were observed only after long periods and then usually only under extreme conditions. Normally, pollutant deposition occurs over similar or longer time periods, so these results warn against the acceptance of short-term findings that suggest a lack of sensitivity. For example, Gunther (1988) treated *Peltigera apthosa* in the field in Alaska with pH 3.3 or 4.4 rain for 21 days. No effect was observed on nitrogenase activity, but it would be dangerous to conclude that a longer period of treatment would be similarly ineffective.

Apart from the experimental application of simulated precipitation, few other manipulative studies of cryptogam communities have been directed towards the impact of acidic deposition. Transplant experiments have often been used to study the response of lichen species to air pollution (Hallingbäck 1990). In a study of the effects of acidic deposition on the epiphytic *Lobarion pulmonariae* community, transplants of all four British *Lobaria* spp were made from Loch Sunart, western Scotland into Seatoller Wood in Borrowdale, Cumbria, an area receiving high levels of wet acidic deposition (Farmer *et al.*, 1992). Table 11.2 shows that after about 31 months in Seatoller Wood two species (*L. scrobiculata* and *L. amplissima*) were severely damaged, while *L. pulmonaria* was less affected and *L. virens* was unharmed. In contrast, the control transplants at Loch Sunart remained healthy.

An indirect manipulative approach for studying the importance of acidic deposition on bryophytes and lichens is to ameliorate the input of acidic

Table 11.2   The final damage score (percentage chlorosis and necrosis) for four *Lobaria* species transplanted from oak trees at Loch Sunart to other oak trees at Loch Sunart or to oaks at Seatoller Wood, Borrowdale, Cumbria. The number of days for which the study was undertaken are given. Values are means ±SEM, $n = 4$. Data are from Farmer *et al.* (1992)

	Loch Sunart	Seatoller Wood
Length of study (days)	943	938
L. amplissima	0.0 ±0.0	97.5 ± 2.5
L. pulmonaria	0.0 ±0.0	52.5 ±20.6
L. scrobiculata	3.8 ±2.4	92.5 ± 8.5
L. virens	0.0 ±0.0	1.7 ± 1.7

pollutants using lime. This has been used successfully to restore plant and animal communities in acidified lakes and rivers (see Farmer 1990). Hallingbäck (personal communication) attached collars containing limestone pieces around the upper trunks of *Quercus robur* trees to neutralize acidity in the stemflow. His study was undertaken in a wood in southern Sweden which receives moderate levels of wet acidic deposition and formerly supported *Lobaria amplissima* and *L. scrobiculata*. Although the liming quickly raised the bark pH, after five years no recolonization of the trees had occurred, but the acid tolerant *Usnea subfloridana* had decreased. This approach may have applications in epiphyte conservation if combined with reintroductions of species.

Effects on physiology

In contrast to the field experiments, laboratory studies show that the physiology of lichens and bryophytes is relatively sensitive to acidic pollutants. Early studies of bisulphite, as an SO_2 analogue, showed that its toxicity was increased when combined with low pH (Turk and Wirth 1975), but it was some time before the effects of varying pH alone were studied. *Umbilicaria mammulata* from Ontario was maintained for 12 weeks under controlled conditions with wetting at different pHs (2, 4, and 6) (Bailey and Larson 1982). A general decline in lichen condition was found for all the material and there was no effect of pH. Similarly Lane and Puckett (1979) had found no detrimental effect of low pH (2.2–8.9) on phosphatase activity in *Cladina rangiferina* and *Lobaria pulmonaria*. In fact enzyme activity was highest at the lower pHs. In contrast Fritz-Sheridan (1985) observed reduced nitrogenase activity in the lichens *Peltigera aphthosa* and *P. polydactyla* in response to low pH treatments. Cessation of nitrogenase activity occurred at pH 2.0 in the former species and at pH 3.0 in the latter. This was seen after only 12 days' treatment, but a reduction was found at all pHs below pH 6.0. This contrasts sharply with Gunther's (1988) study in which the nitrogenase activity of *P. rufescens* was unaffected by incubation in the laboratory for up to 60 days with simulated rain of pH 4.4 and pH 3.4. However, Hallingbäck and Kellner (1992) found that 70 days' incubation of *P. aphthosa* at pH 3.0 also significantly reduced nitrogenase activity.

Lichens of the acid-sensitive *Lobarion* community have also been the subject of laboratory investigations. Sigal and Johnston (1986) studied the effect of acid rain on the physiology of *Lobaria pulmonaria*. The thalli were treated with simulated rain of pH 2.6, 4.2 and 5.6 over a 10 day period. An effect was only seen at pH 2.6, which caused a 100 per cent reduction in nitrogen fixation and a 90 per cent reduction in gross photosynthesis compared to controls. Similarly Dennison *et al* (1977) found that *L. pulmonaria* and *L. oregana* from north-western USA had lower N_2

fixation rates when exposed to sulphuric acid solutions of less than pH 4. Sigal and Johnson (1986) also found that there was no effect of gaseous O_3 on the physiology of *L. pulmonaria*, either singly or in combination with the low pH treatments.

Earlier studies of SO_2 toxicity used electrolyte or K leakage as a measure of physiological damage to lichens. This has not, however, been done previously in acid rain studies on lichens and bryophytes. Table 11.3 presents results based on K leakage in acid solutions for 10 lichen and two bryophyte epiphytes which are common in oak woodlands in NW Britain. There is a considerable range in sensitivity, with *Lobaria* species being particularly sensitive, and *Usnea subfloridana* and *Parmelia laevigata* least sensitive. The latter species is believed to be spreading on acidified bark as

Table 11.3 Sensitivity to low pH of ten lichens and two bryophytes which occur as epiphytes in oak woodlands in north-west Britain. The values are derived from the percentage increase in total tissue potassium which is extracellular in location after two 1 h incubations in solutions at either pH 2.5 or 3.0 in comparison to controls at pH 5.0. An increase in extracellular K is considered to be an indication of membrane stress. Extracellular location was determined following the sequential elution technique of Bates (1989). Data are from Farmer *et al.* (unpublished results)

Species	pH of incubation medium	
	2.5	3.0
Lichens		
Lobaria scrobiculata	56.7	5.2
L. virens	28.5	3.8
Peltigera membranacea	19.5	2.1
Lobaria amplissima	18.2	0.2
L. pulmonaria	17.8	1.2
Hypogymnia physodes	14.0	0.0
Parmeliella atlantica	8.4	1.9
Parmelia caperata	6.4	2.4
Usnea subfloridana	0.0	0.0
Parmelia laevigata	0.0	0.0
Bryophytes		
Isothecium myosuroides	28.5	0.4
Frullania tamarisci	11.5	3.1

the *Lobarion* declines (Seaward and Hitch 1982; Looney and James 1988).

The photosynthetic activity of *Cladina stellaris* was reduced by 27 per cent following wetting with solutions of pH 4 (Lechowicz 1982) and metabolic recovery on resaturation took longer after low pH treatment. Scott and Hutchinson (1987) also found a reduction in net photosynthesis in *C. stellaris* and *C. rangiferina* at pH 2.5 and 3.0 after only three days' treatment in a 2:1 sulphuric/nitric acid mixture. This contrasts sharply with the long exposure periods necessary to produce a response in the field. On return to normal conditions both species recovered six days after treatment and *C. rangiferina* showed a stimulation of photosynthesis. Scott and Hutchinson (1987) suggested that this was due to the fertilizing effect of the nitrate.

Laboratory work on the effects of acid solutions on bryophytes has been very limited. Sheridan and Rosentreter (1973) found that simulated rain (down to pH 1.0) reduced the total chlorophyll and chlorophyll *a:b* ratio of *Tortula ruralis*. Pitkin (1975) grew the common epiphyte *Isothecium myosuroides* at a range of acidities and found that at pH 4.0 the gametophytes degenerated into a protonemal-like growth form, while at pH 3.0 most of the shoots died. Woollon (1975) observed maximum extension growth of the calcicole moss *Fissidens cristatus* at pH 8.0. Growth declined sharply at lower pHs and ceased at pH 3.0. Klein and Bliss (1984) cultured *Polytrichum ohioensis* from spores for three weeks in a range of pHs. They detected a slight decline in growth at pH 4.0 and a larger reduction at pH 3.0. Growth reduction in *P. ohioensis* was enhanced by the presence of Al, a metal that is increasingly mobilized in acidic soils and toxic to many vascular plant species (Anderson 1988).

In contrast to the growth stimulation of *Tomenthypnum nitens* observed in fen plots treated with an H_2SO_4:HNO_3 mixture, Rochefort and Vitt (1988) observed growth reductions when *T. nitens* and *Scorpidium scorioides* were cultured in the laboratory in 1:1 sulphuric/nitric acid solutions. *S. scorpioides* showed a reduction in growth at pH 3.5 or lower and this was even more marked for *T. nitens*, where reduction occurred between pH 3.5 and 3.0. This contrast between field and laboratory results probably reflects the isolation of the laboratory grown plants from the ground water system in the field, which may ameliorate the effects of the low pH treatments.

11.5.4 Acidification of substrates

It is highly likely that, besides direct effects on physiological processes, wet acidic deposition may influence cryptogam communities indirectly by altering substrate chemistry. Many lichen and bryophyte species are closely associated with their substrates. Indeed, Brodo (1973) details how the hyphae of epiphytic lichens penetrate the outer layers of tree bark. Terricolous cryptogams, however, may often be in less intimate contact with their substrates.

Ecological studies of epiphyte communities have shown that, even in unpolluted areas, natural variations in bark chemistry have a major impact on which species are present (Barkman 1958; James *et al.* 1977; Bates and

Brown 1981; Gauslaa 1985). In areas impoverished by acidic pollutants, epiphytes often persist longest on the tree species or individuals with the highest bark buffering capacity (Barkman 1958; Robitaille *et al.* 1977; Gilbert 1986). At several UK sites receiving acid deposition it has been noted that *Lobarion* species persist longer on *Fraxinus excelsior* than *Quercus* spp. (Gilbert 1986; Looney and James 1990). It is usually assumed that this is a function of the higher cation status and buffering capacity (notably K and Mg; Bates and Brown 1981) of *F. excelsior*.

An interesting observation, made by several authors, is the occurrence together of individuals of a single tree species with markedly different bark chemistries and epiphyte communities (Gauslaa 1985; Rose 1988). Rose (1988) suggested that this results from genotypic variation in the phoro-phytes; however, variations in soil chemistry may be more important. In *Quercus*-dominated forests in south-west Norway the occurrence of relict *Lobarion pulmonariae* communities correlated strongly with high bark pH and Ca + Mg content (Gauslaa 1985). The same was found for *Q. petraea* and *Fraxinus excelsior* at sites experiencing high (Borrowdale) and low (Loch Sunart) acid deposition in north-west Britain (Farmer *et al.* 1991*a*; Table 11.4). Norwegian oaks with *Lobarion* grew mainly on nutrient rich soils, and Gauslaa (1985) suggested that variations in bark chemistry were a result of variations in soil cation availability. In woodlands around Loch Sunart, Scotland, a weak positive correlation was found between levels of bark Ca and soil exchangeable Ca in both *Q. petraea* and *F. excelsior* (Bates 1992). Using canonical correspondence analysis, bark pH was found to be a major factor determining epiphyte community composition on both tree species at Loch Sunart. However, bark pH was itself influenced by different features in each phorophyte. In *Q. petraea* it was most strongly correlated with the level of bark Ca and appeared to be influenced by soil Ca availability. In *F. excelsior*, bark pH increased with tree girth, but no correlation was found with bark cation status. The relationship between soil and bark chemistry is further complicated by the ability of trees with acidic stemflow to acidify the surrounding soil, as demonstrated for *Fagus sylvatica* in Sweden (Falkengren-Grerup 1989).

Changes in bark chemistry arising from acid deposition inevitably affect the ionic environment of epiphytes. Table 11.5 presents data from Farmer *et al.* (1991*b*) on stemflow chemistry of *Q. petraea* trees in the woodlands in north-west Britain described above. Trees at Seatoller Wood, Borrowdale, whose bark has been acidified by a history of high acid deposition, further acidified the incoming rainfall as it flowed down their trunks. However, at unpolluted sites the pH of the stemflow was above that of the bulk precipitation. Epiphytes on the better buffered trees in unpolluted sites are, therefore, less likely to be damaged by occasional acidic rainfall episodes.

Table 11.4   Bark chemical characteristics of *Quercus petraea* and *Fraxinus excelsior*, with or without *Lobaria* spp., in north-west Britain. All values are means ± SEM, $n = 4$. After Farmer *et al.* (1991*a*).

Site and tree spp.	*Lobaria* present	Bark pH	Total bark cation content μmol g^{-1} dry wt $^{-1}$		
			K	Ca	Mg
Loch Sunart					
Q. petraea	+	4.20	143.8	659.5	141.7
		±0.19	±29.6	±123.2	±74.9
Q. petraea	−	3.65	109.5	281.4	57.4
		±0.10	±23.7	±47.2	±5.7
Glenlee					
Q. petraea	+	5.98	71.9	789.7	22.6
		±0.16	±13.5	±168.4	±2.0
Q. petraea	−	4.81	46.1	433.1	21.3
		±0.35	±4.8	±192.6	±21.3
F. excelsior	+	5.21	86.1	280.5	54.8
		±0.40	±42.3	±41.6	±5.7
F. excelsior	−	4.91	79.6	286.6	46.2
		±0.14	±36.8	±50.2	±21.5
Borrowdale, Seatoller Wood					
F. excelsior	+	6.02	57.7	698.6	44.2
		±0.19	±21.0	±142.7	±5.4
F. excelsior	−	4.46	48.3	366.3	31.1
		±0.08	±14.8	±79.4	±2.2
Borrowdale, Great Wood					
Q. petraea	+	5.84	23.2	962.9	24.7
		±0.28	±1.9	±108.6	±2.4
Q. petraea	−	4.72	14.9	319.7	13.5
		±0.26	±1.6	±102.8	±2.5

Conversely, it can be expected that acidified bark will continue to acidify stemflow even after reduced acid loading arising from emission reductions. Consequently, the return to a non-acidiphilous flora is likely to be slow. This lag is likely to be greatest in phorophytes which, like *Quercus*, have bark which is not shed rapidly and is, therefore, not quickly replaced by underlying non-acidified tissues (Bates *et al.* 1990). One possible means of reversal of acidification has been noted in The Netherlands, where NH_3 deposition has caused a rise in the bark pH of the previously acidified *Quercus* together with a resultant change in the lichen flora (de Bakker 1989).

Responses of epiphytes to natural fluctuations in bark, stemflow, and rainfall chemistry were also studied by Farmer *et al.* (1991*b*). Cation levels

Table 11.5 The annual mean pH and range of pH for bulk rainfall and stemflow for *Quercus petraea* trees at three sites in north-west Britain. After Farmer *et al.* (1991*b*).

Site	Bulk rainfall	Stemflow
Loch Sunart	5.23	5.66
	4.91–5.79	4.67–6.15
Glenlee	5.36	5.37
	4.75–6.10	4.70–6.39
Borrowdale,	5.22	4.27
Seatoller Wood	3.93–6.12	3.70–4.69

in the epiphytes *Lobaria pulmonaria* and *Isothecium myosuroides* appeared to be influenced by pH fluctuations in the stemflow. During periods of low stemflow pH (e.g. in the autumn) stemflow K levels were high due to leaching from senescing tree leaves, but the tissue K levels of the two epiphytes remained low. Possibly K was also leaking from the epiphyte tissues due to membrane stress occurring at the low pH.

In contrast to epiphytes, terricolous feather mosses and reindeer lichens frequently develop a mat of litter beneath the living shoots or thalli and have rather poor contact with the underlying mineral soil. This has led many workers to emphasize the importance of mineral supply from the atmosphere (e.g. Tamm 1953; Rieley *et al.* 1979; Crittenden 1983, 1989; Brown 1982; Brown and Bates 1990). However, terricolous bryophytes and lichens are also demonstrably sensitive to the chemical nature of their substratum. Calcareous and noncalcareous soils support markedly different cryptogam floras and tissue concentrations of elements often reflect those in the soil. Bates and Farmer (1990) demonstrated significant upward movement of Ca to the growing tips of *Pleurozium schreberi* from a layer of $CaCO_3$ applied to the mineral soil beneath the moss litter, so it is possible that uptake of soil ions is more important in terricolous cryptogams than has been recognized previously. Relict distributions of neutrocline or calcicole cryptogams (e.g. *Rhytidiadelphus triquetrus*) on calcareous soils may arise because the latter afford a source of Ca which is continually being displaced from the shoots during wet periods by acid precipitation (Bates 1989, 1990). Alternatively, the upward supply of Al^{3+} from the soil may become intolerable for many species under an acidifying regime, except on calcareous soils. Relict distributions on calcareous soils may also have arisen from high SO_2 conditions occurring at an earlier period. However, while certain cations can ameliorate the effects of SO_2 (Richardson *et al.* 1979; Baxter *et al.* 1989), it has not been shown that the buffer systems which detoxify SO_2 on the surfaces of calcareous rocks also function in the tissue free space of cryptogams suspended above calcareous soil.

11.6 Concluding remarks

Bark and soil acidification is caused by the total H^+ ion input to the system. For epiphytes, extreme events are evened out by the buffering capacity of bark, so that the best predictions of epiphyte response are probably made on annual averages of pollutant loading. For terricolous cryptogams a similar situation may hold, depending on the intimacy of contact with the substrate. This is in contrast to SO_2 effects, where episodic pollutant peaks may be of overriding importance. As each tree (individual or species) has finite buffering capacity, it may be possible to estimate the additional loading of H^+ ions that will lead to acidification, the so-called *critical load* for the system. Many terricolous species occur on soils whose critical load for H^+ is exceeded and acidification is occurring. Although some species seem to be resistant to field experimental application of acid rain, the decline of others has been described and urgent action to reduce emissions is necessary for their conservation.

The problem of defining critical loads for N has not been resolved. This is determined by the biological response of the system (whereas for H^+ it is the measurable chemical property of buffering capacity), which is very difficult to quantify. Nevertheless, N inputs are causing changes to cryptogam populations and work is, therefore, necessary to discover the mechanism of change.

Future studies of acid rain may need to encompass the effects of predicted climatic change and CO_2 rise. For instance, global climate change may lead to an increase in rainfall for parts of north-west Europe. This may lead to an increase in the total deposition of pollutants onto cryptogam communities, although pollutant concentrations may fall. It is important, therefore, that emission reduction policies take account of possible further climatic scenarios and their relationship with air pollution.

References

Anderson, M. (1988). Toxicity and tolerance of aluminium in vascular plants. *Water, Air and Soil Pollution*, **39**, 439–62.

Arb, C. von, Mueller, C., Amman, K., and Brunold, C. (1990). Lichen physiology and air pollution. II: Statistical analysis of the correlation between SO_2, NO_2, NO and O_3, and chlorophyll content, net photosynthesis, sulphate uptake and protein synthesis of *Parmelia sulcata* Taylor. *New Phytologist*, **115**, 431–8.

Arup, U., Ekman, S., Froberg, L., Knutsson, T., and Mattsson, J.-E. (1989). Changes in the lichen flora on Romeleklint, S. Sweden, over a 50-year period. *Graphis Scripta*, **2**, 148–55.

Ashenden, T. W., Rafarel, C. R., and Bell, S. A. (1989). *Effects of nitrogen dioxide on ferns and bryophytes*. CSD Report Number 993, Nature Conservancy Council, Peterborough, UK.

Ashmore, M. R., Bell, J. N. B., and Mimmack, A. (1988). Crop growth along a gradient of air pollution. *Environmental Pollution*, **53**, 99–121.

Bailey, C. and Larson, D. W. (1982). Water quality and pH effects on *Umbilicaria mammulata* (Ach.) Tuck. *The Bryologist*, **85**, 431–7.

Bakker, A. J. de (1989). Effects of ammonia emission on epiphytic lichen vegetation. *Acta Botanica Neerlandica*, **38**, 337–42.

Barkman, J. J. (1958). *Phytosociology and ecology of cryptogamic epiphytes*. Van Gorcum, Assen, The Netherlands.

Bates, J. W. (1978). The influence of metal availability on the bryophyte and macrolichen vegetation of four rock types on Skye and Rhum. *Journal of Ecology*, **66**, 457–82.

Bates, J. W. (1989). Retention of added K, Ca and P by *Pseudoscleropodium purum* growing under an oak canopy. *Journal of Bryology*, **15**, 589–605.

Bates, J. W. (1990). Interception of nutrients in wet deposition by *Pseudoscleropodium purum*: an experimental study of uptake and retention of potassium and phosphorus. *Lindbergia*, **15**, 93–8.

Bates, J. W. (1992). Influence of chemical and physical factors on *Quercus* and *Fraxinus* epiphytes at Loch Sunart, western Scotland: a multivariate analysis. *Journal of Ecology*, **80**. (In press)

Bates, J. W., Bell, J. N. B., and Farmer, A. M. (1990). Epiphyte recolonization of oaks along a gradient of air pollution in south-east England, 1979–1990. *Environmental Pollution*, **68**, 81–99.

Bates, J. W. and Brown, D. H. (1981). Epiphyte differentiation between *Quercus petraea* and *Fraxinus excelsior* trees in a maritime area of South West England. *Vegetatio*, **48**, 61–70.

Baxter, R., Emes, M. J., and Lee, J. A. (1989). The relationship between extracellular metal accumulation and bisulphite tolerance in *Sphagnum cuspidatum* Hoffm. *New Phytologist*, **111**, 463–77.

Brodo, I. M. (1973). Substrate ecology. In *The lichens*, (eds. V. Ahmadjian and M. E. Hale), pp. 401–42. Academic Press, New York.

Brown, D. H. (1980). Notes on the instability of extracted chlorophyll and a reported effect of ozone on lichen algae. *Lichenologist*, **12**, 151–4.

Brown, D. H. (1982). Mineral nutrition. In *Bryophyte ecology*, (ed. A. J. E. Smith), pp. 229–55. Chapman and Hall, London.

Brown, D. H. and Bates, J. W. (1990). Bryophytes and nutrient cycling. *Botanical Journal of the Linnean Society*, **104**, 129–47.

Brown, D. H. and Smirnoff, N. (1978). Observations on the effect of ozone on *Cladonia rangiformis*. *Lichenologist*, **10**, 91–4.

Büscher, P., Koedam, N., and Van Speybroek, D. (1990). Cation-exchange properties and adaptation to soil acidity in bryophytes. *New Phytologist*, **115**, 177–86.

Comeau, G. and Leblanc, F. (1971). Influence de l'ozone et de l'anhydride sulfureux sur la regeneration des feuilles de *Funaria hygometrica* Hedw. *Naturaliste Canadian. Quebec*, **98**, 347–58.

Crittenden, P. D. (1983). The role of lichens in the nitrogen economy of subarctic woodlands: nitrogen loss from the nitrogen-fixing lichen *Stereocaulon paschale* during rainfall. In *Nitrogen as an ecological factor*, (eds. J. A. Lee, S. McNeill, and I. H. Rorison), pp. 43–68, Blackwell Scientific Publications, Oxford.

Crittenden, P. D. (1989). Nitrogen relations of mat-forming lichens. In *Nitrogen, phosphorus and sulphur utilization by fungi*, (eds. L. Boddy, R. Marchant, and D. J. Read), pp. 243–68. Cambridge University Press, Cambridge.

Day, I. P. (1985). *Lichens in Borrowdale and pollution*. Report to the Nature Conservancy Council, Peterborough.

Dennison, R., Caldwell, B., Bormann, B., Eldred, L., Swanberg, C., and Anderson, S. (1977). The effects of acid rain on nitrogen fixation in western Washington coniferous forests. *Water, Air and Soil Pollution*, **8**, 21–34.

Dobben, H. F. van, de Wit, T., and Dam, D. van (1983). Effects of acid deposition on vegetation in The Netherlands. *VDI-Berichte*, **500**, 225–9.

Eversman, S. and Sigal, L. L. (1984). Ultrastructural effects of peroxyacetylnitrate (PAN) on two lichen species. *The Bryologist*, **87**, 112–18.

Eversman, S. and Sigal, L. L. (1987). Effects of SO_2, O_3 and SO_2 and O_3 in combination on photosynthesis and ultrastructure of two lichen species. *Canadian Journal of Botany*, **65**, 1806–18.

Fairfax, J. A. W. and Lepp, N. W. (1976). The effect of some atmospheric pollutants on the cation status of two woodland mosses. In *Proceedings of the Kuopio Meeting on plant damage caused by air pollution*, (ed. L. Karenlampi), pp. 26–36. University of Kuopio, Kuopio.

Falkengren-Grerup, U. (1986). Soil acidification and vegetation changes in deciduous forest in southern Sweden. *Oecologia*, **70**, 339–47.

Falkengren-Grerup, U. (1989). Effect of stemflow on beech forest soils and vegetation in southern Sweden. *Journal of Applied Ecology*, **26**, 341–52.

Farmer, A. M. (1990). The effects of lake acidification on aquatic macrophytes—a review. *Environmental Pollution*, **65**, 219–40.

Farmer, A. M., Bates, J. W., and Bell, J. N. B. (1991*a*). Comparisons of three woodland sites in NW Britain differing in richness of the epiphytic *Lobarion pulmonariae* community and levels of wet acidic deposition. *Holarctic Ecology*, **94**, 85–91.

Farmer, A. M., Bates, J. W., and Bell, J. N. B. (1991*b*). Seasonal variations in acidic pollutant inputs and their effects on the chemistry of stemflow, bark and epiphyte tissues in three oak woodlands in NW Britain. *New Phytologist*, **118**, 441–51.

Farmer, A. M., Bates, J. W., and Bell, J. N. B. (1992). The transplantation of four species of *Lobaria* lichens to demonstrate a field acid rain effect. In *Acidification research: evaluation and policy applications*, (ed. T. Schneider). Elsevier, Amsterdam (in press).

Fields, R. F. (1988). Physiological responses of lichens to air pollutant fumigations. In *Lichens, bryophytes and air quality*, (eds T. H. Nash and V. Wirth), pp. 175–200. Cramer, Berlin.

Flatberg, K. I., Foss, B., Loken, A., and Saasted, S. M. (1990). *Moss damage in coniferous forests*. Directorate for Nature Management (DN), Trondheim.

Fritz-Sheridan, R. P. (1985). Impact of simulated acid rains on nitrogenase activity in *Peltigera apthosa* and *P. polydactyla*. *The Lichenologist*, **17**, 27–31.

Gardiner, J. C. (1981). A bryophyte flora of Surrey. *Journal of Bryology*, **15**, 377–491.

Gauslaa, Y. (1985). The ecology of *Lobarion pulmonariae* and *Parmelion caperatae* in *Quercus* dominated forests in south-west Norway. *The Lichenologist*, **17**, 117–40.

Gilbert, O. L. (1986). Field evidence for an acid rain effect on lichens. *Environmental Pollution (Series A)*, **40**, 227–231.

Gunther, A. J. (1988). Effect of simulated acid rain on nitrogenase activity in the lichen genus *Peltigera* under field and laboratory conditions. *Water, Air and Soil Pollution*, **38**, 379–85.

Hallingbäck, T. (1986). Lunglarvarna, *Lobaria*, på reträtt i Sverige. *Svensk Botanisk Tidskift*, **80**, 373–81.

Hallingbäck, T. (1989). Bokfjädermossa, *Neckera pumila*, en försurningshotad mossa. *Svensk Botanisk Tidskrift*, **83**, 161–73.

Hallingbäck, T. (1990). Transplanting *Lobaria pulmonaria* to new localities and a review on the transplanting of lichens. *Windahlia*, **18**, 57–64.

Hallingbäck, T. and Kellner, O. (1992). Effects of simulated nitrogen rich and acid rain on the nitrogen-fixing lichen *Peltigera aphthosa* (L.) Willd. *New Phytologist*, **120**, 99–103.

Hallingbäck, T. and Martinsson, P.-O. (1987). The retreat of two lichens, *Lobaria pulmonaria* and *L. scrobiculata* in the district of Gäsene (SW Sweden). *Windahlia*, **17**, 27–32.

Hallingbäck, T. and Olsen, K. (1987). Lunglavens Tillbakagång i Skåne. *Svensk Botanisk Tidskrift*, **81**, 103–8.

Hallingbäck, T. and Thor, G. (1988). Jättelav, *Lobaria amplissima*, i Sverige. *Svensk Botanisk Tidskrift*, **82**, 125–39.

Hawksworth, D. L. (1990). The long-term effects of air pollutants on lichen communities in Europe and North America. In *The Earth in transition. Patterns and processes of biotic impoverishment*, (ed. G. M. Woodwell), pp. 45–64. Cambridge University Press, Cambridge.

Hawksworth, D. L. and Rose, F. (1970). Qualitative scale for estimating sulphur dioxide air pollution in England and Wales using epiphytic lichens. *Nature*, **227**, 145–8.

Hawksworth, D. L. and Rose, F. (1976). *Lichens as pollution monitors*, Edward Arnold, London.

Hutchinson, T. C., Dixon, M. and Scott, M. (1986). The effect of simulated acid rain on feather mosses and lichens of the boreal forest. *Water, Air and Soil Pollution*, **31**, 409–16.

Hutchinson, T. C. and Scott, M. (1988). The response of the feather moss *Pleurozium schreberi* (Brid.) Mitt. to five years of simulated acid precipitation in the Canadian boreal forest. *Canadian Journal of Botany*, **66**, 82–8.

James, P. W., Hawksworth, D. L., and Rose, F. (1977). Lichen communities in the British Isles: a preliminary conspectus. In *Lichen ecology*, (ed. M. R. D. Seaward), pp. 295–413. Academic Press, London.

Jones, E. W. (1953). A bryophyte flora of Berkshire and Oxfordshire. II: Musci.

Transactions of the British Bryological Society, **2**, 220–82.

Jones, E. W. (1955). A bryophyte flora of Berkshire and Oxfordshire. III: Addenda. *Transactions of the British Bryological Society*, **2**, 537–8.

Jones, E. W. (1991). The changing bryophyte flora of Oxfordshire. *Journal of Bryology*, **16**, 513–49.

Kinzel, H. (1983). Influence of limestone, silicates and soil pH on vegetation. In *Physiological plant ecology. III: Responses to the chemical and biological environment*, (eds. O. L. Lange, P. S. Nobel, C. B. Osmond, and H. Ziegler), pp. 201–44. Springer, Berlin.

Klein, R. M. and Bliss, M. (1984). Decline in surface coverage by mosses on Camels Hump Mountain, Vermont: possible relationship to acidic deposition. *The Bryologist*, **87**, 128–31.

Lane, I. and Puckett, K. J. (1979). Responses of the phosphatase activity of the lichen *Cladina rangiferina* to various environmental factors. *Canadian Journal of Botany*, **57**, 1534–40.

Laxen, D. P. H. and Thompson, M. A. (1987). Sulphur dioxide in Greater London, 1931–1985. *Environmental Pollution*, **43**, 103–14.

Lechowicz, M. J. (1982). The effect of simulated acid precipitation on photosynthesis in the caribou lichen, *Cladina stellaris* (Opiz.) Brodo. *Water, Air and Soil Pollution*, **18**, 421–30.

Lechowicz, M. J. (1987). Resistance of the caribou lichen *Cladina stellaris* (Opiz.) Brodo to growth reduction by simulated acidic rain. *Water, Air and Soil Pollution*, **34**, 71–7.

Looney, J. H. H. (1991). Effects of acidification on lichens. In *The effects of acid deposition on nature conservation in Great Britain*, (eds. S. J. Woodin and A. M. Farmer), pp. 45–55, Focus on Nature Conservation No. 26, Nature Conservancy Council, Peterborough, UK.

Looney, J. H. H. and James, P. W. (1990). *The effects of acidification on lichens*. CSD Report 1057, Nature Conservancy Council, Peterborough, UK.

Marti, J. (1983). Sensitivity of lichen phycobionts to dissolved air pollutants. *Canadian Journal of Botany*, **61**, 1647–53.

McCune, B. (1988). Lichen communities along O_3 and SO_2 gradients in Indianapolis. *The Bryologist*, **91**, 223–8.

Morgan, S. M., Lee, J. A., and Ashenden, T. W., (1992). Effects of nitrogen oxides on nitrate assimilation in bryophytes. **120**, 89–97.

Moser, T. J., Nash, T. H., and Clark, W. D. (1980). Effects of long-term field sulphur dioxide fumigation on arctic caribou forage lichens. *Canadian Journal of Botany*, **58**, 2235–40.

Nash, T. H., III (1976). Sensitivity of lichens to nitrogen dioxide fumigations. *The Bryologist*, **79**, 103–6.

Nash, T. H., III and Sigal, L. L. (1979). Gross photosynthetic response of lichens to short-term ozone fumigation. *The Bryologist*, **82**, 280–5.

Pitcairn, C. E. R., Fowler, D., and Grace, J. (1991). *Changes in species composition of semi-natural vegetation associated with the increase in atmospheric inputs of nitrogen*. CSD Report 1246, Nature Conservancy Council, Peterborough, UK.

Pitkin, P. H. (1975). Aspects of the ecology and distribution of some widespread corticolous bryophytes. Unpublished D.Phil thesis. University of Oxford.

PORG (1990). *Oxides of nitrogen in the United Kingdom*. The second report of the United Kingdom Photochemical Oxidants Review Group. Department of the Environment and Department of Transport Publications, South Ruislip.

Raeymaekers, G. (1987). Effects of simulated acidic rain and lead on the biomass, nutrient status, and heavy metal content of *Pleurozium schreberi* (Brid.) Mitt. *Journal of the Hattori Botanical Laboratory*, **63**, 219–30.

Raeymaekers, G. and Glime, J. M. (1986). Effects of simulated acidic rain and lead interaction on the phenology and chlorophyll content of *Pleurozium schreberi* (Brid.) Mitt. *Journal of the Hattori Botanical Laboratory*, **61**, 525–41.

Rafarel, C. R. and Ashenden, T. W. (1991). A facility for the large-scale exposure of plants to gaseous atmospheric pollutants. *New Phytologist*, **117**, 345–9.

Richardson, D. H. S. (1988). Understanding the pollution sensitivity of lichens. *Botanical Journal of the Linnean Society*, **96**, 31–43.

Richardson, D. H. S. and Nieboer, E. (1983). Ecophysiological responses of lichens to sulphur dioxide. *Journal of the Hattori Botanical Laboratory*, **54**, 331–51.

Richardson, D. H. S., Nieboer, E., Lavoie, P., and Padovan, D. (1976). The role of metal-ion binding in modifying the toxic effects of sulphur dioxide on the lichen *Umbilicaria muhlenbergii*. II: ^{14}C-fixation studies. *New Phytologist*, **82**, 633–43.

Rieley, J. O., Richards, P. W., and Bebbington, A. D. L. (1979). The ecological role of bryophytes in a north Wales woodland. *Journal of Ecology*, **67**, 497–527.

Robitaille, G., LeBlanc, F., and Rao, D. N. (1977). Acid rain: a factor contributing to the paucity of epiphytic cryptogams in the vicinity of a copper smelter. *Revue Bryologique et Lichénologique*, **43**, 53–66.

Rochefort, L. and Vitt, D. H. (1988). Effects of simulated acid rain on *Tomenthypnum nitens* and *Scorpidium scorpioides* in a rich fen. *The Bryologist*, **9**, 121–9.

Ronen, R. (1986). The effect of air pollution on physiological parameters of the lichen *Ramalina duriaei*. Unpublished Ph.D thesis, University of Tel Aviv.

Rorison, I. H. and Robinson, D. (1984). Calcium as an environmental variable. *Plant, Cell and Environment*, **7**, 381–90.

Rose, F. (1974). The epiphytes of oak. In *The British oak*, (eds. M. G. Morris and F. H. Perring), pp. 250–73. Classey, Faringdon, UK.

Rose, F. (1988). Phytogeographical and ecological aspects of *Lobarion* communities in Europe. *Botanical Journal of the Linnean Society*, **96**, 69–79.

Rosentreter, R. and Ahmadjian, V. (1977). Effect of ozone on the lichen *Cladonia arbuscula* and the *Trebouxia* phycobiont of *Cladina stellaris*. *The Bryologist*, **80**, 600–5.

Ross, L. J. and Nash, T. H., III (1983). Effect of ozone on gross photosynthesis of lichens. *Environmental and Experimental Botany*, **23**, 71–7.

Schmid, M. L. and Kreeb, K. (1975). Enzymatische Indikation gasgeschädigter Flechten. *Angewandte Botanik*, **49**, 141–54.

Scott, M. G. and Hutchinson, T. C. (1987). Effects of a simulated acid rain episode on photosynthesis and recovery in the caribou-forage lichens, *Cladina stellaris* (Opiz.) Brodo and *Cladina rangiferina* (L.) Wigg. *New Phytologist*, **107**, 567–75.

Scott, M. G., Hutchinson, T. C., and Feth, M. J. (1989). A comparison of the effects on Canadian boreal forest lichens of nitric and sulphuric acids as sources of rain acidity. *New Phytologist*, **111**, 663–71.

Seaward, M. R. D. (1989). Lichens as monitors of recent changes in air pollution. *Plants Today*, March/April 1989, 64–9.

Seaward, M. R. D. and Hitch, C. B. (1982). *Atlas of the lichens of the British Isles*, Vol. 1. Natural Environment Research Council, Cambridge.

Sheridan, R. P. and Rosentreter, R. (1973). The effect of hydrogen ion concentrations in simulated rain on the moss *Tortula ruralis* (Hedw.) Sm. *The Bryologist*, **76**, 168–73.

Sigal, L. L. and Johnston, J. W. (1986). Effects of acidic rain and ozone on nitrogen fixation and photosynthesis in the lichen *Lobaria pulmonaria* (L.) Hoffm. *Environmental and Experimental Botany*, **26**, 59–64.

Sigal, L. L. and Nash, T. H., III (1983). Lichen communities on conifers in southern California mountains: an ecological survey relative to oxidant air pollution. *Ecology*, **64**, 1343–54.

Sigal, L. L. and Taylor, O. C. (1979). Preliminary studies of the gross photosynthetic response of lichens to peroxyacetylnitrate fumigations. *The Bryologist*, **82**, 564–75.

Sjors, H. (1950). On the relation between vegetation and electrolytes in north Swedish mire waters. *Oikos*, **2**, 241–58.

Søchting, U. (1987). Injured reindeer lichens in Danish lichen heaths. *Graphis Scripta*, **1**, 103–6.

Søchting, U. (1990). Reindeer lichens injured in Denmark. *Bulletin of the British Lichen Society*, **67**, 1–4.

Sparling, J. H. (1967). The occurrence of *Schoenus nigricans* L. in blanket bogs. II: Experiments on the growth of *S. nigricans* under controlled conditions. *Journal of Ecology*, **55**, 15–31.

Stanosz, G. R., Smith, V. L., and Bruck, R. I. (1987). Effect of ozone on colonization of disturbed forest soil by moss. *Phytopathology*, **77**, 1727–8.

Storm, C. (1990). Relationships between forest decline and forest undergrowth—a case study of the southeastern Black Forest. *Angewandte Botanik*, **64**, 51–68.

Tamm, C. O. (1953). Growth, yield and nutrition in carpets of a forest moss (*Hylocomium splendens*). *Meddelanden fran Statens Skogsforskningsinstitut*, **43**, 1–140.

Thompson, D. B. A. and Baddeley, J. (1991). Some effects of acidic deposition on montane *Racomitrium lanuginosum* heaths. In *The effects of acid deposition on nature conservation in Great Britain*, (eds S. J. Woodin and A. M. Farmer), pp. 17–28. Focus on Nature Conservation No. 26. Nature Conservancy Council, Peterborough, UK.

Turk, R. and Wirth, V. (1975). Uber der SO_2-Empfinol-lichkeit einiger Moose. *The Bryologist*, **78**, 187–93.

UKRGAR (1990). *Acid deposition in the United Kingdom 1986–1988*. Third report of the UK Review Group on Acid Rain, Department of the Environment, South Ruislip.

Wellburn, A. R. (1990). Why are atmospheric oxides of nitrogen usually phytotoxic and not alternative fertilizers? *New Phytologist*, **115**, 395–429.

Williams, M. L., Atkins, D. H. F., Bower, J. S., Campbell, W., Irwin, J. G., and Simpson, D. (1989). *A preliminary assessment of the air pollution climate of the UK*. Warren Spring Laboratory Report LR723(AP), Warren Spring Laboratory, Stevenage, UK.

Winner, W. E. (1988). Responses of bryophytes to air pollution. In *Lichens, bryophytes and air quality*, (eds T. H. Nash and V. Wirth), pp. 141–73. Cramer, Berlin.

Winner, W. E. and Bewley, J. D. (1983). Photosynthesis and respiration of feather mosses fumigated at different hydration levels with SO_2. *Canadian Journal of Botany*, **61**, 1456–61.

Wodzinski, R. S., Labeda, D. P., and Alexander, M. (1977). Toxicity of SO_2 and NO_x: selective inhibition of blue–green algae by bisulphite and nitrite. *Journal of the Air Pollution Control Association*, **27**, 891–3.

Wolseley, P. A. and James, P. W. (1991). *The effects of acidification on lichens 1986–90*. CSD Report 1247, Nature Conservancy Council, Peterborough, UK.

Wolseley, P. A. and O'Dare, A. M. (1990). The use of epiphytic lichens as environmental indicators in Exmoor Woodlands. *Ecology in Somerset*, **1**, 3–22.

Woodin, S. J. (1989). Environmental effects of air pollution in Britain. *Journal of Applied Ecology*, **26**, 749–61.

Woollon, F. B. (1975). Mineral relationships and ecological distribution of *Fissidens cristatus* Wils. *Journal of Bryology*, **8**, 455–64.

12

Responses of *Sphagnum* species to polluted environments

JOHN A. LEE AND COLIN J. STUDHOLME

12.1 Introduction

Peatlands cover approximately 3 per cent of the Earth's land surface (Kivinen and Pakarinen 1981) and play a significant role in the global cycles and geochemical balances of carbon, nitrogen, and sulphur (Gorham *et al.* 1984). In many peatlands the major peat-forming plants are *Sphagnum* species, and thus the response of these species to pollutants may influence markedly biogeochemical cycles.

There is every reason to suspect that bryophytes of ombrotrophic communities will be particularly sensitive to atmospheric pollution. First, the photosynthetic cells of most bryophyte species are directly and continuously exposed to their atmospheric environment. Second, the leaves of many species are one cell thick, effectively facilitating the exposure of the photosynthetic cells to pollutants. Third, ombrotrophic ecosystems depend on the atmospheric source of elements. In ombrotrophic peatlands, the surface mire vegetation is particularly dependent on the atmospheric supply of elements because element cycling is very limited by the combination of slow decomposition (Clymo and Haywood 1982) with continuous growth of the bog surface (Damman 1986). Fourth, the atmospheric supply of elements may be small and readily perturbed. Fifth, many ombrotrophic mire bryophytes, notably *Sphagnum* species, form a continuous deposition surface at the interface of the atmosphere with the land. Thus bryophytes are the primary receptors of atmospheric deposition. Sixth, bryophytes have been shown to accumulate atmospheric pollutants (e.g. Goodman and Roberts 1971) and this process may potentially result in the perturbation of their mineral nutrition and metabolism.

Evidence that atmospheric pollution has caused widespread change in *Sphagnum*-dominated ombrotrophic mire communities is however rather restricted, possibly at least in part because these communities have been largely ignored by pollution scientists. The most complete record of change resulting from atmospheric pollution comes from the southern Pennine uplands of England which contain approximately 300 km² of blanket peat. These blanket mires have been subjected to considerable atmospheric pollution over at least the last 200 years (Press *et al.* 1983; Ferguson and Lee 1983). The dominance of *Sphagnum* species in the southern Pennines prior to the Industrial Revolution can be inferred from historical descriptions (e.g. Farey 1813) and from macrofossil remains in the peat (Conway 1947, 1949; Tallis 1964). Historical accounts and macrofossil analyses can also be used to document the decline of *Sphagnum* in the region (e.g. Moss 1913; Tallis 1964). At the present day, the minerotrophic species *S. recurvum* P. Beauv. is present in flushes and pools throughout the southern Pennines, but ombrotrophic species have been eliminated from large areas of peatland. Tallis (1964) was amongst the first to suggest that this decline was due to atmospheric pollution since the loss of *Sphagnum* species from the peat profile coincided with the marked accumulation of soot originating from nearby industrial towns.

Other evidence that atmospheric pollution causes the decline of *Sphagnum* species comes from the Sudbury region of Canada. Metal smelting at Sudbury has resulted in the release of metal-containing particulates and large quantities of SO_2 into the environment. Gignac and Beckett (1986) showed that the smelting had resulted in the loss of all *Sphagnum* species from peatlands closest to the smelter. Peatlands further away showed a marked reduction in species number, with *S. fallax* (Klinggr.) Klinggr. being amongst the most abundant and widespread species. This is closely analogous to the changes observed in the southern Pennines, but from field observations alone it is not possible to ascribe the changes to a particular pollutant or pollutant mixture.

Sphagnum recurvum (*fallax*) has apparently become more important in some habitats during the last century. Thus Tallis (1973) showed that carpets of this species had largely replaced a more species-rich bog surface in recent times in lowland schwingmoor vegetation in Cheshire, England, and there are similar reports from lowland mires in polluted parts of central Europe (e.g. in Poland). The apparent success of this minerotrophic species may result from either a greater tolerance of pollutants than other *Sphagnum* species, or from increased growth in response to a greater solute supply as the result of atmospheric pollution, or from changes in the catchments enriching the groundwaters, or from any combination of these factors. A further possibility is that this change in the relative abundance of species is caused by some subtle response to climatic change.

Sphagnum imbricatum Hornsch. ex Russ. shows a contrasting history in

Britain to that of *S. recurvum*. This species has declined much more markedly than other ombrotrophic species in recent centuries (Green 1968). This species was a major peat-former in the southern Pennines, but is now absent from the region and rare in northern England and Wales. This may indicate extreme sensitivity to atmospheric pollutants or a response to climatic and/or land-use changes resulting in a widespread drying of the mire surfaces.

Documented changes in species distribution and abundance provide a basis from which experimentation and observation can be used to elucidate the likely causes of past vegetation changes. Attempts to elucidate the role of atmospheric pollutants in the changed and changing status of *Sphagnum* species are described below.

12.2 The nature of polluted environments

Gaseous pollutants and particles may directly reach the vegetation surface as dry deposition, or may be deposited in or on hydrometeors as wet deposition. Dry deposition greatly predominates close to pollution sources, but wet deposition predominates at sites remote from sources where the concentrations of gaseous pollutants are low. At these latter sites, particularly where overall pollutant deposition is very small, it can be shown that ombrotrophic *Sphagnum* species are closely coupled to the atmospheric supply of solutes. Figure 12.1 shows that nitrate deposition rapidly induces the activity of nitrate reductase, the initial and rate-limiting enzyme of nitrate assimilation. Activity of the enzyme declines following each deposition event, presumably following the utilization of the available substrate nitrate. Thus wet deposition is an important source of solutes for ombrotrophic bryophytes.

A further important deposition process is the impaction of wind-driven cloud and fog particles. Occult deposition, as this is known, is of particular importance in mountainous districts where vegetation may be in cloud for an average of several hours each day. Cloud and fog droplets commonly contain major ions at concentrations of 50–200 μEq. l^{-1}, up to an order of magnitude greater than their concentration in rain, thus they can provide directly a relatively concentrated supply of pollutants. Cap cloud on summits also provides the explanation of increased concentrations of solutes in precipitation with altitude. Precipitation emanating from frontal clouds efficiently collects fog droplets (3–15 μm in radius) as it falls through the cap cloud, and thus increases in solute concentration, whereas the sub-micron aerosol particles encountered by precipitation falling at lower altitudes are not efficiently collected.

The oxides of sulphur (SO_x) and nitrogen (NO_x) are amongst the most

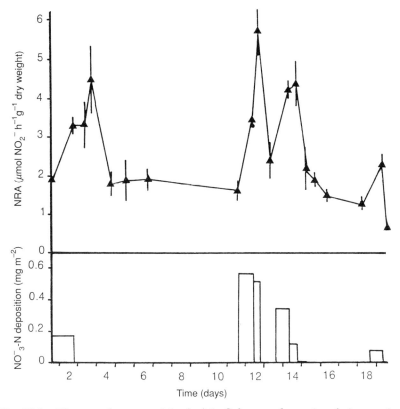

Fig. 12.1 Nitrate reductase activity (*top*) in *Sphagnum fuscum* in relation to nitrate deposition from the atmosphere (*bottom*) at an unpolluted mire at Abisko, north Sweden in July 1985. Vertical bars represent ±SEM, *n* = 4.

important gaseous pollutants, and are the most important precursors of precipitation acidity resulting in the formation of sulphuric and nitric acids. A variety of other emissions are also potentially toxic and influence the composition of precipitation, e.g. ammonia, hydrogen chloride, and hydrogen fluoride. The secondary gaseous pollutant, ozone, derived from the action of sunlight on mixtures of NO_x and volatile hydrocarbons, is potentially highly phytotoxic. In lowland regions the concentration of ozone shows marked diurnal fluctuation, but at high altitudes high concentrations of ozone persist at night, potentially increasing the exposure of vegetation to the gas. Marked fluctuation in the concentrations of other gaseous pollutants occurs giving a wide range of pollution episodes. These fluctuations in polluted regions are conditioned for example by seasonal and diurnal variations in fossil fuel usage and by variation in climate (e.g. temperature inversions and wind speed).

The relative importance of the major atmospheric pollutants has changed over the years. For much of the last two centuries SO_2 has been the major gaseous atmospheric pollutant in Britain. But in recent decades the concentration of this gas has fallen markedly with changes in fuel usage and combustion processes. A very large fall has also been observed in particulate pollution. These falls have been to a degree offset by a marked increase in the emission of nitrogen oxides. Brimblecome and Stedman (1982) showed a doubling of the nitrate concentration in wet deposition at Rothamsted, England over the last century, and in the southern Pennine region nitrate deposition may have quadrupled since 1868 (Lee *et al.* 1987). Similar large increases for the concentration of nitrate in precipitation this century have been demonstrated in remote parts of the northern Hemisphere, e.g. Greenland (Delmas 1986). The increase in emission of nitrogen oxides has also resulted in the increasing importance of ozone as an atmospheric pollutant.

Thus, over the period that changes in the distribution and abundance of *Sphagnum* species have been observed, there have been marked changes in the pollution climate. Which of these changes has exerted the major effect on *Sphagnum* communities in the past, and what are the effects of the present pollution climate on them?

12.3 The effects of SO_2 on *Sphagnum* species

The marked decline in *Sphagnum* in the southern Pennines during the 19th and early 20th centuries was correlated with high concentrations of SO_2 in the region (Ferguson and Lee 1983). Thus SO_2 and its solution products must be prime candidates as potential causal agents. Figure 12.2 shows the effect of fumigation of *Sphagnum* species with 131 μg m^{-3} SO_2. The growth of all species was reduced relative to that in the clean air control (9 μm^{-3} SO_2), but some species were much more severely affected than others. One of the most tolerant species was *S. recurvum* and one of the least was *S. imbricatum*. Ferguson *et al.* (1978) in a series of experiments with SO_2 and its solution products showed that *S. recurvum* was the most tolerant of the tested *Sphagnum* species to the treatments, and this is consistent with its survival in polluted districts. Sulphate ions were much less toxic than SO_2 or bisulphite ions. These workers also showed (Ferguson and Lee 1980) that in a field experiment in an 'unpolluted' mire, it was possible using bisulphite ions to mimic SO_2 concentrations to kill the *Sphagnum* carpet with no observable effect on the associated vascular plants and little effect on some other bryophyte species (e.g. *Polytrichum commune*). Thus it was possible to produce experimentally a similar effect with an SO_2 solution product to that observed under 'natural' pollution climates in the southern

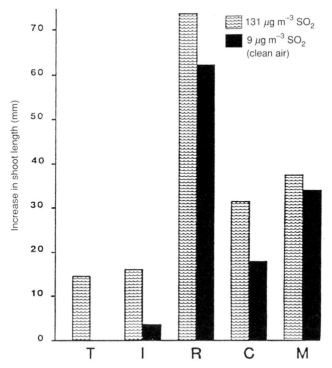

Fig. 12.2 Effects of SO_2 on the growth (increase in length) of *Sphagnum* species. T: *Sphagnum tenellum*; I: *S. imbricatum*; R: *S. recurvum*; C: *S. capillifolium*; M: *S. magellanicum* (data from Ferguson *et al.* 1978).

Pennines. There can be little doubt that phytotoxic concentrations of SO_2 were widespread in the southern Pennines for over a century (Ferguson and Lee 1983), and therefore that this gas probably played a major role in the demise of ombrotrophic *Sphagnum* species in the region.

However, a number of ombrotrophic *Sphagnum* species occur very locally at a few sites in the southern Pennines today. These could represent either relict populations of the formerly more or less continuous *Sphagnum* carpet resistant to atmospheric pollutants or recent re-invasions of the mire surfaces in response to the marked reduction in sulphur pollution in recent decades. Figure 12.3 shows the response of three southern Pennine populations of *Sphagnum cuspidatum* Hoffm. (Alport Moor, Holme Moss, and Ringinglow) to a SO_2 solution product (bisulphite ions) under laboratory conditions, and contrasts this with the response of populations from less polluted regions of northern Britain (Migneint, Berwyns, and Butterburn). These latter populations showed no growth in the highest bisulphite concentration (0.3mM) and quickly succumbed to the treatment, whereas

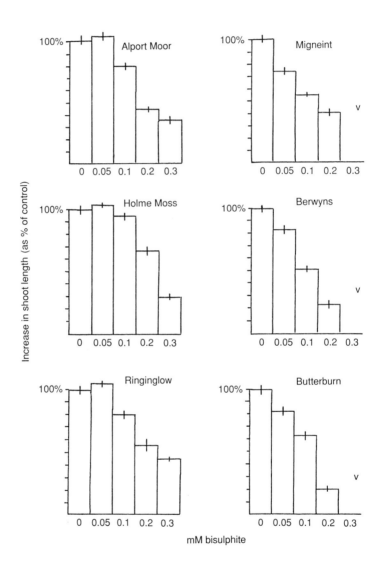

Fig. 12.3   The growth (increase in length as a percentage of control) of *Sphagnum cuspidatum* from six populations following 34 days in different concentrations of bisulphite ions. The polluted southern Pennine populations are Alport Moor, Holme Moss, and Ringinglow. Vertical bars represent ± SEM, *n* = 4. Plants in treatments indicated v made no growth (from Studholme 1989).

the southern Pennine plants survived to the end of the experiment (34 days) and showed some growth (25–45 per cent of the control treatment). (Full details of the experiment can be found in Studholme (1989).) These data suggest that the southern Pennine populations may be relict in that they are more tolerant of the previously prevailing high SO_2 concentrations. Studholme (1989), in a study of the genetic variation based on isozymes of peroxidase, acid phosphatase, and superoxide dismutase, showed that plants from southern Pennine populations were less genetically variable (polymorphism index of 0.023) than plants from populations more remote from sources of population (polymorphism index of 0.055). However, there was no evidence to suggest that a selection process had occurred to reduce the numbers of genotypes present and a genetic basis for the difference in bisulphite tolerance was not found (Studholme 1989).

This study initiated a detailed consideration of SO_2 tolerance mechanisms in *Sphagnum* species. Baxter *et al.* (1989) were able to show that SO_2 tolerance in southern Pennine populations of *S. cuspidatum* was the result of an avoidance mechanism linked to the pollution history of the mires. Livett *et al.* (1979) showed that the surface of southern Pennine blanket peat had high concentrations of metals as the result of a long history of particulate air pollution, and Ferguson *et al.* (1984) showed that these high concentrations were correlated with high concentrations of metals in living *Sphagnum* species. Baxter *et al.* (1989) demonstrated large differences (> four-fold) in the iron concentrations of apical lengths of *S. cuspidatum* in populations shown by Studholme (1989) to differ in tolerance to bisulphite ions. These workers also demonstrated that under laboratory conditions bisulphite ions disappeared much more rapidly from solution surrounding plants from southern Pennine populations of *S. cuspidatum* (see Fig. 12.4(*b*)) than from solutions surrounding plants from less polluted localities (Fig. 12.4(*a*)). This phenomenon was unaltered by killing the moss by boiling, and thus was caused by a physical feature of the plants. Experimental modification of the metal concentration on the cation exchange sites of the moss cell walls had a major effect on the disappearance of bisulphite ions from solution. Figure 12.5 shows the effect of washing southern Pennine plants of *S. cuspidatum* in distilled water or 5 mM EDTA, prior to exposure to bisulphite ions, on the subsequent rate of bisulphite ion disappearance. Washing in EDTA reduced considerably the rate of bisulphite disappearance from solution, and Baxter *et al.* (1989) also showed that this was associated with a marked increase in sensitivity of the moss to bisulphite ions. (The EDTA treatment on its own had no effect on the subsequent growth and physiology of the moss in an artificial rainwater solution.) When *S. cuspidatum* from an unpolluted site was grown in an artificial rainwater solution amended with iron (as $FeCl_3$ 1 μg cm^{-3}) for 14

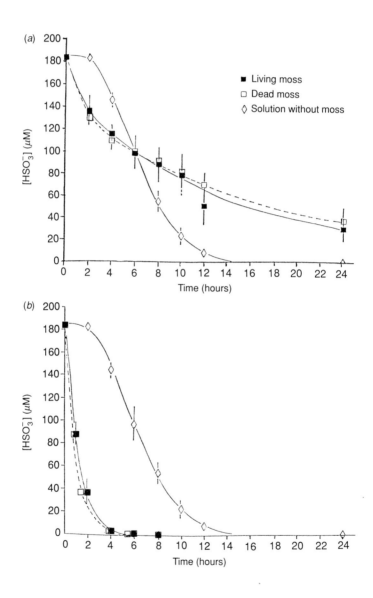

Fig. 12.4 Effects of *Sphagnum cuspidatum* (*a*) from an unpolluted mire, (*b*) from the polluted southern Pennines, on the disappearance of bisulphite ions in artificial rainwater. Vertical bars are ± SEM, *n* = 3 (from Baxter 1989).

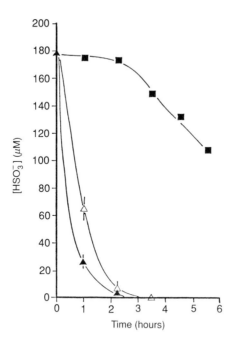

Fig. 12.5 Effects of *Sphagnum cuspidatum* from the polluted southern Pennines on the oxidation of bisulphite ions in artificial rainwater. Plants were washed in distilled water ($\triangle$) or 5 mM EDTA ($\blacksquare$) for 30 minutes prior to treatment. $\blacktriangle$ represents no prior treatment. Vertical bars are $\pm$ SEM, $n = 3$ (from Baxter 1989).

days prior to exposure to bisulphite ions, a greatly improved tolerance of the plants was observed together with rapid disappearance of bisulphite from the bathing solution. Baxter (1989) was able to show that exchangeable iron concentration was largely responsible for the difference in bisulphite tolerance in populations of *S. cuspidatum* and that the tolerance mechanism was essentially one of avoidance through Fe III promoting the oxidation of sulphite to the less toxic sulphate.

Baxter *et al.* (1991) also demonstrated that the bisulphite tolerance of *S. recurvum* could be markedly influenced by the manipulation of the cell wall iron content. This minerotrophic species usually has a higher iron content than that of ombrotrophic *Sphagnum* species, and a higher tolerance of SO_2 (see above). However, removal of cell wall exchangeable iron by EDTA increased the susceptibility of this species to bisulphite ions. Thus it may be concluded that much, if not all, of the apparent difference in tolerance to SO_2 and its toxic solution product bisulphite within *Sphagnum* is habitat-induced and is not an intrinsic property of the moss species or population. In this regard, it is interesting to note that Ferguson and Lee

(1979) were unable to detect any differences between species in terms of the initial response (< 5 min) of photosynthesis to bisulphite ions. These experiments were carried out by the injection of bisulphite into aqueous medium in a well-stirred oxygen electrode chamber. Under these conditions, the exposure of the chlorophyllose cells to bisulphite ions may not have been influenced appreciably by the 'normal' cell wall phenomenon, and thus species differences were not apparent. Thus the survival of *S. recurvum* and other minerotrophic species in regions grossly polluted with SO_2 is largely a function of habitat, and a similar induced SO_2 avoidance mechanism may prove widespread in other apparently tolerant bryophyte species.

12.4 Responses of *Sphagnum* species to nitrogen deposition

Woodin *et al.* (1985) demonstrated the close coupling between atmospheric nitrate deposition and nitrate assimilation in the ombrotrophic species, *S. fuscum*, (Schimp.) Klinggr. (see also Fig. 12.1) and this relationship is probably typical for many ombrotrophic Sphagna in environments remote from major sources of pollution. Woodin and Lee (1987) showed that in a subarctic mire this species was also able effectively to immobilize the atmospheric combined nitrogen supply, and this reflects the importance of nitrogen as a mineral nutrient and its scarcity of supply in 'unpolluted' environments. The conclusion from these studies must be that the scarce atmospheric nitrogen supply is rapidly absorbed and utilized by the moss. If nitrogen supply is limiting in these environments, then an increase in atmospheric nitrogen supply will stimulate the growth of *Sphagnum* species and there will be little or no change in the total tissue nitrogen content of the plants (Malmer 1990). Thus a modest increase in atmospheric nitrogen supply need not be detrimental to ombrotrophic Sphagna.

In environments with high atmospheric nitrogen pollution, such as the southern Pennines of England, growth of *Sphagnum* species is usually slower than in less polluted regions and appreciable nitrogen accumulation occurs. Figure 12.6 shows the accumulation of total tissue nitrogen in *S. cuspidatum* transplanted from an 'unpolluted' site and floated in an artificial bog pool in the southern Pennines receiving only the atmospheric supply of elements. This rapid increase in tissue nitrogen is also associated with the progressive decline in the ability of the moss to assimilate further nitrogen as demonstrated by a loss of inducibility of nitrate reductase (Lee *et al.* 1987) and by a reduced ability to take up NO_3^- and NH_4^+ from the bathing solution (Press *et al.* 1986). The exact mechanism by which nitrate reductase inducibility is lost in these *S. cuspidatum* transplants is unknown, but Baxter (1989) showed that moss from 'unpolluted' sites underwent

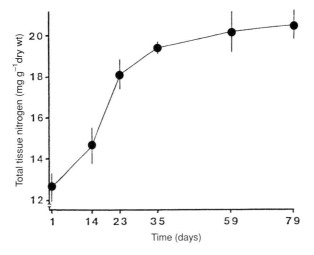

Fig. 12.6   Increase in tissue nitrogen content of *Sphagnum cuspidatum* plants following transplantation from an unpolluted mire to the polluted southern Pennines. Plants received only the atmospheric nitrogen supply. Vertical bars are ± SEM, *n* = 4 (from Press *et al.* 1988).

large initial increases in tissue amino acid pools (notably asparagine, glutamine, and arginine) when grown under laboratory conditions with elevated nitrogen supply (0.1 mM NH$_4^+$), and therefore some form of end production inhibition of the enzyme activity may occur. The net result of this effect on uptake and assimilation processes is that the moss is unable totally to sequester the atmospheric nitrogen supply (Woodin and Lee 1987) and a proportion of this becomes available to the associated angiosperm species and probably also to the microbial flora in the acrotelm. Even if there is no adverse effect of the high nitrogen deposition on the growth of *Sphagnum* species (but see Press *et al.* 1986), it is plausible that the excess nitrogen supply may alter the competitive balance between the angiosperm and bryophyte components of mire vegetation to the detriment of the latter. There are, however, no field data currently available to support this view.

A further possible effect of increased tissue nitrogen content is an acceleration of the rate of decay (e.g. Clymo and Hayward 1982). This could also potentially affect the competitive balance between angiosperms and *Sphagnum* species in favour of the former by effectively decreasing the thickness of the bryophyte mat and its element and water store, thus increasing the rate of element cycling. There is no direct evidence to support this, but *Sphagnum* species in areas of high nitrogen deposition often have short stems, and Baddeley (1991) has shown that increasing the nitrogen supply to *Racomitrium lanuginosum* mats causes an increase in the respiration rate of decaying stem sections.

Many ombrotrophic mire systems occur in regions of high precipitation remote from major pollution sources where wet and occult deposition represent the major pollutant inputs. Even in the grossly polluted southern Pennines the measured concentrations of NO_x and NH_x are rather small. A series of diffusion tube studies on southern Pennines mires (Press *et al.* 1986; D. E. Conlan, unpublished data) showed that mean annual concentrations of NO_2 did not exceed 8 p.p.b. and NH_3 concentrations were always below the limits of detection ($<$ 1 p.p.b.). Other studies from the southern Pennines suggest mean annual NH_3 concentration of c. 3 p.p.b. (UKRG 1990). Although fumigation studies with *Sphagnum* species are generally lacking, Morgan *et al.* (1992) showed no marked toxicity of a number of bryophyte species to fumigation with 40 p.p.b. NO_x (35 p.p.b. NO_2 and 5 p.p.b. NO). However, there were effects on nitrate assimilation, an initial stimulation of nitrate reductase activity followed by a loss of inducibility of the enzyme. A lower concentration (*c.* 25 p.p.b. NO_x) had little consistent effect on the mosses, and although appropriate experimentation on *Sphagnum* is lacking it is doubtful whether there is a significant toxic effect of NO_x on ombrotrophic mire bryophytes in the British uplands. NH_x is potentially very toxic to bryophytes, but the threat is perhaps greatest in the lowlands, where large sources as the result of animal husbandry may abut on to mire systems. Major changes in the floristic composition of Dutch heathlands in recent decades have been ascribed to this cause (Heil and Diemont 1983), and despite the generally low mean annual NH_3 concentrations observed in upland Britain (mostly $<$ 5 p.p.b.) it has been stated that dry deposition of the gas dominates total nitrogen deposition in the United Kingdom (UKRG 1990). Thus the ecological importance of NH_x deposition may be greater than has been realized hitherto.

12.5 Other responses to polluted environments

A major problem in understanding the response of *Sphagnum* species and other bryophytes to gaseous air pollutants is the difficulty of performing fumigation experiments. In these experiments this is primarily caused by the poikilohydric nature of the bryophytes. Attempts to maintain the water status by spray systems or slow air flow rates may be inimical to the maintenance of gas concentrations within the fumigation chamber. These problems, compounded by the need to maintain adequate mineral nutrition over perhaps several months in order to detect appreciable growth responses, have limited severely the number of fumigation experiments attempted on *Sphagnum* species. Thus we have a very incomplete knowledge of responses of these species to NO_x, NH_x (see above), and to ozone. Similarly, we have little knowledge of responses of *Sphagnum* to pollutant mixtures. The lack of knowledge of the responses of *Sphagnum* species to ozone is particularly unfortunate given the increasing importance of this gas as a pollutant in at least some parts of the Northern Hemisphere, and evidence that diurnal fluctuation in concentration of ozone may be less marked in the uplands where many extensive mire systems exist (e.g. Grace and Unsworth 1988).

Attack by a number of gaseous pollutants, including ozone, may result in the formation of hydrogen peroxide, superoxide, and hydroxyl radicals within plant tissues. A number of antioxidant and free radical scavenging mechanisms have been proposed to operate in plants and these include superoxide dismutase and glutathione peroxidase (see e.g. Wellburn (1988) for a more detailed description). For example, glutathione peroxidase may remove hydrogen peroxide and other organic peroxides by catalysing the conversion of two molecules of glutathione (GSH) to one oxidized molecule of glutathione (GSSG).

Many higher plant studies have documented that atmospheric pollutants (including O_3, SO_2 and NO_2) give rise to increase in peroxidase activity (e.g. Patton and Garraway 1986; Horsman and Wellburn 1975), which have often been attributed to the enzyme's detoxifying capabilities. A similar response for superoxide dismutase has also been reported (e.g. Tanaka and Sugahara 1980). However, Studholme (1989) in a study of *S. cuspidatum* plants from populations in grossly polluted and largely unpolluted regions, showed much lower peroxidase in plants from the former populations. Fig. 12.7 shows this response for two populations, one from the polluted southern Pennines and one from 'unpolluted' North Wales. Peroxidase activity was lower in plants from the former population throughout the year and both populations showed significant seasonal variation in activity. Studholme, in a detailed study of the response of peroxidase isozymes to a wide range of pollutants under controlled environment conditions, was unable to reproduce the field response, although some treatments, such as elevated bisulphite and heavy metal ions, resulted in some overall reduction in activity. (It is perhaps of some significance that the major pollutant missing from these experiments was ozone.) Superoxide dismutase also showed a lower activity in *S. cuspidatum* plants from polluted environments, although the effect was much less marked than in the case of peroxidase. These observations suggest that these enzymes are not primarily involved in protecting *S. cuspidatum* against toxic products resulting from the pollutants encountered in this study. Interestingly, Studholme (1989) reports a study in which a number of moss species (including *S. magellanicum*) and vascular plants (including *Eriopho-*

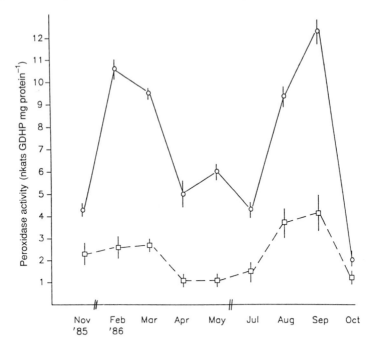

Fig. 12.7 Seasonal variation in peroxidase activity in *Sphagnum cuspidatum* plants in the polluted southern Pennines (□) or in an 'unpolluted' ombrotrophic mire in North Wales (○). Vertical bars are ± SEM, *n* = 3 (from Studholme 1989).

rum vaginatum) were transplanted from North Wales and exposed to ambient atmospheric pollution in Manchester under winter conditions. The bryophytes showed a significant reduction in peroxidase activity within seven days, whereas the vascular plants showed an increase. Thus it is possible that a reduction in peroxidase activity under polluted conditions is a general feature of bryophytes as distinct from the typical response of vascular plants.

Another feature which *Sphagnum* species share with other bryophytes is the ability to accumulate metal ions from the environment to concentrations, in the case of micronutrients, far in excess of demand for growth. Much of the accumulation has been shown to be a physical process associated with cation exchange sites on the cell walls, and to occur whether the moss is alive or dead. This property of *Sphagnum* has been employed to characterize atmospheric metal loading in the environment using the so-called moss bag technique. Acid-washed dead *Sphagnum* material is contained within a small hair net and exposed to the atmosphere for appropriate periods of time (e.g. Little and Martin 1974). The *Sphagnum* plants are then digested in acid and the total metal content subsequently determined. This provides a cheap and accurate integration of atmospheric metal deposition. This property of mosses is undoubtedly an important feature of element acquisition and retention as part of normal growth and metabolism. In polluted environments the accumulation of non-essential elements such as lead may not only interfere with metabolism in the cells directly, but may also appreciably influence the acquisition and retention of essential elements through competition for exchange sites. The importance of this potential cause of growth reduction in *Sphagnum* and other bryophyte species is little understood.

12.6 Conclusion

Despite the importance of *Sphagnum* as a major peat-forming plant, dominant in many mire ecosystems, relatively little attention has been paid to its response to pollutants. This is all the more surprising given the fact that the macrofossil record in peat profiles allows to some extent the reconstruction of vegetation responses to pollution history. Most investigations have focused on descriptive accounts of mire vegetation in response to pollution sources combined with measurements of element composition.

A number of factors have combined to produce this state of affairs. First, *Sphagnum* does not provide easy experimental material when compared with many vascular plants. Second, with the exception of aquatic species, it is difficult to maintain an appropriate hydrological regime. Third, reproducible fumigation experimentation is both difficult and expensive. Fourth, most interest amongst experimenters has been in understanding crop response to pollutants. Fifth, physiological techniques produced for the study of crops cannot always be readily applied to studies of mosses.

From those studies which have been undertaken, there is little doubt that *Sphagnum* species are amongst the most sensitive to atmospheric pollution, and that SO_2 and its solution products have had a profound effect on at least some mire ecosystems, with the effective elimination of ombrotrophic species over large areas.

Experimentation on apparently 'simple' plants does have its advantages however. The demonstration of the importance of extracellular oxidation of sulphite to sulphate catalysed by Fe III as a SO_2 tolerance mechanism would not have been so easy in higher plants. Also, a demonstration of the coupling of nitrogen assimilation with the atmospheric combined nitrogen supply, if present in higher plants, would also be difficult to disassociate from soil factors.

The changing nature of pollution climates poses a number of important questions as to how *Sphagnum*-dominated ecosystems will respond. To what extent is the increased atmospheric nitrogen deposition of recent decades affecting mires in regions remote from pollution sources? Does ozone alone or in combination with other gaseous pollutants have a marked effect on *Sphagnum* species at ambient concentrations? To what extent does the growth of *Sphagnum* species respond to elevated CO_2 concentrations, and is this response at all influenced by the atmospheric combined nitrogen supply? The answers to these and other important related questions must await further detailed experimental studies.

References

Baddeley, J. A. (1991). The physiological ecology of *Racomitrium lanuginosum* with respect to atmospheric nitrogen deposition. Unpublished Ph.D thesis. University of Manchester.

Baxter, R. (1989). Physiological responses of *Sphagnum cuspidatum* Ehr. (ex Hoffm.) to acidic deposition. Unpublished Ph.D thesis. University of Manchester.

Baxter, R., Emes, M. J., and Lee, J. A. (1989). The relationship between extracellular metal accumulation and bisulphite tolerance in *Sphagnum cuspidatum* Hoffm. *New Phytologist*, **111**, 463–72.

Baxter, R., Emes, M. J., and Lee, J. A. (1991). Transition metals and the ability of *Sphagnum* species to withstand phytotoxic effects of the bisulphite ion. *New Phytologist*, **118**, 433–9.

Brimblecombe, P. and Stedman, D. H. (1982). Historical evidence for a dramatic increase in the nitrate component of acid rain. *Nature*, **298**, 460–2.

Clymo, R. S. and Hayward, P. M. (1982). The ecology of *Sphagnum*. In *Bryophyte ecology*, (ed. A. J. E. Smith), pp. 229–89. Chapman and Hall, London.

Conway, V. M. (1947). Ringinglow Bog near Sheffield. Part I: Historical. *Journal of Ecology*, **34**, 149–81.

Conway, V. M. (1949). Ringinglow Bog near Sheffield. Part II: The present surface. *Journal of Ecology*, **37**, 148–70.

Damman, A. W. H. (1986). Hydrology, development and biogeochemistry of ombrogenous peat bogs with special reference to nutrient relocation in a West Newfoundland bog. *Canadian Journal of Botany*, **64**, 384–94.

Delmas, R. J. (1986). Past and present chemistry of north and south polar snow. In *Arctic air pollution*, (ed. B. Stonehouse), pp. 175–86. Cambridge University Press, Cambridge.

Farey, J. (1813). *A general view of the agriculture and minerals of Derbyshire, with observations on the means of their improvement*. Board of Agriculture, London.

Ferguson, P., Lee, J. A., and Bell, J. N. B. (1978). Effects of sulphur pollutants on the growth of *Sphagnum* species. *Environmental Pollution (Series A)*, **16**, 151–62.

Ferguson, P. and Lee, J. A. (1979). The effects of bisulphite and sulphate upon photosynthesis in *Sphagnum*. *New Phytologist*, **82**, 703–12.

Ferguson, P. and Lee, J. A. (1980). Some effects of bisulphite and sulphate on the growth of *Sphagnum* species in the field. *Environmental Pollution (Series A)*, **21**, 59–71.

Ferguson, P. and Lee, J. A. (1983). Past and present sulphur pollution in the southern Pennines. *Atmospheric Environment*, **17**, 1131–7.

Ferguson, P., Robinson, R. N., Press, M. C., and Lee, J. A. (1984). Element concentrations in five *Sphagnum* species in relation to atmospheric pollution. *Journal of Bryology*, **13**, 107–14.

Gignac, L. D. and Beckett, P. (1986). The effect of smelting operations on peatlands near Sudbury, Ontario, Canada. *Canadian Journal of Botany*, **64**, 1138–47.

Goodman, G. T. and Roberts, T. M. (1971). Plants and soils as indicators of airborne heavy metals. *Nature*, **231**, 287–92.

Gorham, E., Bayley, S. E., and Schindler, D. W. (1984). Ecological effects of acid deposition upon peatlands: a neglected field in acid rain research. *Canadian Journal of Fisheries and Aquatic Sciences*, **41**, 1256–68.

Grace, J. and Unsworth, M. H. (1988). Climate and microclimate of the uplands. In *Ecological change in the uplands*, (eds M. B. Usher and D. B. A. Thompson), pp. 137–50. Blackwell Scientific Publications, Oxford.

Green, B. H. (1968). Factors influencing the spatial and temporal distribution of *Sphagnum imbricatum* Hornsch, ex Russ. in the British Isles. *Journal of Ecology*, **56**, 47–58.

Heil, G. W. and Diemont, W. H. (1983). Raised nutrient levels change heathland into grassland. *Vegetatio*, **53**, 113–20.

Horsman, D. C. and Wellburn, A. R. (1975). Synergistic effect of SO_2 and NO_2 polluted air upon enzyme activity in pea seedlings. *Environmental Pollution*, **8**, 123–33.

Kivinen, E. and Pakarinen, P. (1981). Geographical distribution of peat resources and major peatland complex types in the world. *Annales. Academiae Scientiarum Fennicae A. III: Geologica–Geographia*, **132**, 1–28.

Lee, J. A., Press, M. C., Woodin, S., and Ferguson, P. (1987). Responses to acidic deposition in ombrotrophic mires in the UK. In *Effects of atmospheric pollutants on forests, wetlands and agricultural ecosystems*, (eds T. C. Hutchinson and K. M. Meema), pp. 549–60. NATO ASI Series G 16, Berlin.

Little, P. and Martin, M. H. (1974). Biological monitoring of heavy metal pollution. *Environmental Pollution*, **6**, 1–19.

Livett, E. A., Lee, J. A., and Tallis, J. H. (1979). Lead, zinc and copper analyses of British blanket peats. *Journal of Ecology*, **67**, 865–91.

Malmer, N. (1990). Constant or increasing nitrogen concentrations in *Sphagnum* mosses on mires in Southern Sweden during the last few decades. *Aquilo Series Botanique*, **28**, 57–65.

Morgan, S. M., Lee, J. A., and Ashenden, T. W. (1992). Effects of nitrogen oxides on nitrate assimilation in bryophytes. *New Phytologist*, **120**, 89–97.

Moss, C. E. (1913). *The vegetation of the Peak District*. Cambridge University Press, Cambridge.

Patton, R. L. and Garraway, M. O. (1986). Ozone-induced necrosis and increased peroxidase activity in hybrid poplar leaves. *Environmental and Experimental Botany*, **26**, 137–41.

Press, M. C., Ferguson, P., and Lee, J. A. (1983). Two hundred years of acid rain. *Naturalist*, **108**, 125–9.

Press, M. C., Woodin, S. J., and Lee, J. A. (1986). The potential importance of an increased atmospheric nitrogen supply to the growth of ombrotrophic *Sphagnum* species. *New Phytologist*, **103**, 45–55.

Studholme, C. J. (1989). Isozyme variation, physiology and growth of *Sphagnum cuspidatum* Hoffm. in a polluted environment. Unpublished Ph.D thesis. University of Manchester.

Tallis, J. H. (1964). Studies on the Southern Pennine peats. III: The behaviour of *Sphagnum*. *Journal of Ecology*, **52**, 345–53.

Tallis, J. H. (1973). The terrestrialization of lake basins in North Cheshire, with special reference to the development of a Schwingmoor structure. *Journal of Ecology*, **62**, 537–67.

Tanaka, K. and Sugahara, K. (1980). Role of superoxide dismutase in defence against SO_2 and an increase in superoxide dismutase activity with SO_2 fumigation. *Plant and Cell Physiology*, **21**, 601–11.

UKRG (United Kingdom Review Group on Acid Rain) (1990). *Acid deposition in the United Kingdom 1986–88*. Department of the Environment, London.

Woodin, S. J. and Lee, J. A. (1987). The fate of some components of acidic deposition in ombrotrophic mires. *Environmental Pollution*, **45**, 61–72.

Woodin, S. J., Press, M. C., and Lee, J. A. (1985). Nitrate reductase activity in *Sphagnum fuscum* in relation to wet deposition of nitrate from the atmosphere. *New Phytologist*, **99**, 381–8.

13

Effects of pollutants on aquatic species

JANICE M. GLIME

13.1 Introduction

Aquatic systems are likely to experience a much greater variety of changes, at least through human activity, than terrestrial systems. The increase of nutrients has created the most visible problems, resulting in algal blooms, loss of oxygen, decreasing fish populations, and ultimately depriving bryophytes and lichens of light. As in the terrestrial system, the decrease of pH and increase of heavy metals have caused more subtle changes, often favouring bryophytes over their vascular competitors. But in addition to those problems that are also experienced by terrestrial situations, the aquatic system can suffer from unique problems. Construction of dams alters flow regimes and can alter the dissolved substances as a result of oxygen depletion in the reservoir. Temperatures are likely to rise, not only as the result of the greenhouse effect, but also from deforestation, reservoirs, and heated effluents, creating problems for those organisms that are adapted to the previously existing narrow temperature range. Deforestation, agriculture, and construction increase turbidity, which may affect light intensities or cause abrasion. Organic pollutants, much more common in aquatic systems, can be toxic to selected members of the community.

Thus, as the water quality changes, the communities of bryophytes will change, and ultimately we might expect the communities of animals dependent upon them or on their absence to change (Glime 1978). Although lichens in the aquatic habitat usually lack the abundance necessary to be of practical use, bryophytes, because of the sensitivity of some taxa and the tolerance or requirements of others, can often serve as indicators of changing water conditions and impending danger to other organisms.

In a Swedish river, the high concentration of cobalt in *Fontinalis* apparently was responsible for the great reduction in the mayfly *Ephermer-ella ignita* that ate *Fontinalis* (Södergren 1976). In California, construction dams on the Sacramento River seemed to be responsible for increased growth of *Hygrohypnum ochraceum*, which may have been sufficient to hinder the spawning activities of migratory salmonid fish (Balch 1970). On the other hand, the disappearance of most water mites from several river stations in Belgium is attributed to the disappearance of the moss vegetation, particularly *Cratoneuron filicinum* and *Amblystegium riparium* (Bolle *et al.* 1977). Thus, aquatic community structure, especially in streams and rivers, is often highly dependent upon the cryptogamic components, and cryptogams, especially bryophytes, often respond quite differently from their vascular plant associates.

13.2 Nutrient enrichment

The C:N:P ratio of aquatic vascular plants is essentially the same as in terrestrial higher plants (145:14.5:1), but in aquatic mosses the different balance (44:3.7:1), results in a C:N ratio of 12:1 instead of 14.5:1 (Dietz 1973). The K content, on the other hand, is much lower in aquatic mosses, averaging 7–11 g kg^{-1} dry weight against 20–50 in the vascular plants. This difference may reflect the greater stem to leaf weight ratio in mosses that have both a greater storage in leaves and a greater need to maintain leaves. The K content in both vascular plants and mosses varies little even though it varies in the medium, suggesting that macronutrient uptake is largely independent of the water quality.

Nevertheless, eutrophication can benefit some bryophytes, especially floating taxa such as *Riccia fluitans* and *Ricciocarpus natans* (Uotila 1971). *Riccia fluitans* can increase in total biomass in non-acidified waters as a result of phosphate enrichment of both sediment and water column, whereas sediment enrichment alone benefits only the vascular macrophytes (Roelofs *et al.* 1984).

Frahm (1975) examined lethal levels of several ions and found that *Fontinalis antipyretica* (least tolerant) and *Amblystegium riparium* (most tolerant) among five species had high tolerances for Na$^+$ ($>$ 950 mg l^{-1}) and Cl$^-$ ($>$ 1550 mg l^{-1}), but low tolerances for NH$_4^+$ (10 and $>$ 25 mg l^{-1} respectively), Fe^{3+} ($>$2 and $>$12), and PO$_4^{3-}$ ($>$ 5 and $>$ 25). Frahm (1976) also found that tolerance levels for phosphates decreased with time of exposure (Fig. 13.1). Baudo *et al.* (1981) found that *Fontinalis* had relatively low concentrations of Na, perhaps indicating that exclusion may play a role in protecting the moss from Na. Contrary to the other elements, concentrations of Na seem to be greater in aquatic vascular plants than in bryophytes (Satake 1983).

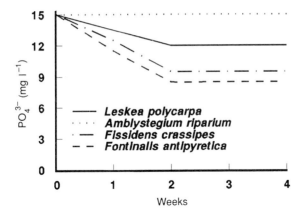

Fig. 13.1 Toxicity tolerance to levels of phosphates in four bryophytes for increasing periods of exposure of 0–4 weeks. Above the levels indicated, the mosses cannot grow (redrawn from Frahm 1976).

Say *et al.* (1981) found that concentrations of manganese in mosses living in clean water downstream were greater than in polluted upstream water, although there was no elevation in Mn levels in the downstream water. It is possible that competitive inhibition by the elevated heavy metals (Zn, Pb, Cu, Cd) upstream had lowered the ability of Mn to enter those mosses.

Inorganic fertilizers seem to be particularly detrimental to most aquatic lichens, largely because the free-living cyanobacteria become extremely abundant and overgrow the crustose lichens (James *et al.* 1977). We might expect *Physcia caesia* to increase in the splash zone on concrete walls of water channels enriched with N and P, since this species occurs on rocks used as bird perches.

13.3 Turbidity

The same effluents that bring high nutrient levels can cause bryophyte populations to degrade due to turbidity. Decrease of light can eliminate light-loving taxa such as *Amblystegium riparium* and *A. kochii* (Uotila 1971). Black plastic caused a rapid deterioration of *Brachythecium rivulare* on filter beds, but 18 months after removal of the plastic, substantial moss growth had returned (Hussey 1982). Lewis (1973*b*) showed that germination of spores was reduced by 42 per cent in *Eurhynchium riparioides* due to the action of 5000 mg l⁻¹ suspended coal particles. However, during periods of low flow and high light intensity, establishment of protonemata occurred, and Lewis suggested that the real effect of the coal particles might be to limit light as they settle on the surfaces of the substrata. She found that chlorophyll *a* was reduced by 50 per cent in *Eurhynchium riparioides* at coal dust concentrations of 5000 mg l^{-1}, apparently the result of reduced light penetration. However, bryophytes are C_3 plants that tend to have low light compensation points, as evidenced by their presence in very deep water (Hasler 1938; Glime 1987*a*). Wetzel *et al.* (1985) measured the light compensation point of *Sphagnum inundatum* to be 3–10 μE m^{-2} s^{-1}, whereas for the shallow water red alga *Batrachospermum* sp. it was >20 μE $m^{-2}s^{-1}$.

Suspended particles with rough edges can cause physical damage to mosses. Lewis (1973*a, b*) found that suspended coal particles caused severe abrasive damage to leaves of *Eurhynchium riparioides* in one week at 500 mg l^{-1} and after three weeks at 100 mg l^{-1}. Loss of lateral branches likewise prevents production of sex organs, further reducing the success of the species. Production of lateral shoots was nearly absent at 500 mg l^{-1}, presumably due to reduced light intensity and/or physical damage to the buds. Rhizoid production seemed to be unaffected even at 5000 mg l^{-1}.

James *et al.* (1977) report that the lichen *Verrucaria praetermissa* occurs in shady habitats of stream or lake banks, and that *Verrucaria hydrela* and *V. margacea* require very shaded, more humid environments. *Verrucaria silicea* grows on siliceous rocks in streams and lakes, preferring rapid water and good light with no mud and silt; disturbances from mining or logging could eliminate these species by changing the light available to them. The lichen *Collema fluviatile* is a member of this association, and because of its rarity could be eliminated altogether following any extensive widespread disturbance that created silt or blocked light.

13.4 Organic pollution

Moss chlorophyll is easily damaged by some kinds of organic pollution, resulting in a reduced chlorophyll/phaeopigment ratio (Peñuelas 1984*a*; Abaigar and Diaz 1987; López and Carballeira 1989), presumably because the leaves are only one cell thick and permit easy entry. Although industrial organic loading on filter beds in England seemed to have no effect on moss growth, dilution of domestic loading caused mosses to grow where they had been absent (Hussey 1982). *Eurhynchium riparioides* seems to be particularly tolerant (Empain 1974, 1976), *Fontinalis antipyretica* relatively tolerant (Peñuelas 1984*a*), and liverworts more sensitive (Abaigar and Diaz 1987) to organic pollution.

One reason for chlorophyll sensitivity in bryophytes is the rapid absorption of organic pollutants. Eidt *et al.* (1984) found that the leafy liverwort *Jungermannia* sp. accumulated fenitrothion to over 100× the water concentration; Morrison and Wells (1981) found that *Fontinalis antipyretica* collected 33×, phytoplankton accumulated only about 8×. The fenitrothion likewise persisted longer in the liverwort, decreasing substantially within seven days after spraying in all other plants, including both mosses and higher plants (Eidt *et al.* 1984).

One of the consequences of many organic effluents is an increase in bacteria and hence an increase in respiration, resulting in elevated CO_2. Sanford *et al.* (1974) found a significant increase in productivity as the ratio of CO_2 to air was increased from 5:5000 to 100:5000 in the culture medium of *Hygrohypnum ochraceum*, and they suggested that such an increase from bacterial respiration in the sediments could enhance productivity of this moss in the field. Vrhovsek *et al.* (1981) noted that *Fissidens crassipes* appeared only in the much polluted part of the river where bacterial respiration was high. Wetzel *et al.* (1985) found that increased CO_2 in the sediments increased production not only of rooted macrophytes that transferred the CO_2 to the leaves through internal passages, but also of bryophytes such as *Sphagnum inundatum* closely associated with the sediments.

13.5 Phenols and chlorinated hydrocarbons

Phenols arrive in aquatic systems from industrial waste and from decomposition of dead plants. These are quickly incorporated into moss tissue (Merlin 1988; Mouvet 1989). At low concentrations in the medium (50 mg phenol dm^{-3}), *Fontinalis antipyretica* can decompose 32–43 per cent of phenol, and *Eurhynchium riparioides* can decompose 20–27 per cent (Samecka-Cymerman 1983). The ability decreases as concentrations increase. At 50 mg dm^{-3}, apical growth is diminished, but lateral growth is still considerable. At lower concentrations, more phenol is decomposed, whereas at higher concentrations increasingly larger percentages of the plants die. Such ability to decompose phenol is already known for the aquatic higher plants *Elodea* and *Ceratophyllum* (Timofiejeva *et al.* 1977).

Frahm (1976) found *Fontinalis antipyretica* to be intolerant of four weeks of exposure to 0.02 mg l^{-1} phenol, whereas *Leskea polycarpa*, *Amblystegium riparium*, and *Fissidens crassipes* were tolerant of 0.08 mg l^{-1} for the same time period.

Cinclidotus danubicus concentrates PCBs and can be used as an indicator of their presence in freshwater when the water concentrations may be too low to detect (Mouvet *et al.* 1985). Mouvet *et al.* (1986) found concentrations in mosses ranging from 90 to 289 000 μg kg^{-1}. Other hydrocarbons (α-HCH, β-HCH, γ-HCH, DDT, DDD, DDE) were in less spectacular concentrations, but reflected water concentrations.

Auerbach *et al.* (1973) found that in *Fontinalis antipyretica* toxicity of phenols and other compounds could be recognized long before lethality by observing CO_2 exchange and time of deplasmolysis of leaves. Deplasmolysis is indicative of membrane damage, permitting water gained under early stages of high osmotic potential and toxicity to leave the cell. Old cells of *F. antipyretica* exhibited plasmolysis more easily than young cells in KCl and $CaCl_2$ because the older cells had lower cytoplasmic viscosity (Fischer 1948).

13.6 Acidification

Acidification results largely from input of sulphurous and nitrogenous compounds in rainwater and from acid mine wastes. One not surprising result in Swedish (Grahn 1976), Adirondack (Roberts *et al.* 1985) and Dutch (Roeloffs *et al.* 1984) lakes is the increase of *Sphagnum* species. Even in streams, *Sphagnum* can occur in dense mats in the deeper parts when acid pollutants acidify the water (Harrison 1958). *Sphagnum* changes the water composition as it increases in biomass. Its strong cation exchange capacity removes biologically important cations (Skene 1915; Ramaut 1954; Anschütz and Gessner 1954; Clymo 1963) and replaces them with H^+. But in Ontario, Canada, succession to *Sphagnum* does not seem to be a consequence of anthropogenic acidification (Yan *et al.* 1985; Stokes 1986), perhaps because the Ontario lakes are less acidic than in the Netherlands (pH of 3.9).

Yan *et al.* (1985) found a negative correlation between pH and aquatic bryophyte richness. One unusual consequence of low pH is the tendency for normally terrestrial taxa to grow submersed. *Anisothecium vaginale, Polytrichum longisetum*, and a species of *Bryum* grow submersed in streams with a water pH less than 4.0 (Sand-Jensen and Rasmussen 1978), presumably because the CO_2 is all present as free CO_2, making it possible for the bryophytes to obtain adequate CO_2 for photosynthesis. Bryophytes can utilize only free CO_2, unlike higher plants that can use bicarbonates (Bain and Proctor 1980; Allen and Spence 1981).

Low pH has not only a direct effect on aquatic organisms, but it also affects the chemistry of all the other water components. Caines *et al.* (1985) found an inverse relationship between H^+ concentration and the ability of *Nardia compressa* (an acid stream liverwort) to accumulate ions of Al, Mn, and Zn (Fig. 13.2). The additional H^+ ions in the water at low pH compete with other ions for the binding sites of bryophytes.

Another pH-mediated relationship is the change from NO_3^--dominated N utilization, with roots being the major uptake sites in the ecosystem, to NH_4^+-dominated N utilization, with leaves as the major uptake sites. Melzer (1980) determined that some aquatic plants are unable to use NO_3^- and may starve in NO_3^--dominated N-poor waters. Thus, when NH_4NO_3

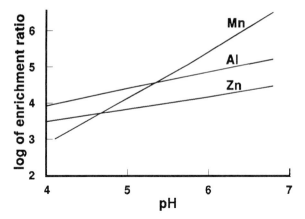

Fig. 13.2   The effect of pH on enrichment ratios of Mn, Al, and Zn in *Nardia compressa* (redrawn from Caines *et al.* 1985).

is supplied to acidified water, communities can shift, favouring the rushes (*Juncus*), *Sphagnum*, and *Drepanocladus fluitans* (Schuurkes *et al.* 1986). *Sphagnum flexuosum* exhibited no NO_3^- utilization in experiments. On the other hand, *Drepanocladus fluitans* had a high NH_4^+ uptake, but its NO_3^- uptake nevertheless exceeded that by *S. flexuosum*. *Sphagnum* species increased in Netherlands streams concomitantly with pH < 4.5, high NH_4^+ concentration, and strongly increased CO_2 levels (Roelofs *et al.* 1984). In the Netherlands, acidification is generally due to $(NH_4)_2SO_4$, whereas in Canada, the UK, and Scandinavia it is primarily due to H_2SO_4 or HNO_3, thus emphasizing the importance of NH_4^+. *Fontinalis antipyretica* likewise uses ammonium N (Schwoerbel and Tillmans 1974) and *Riccia fluitans* occurs in waters that are simultaneously low in NO_3^-, but high in NH_4^+ (Roelofs 1983).

Satake *et al.* (1989*a*), in their study of the acid River Akagawa in Japan, found *Jungermannia vulcanicola* at pH 3.6 to 4.6 and *Scapania undulata* at pH levels down to 3.9, but *Eurhynchium riparioides* was only present when confluent water of pH 5.5 entered the river. *Eurhynchium riparioides* is probably entirely absent from streams with low pH or very soft water (Hargreaves *et al.* 1975; Whitton and Diaz 1981; Wehr and Whitton 1983*b*). Satake *et al.* (1984) have recorded *Jungermannia vulcanicola* in acid springs at a pH as low at 1.9! Concentrations of K in shoots of *J. vulcanicola* and *S. undulata* were high in the acid streams, reaching 5.78 per cent and 2.46 per cent dry weight, respectively. In the same acidic river, concentrations in *Sphagnum* and *Eurhynchium riparioides* were less than 1 per cent.

Webster (1985) found high levels of Fe, Al, Mn, and Zn in *Pohlia nutans* in acid coal seeps in Pennsylvania. These clumps were as deep as 11.5 cm, but only the upper 2–3 mm remained green, suggesting that they may have been growing at that location for some time. Wenerick *et al.* (1989) found that acidified mine water reduced the chlorophyll content in *Pohlia nutans* and *Sphagnum recurvum* in a greater dilution than it did in *Typha latifolia*. However, the chlorophyll *a/b* ratio actually increased in the mosses. The enzyme that facilitates the breakdown of chlorophyll *b* to chlorophyllide *b* is only active in a narrow acidity range, whereas the enzyme that facilitates the breakdown of chlorophyll *a* to phaeophytin *a* is active in a wide range (Malhotra and Hocking 1976). This normally results in the preferential breakdown of chlorophyll *a* over chlorophyll *b* with increasing acidity, thus lowering the *a/b* ratio. Malhotra and Hocking found no change in chlorophyllide *a* or phaeophytin *b* concentrations with increased acidity. Wenerick *et al.* (1989) suggested that their observed increase in the *a/b* ratio may have been due to the high concentration of Fe in solution, interfering with Mn uptake (Somers and Shive 1942; Cheniae and Martin 1969).

The copper mosses may in fact be acid mosses (Persson 1956). *Scopelophila ligulata* is associated with an average pH of less than 3.5 (Persson 1956; Shaw and Anderson 1988). One might explain this linkage by the common occurrence of Cu as $CuSO_4$, which dissociates to form H_2SO_4, thus lowering the pH of the soil.

It is difficult to separate the effects of acidification from effects of sulphur. Frahm (1976) found that several species of aquatic mosses varied considerably in their tolerances to the addition of SO_4^{2-} as Na_2SO_4 (Fig. 13.3). Furthermore, acid mine wastes frequently are laden with high levels of heavy metals that contribute to the toxicity levels, as discussed below.

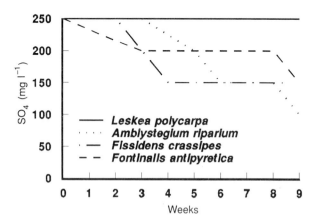

Fig. 13.3 Toxicity tolerance to sulphates (as Na_2SO_4) in four bryophytes for periods of exposure of 0–9 weeks (redrawn from Frahm 1976).

Few lichens seem to occupy aquatic alkaline rocks (James *et al.* 1977), so we would expect acid rain and acid mine run-off to favour more of these communities; however, the basophilic *Verrucaria elaeomeleana* and the rare *Placynthium tantaleum* and *Aspicilio calcareae* would be likely to disappear under acidification.

13.7 Heavy metals

Although bryophytes seem to regulate absorption of macronutrients, they readily absorb excessive quantities of heavy metals. Contrary to vascular plants, their uptake is primarily from the water, whereas vascular plants take up most metals (Pb is an exception) primarily via their roots (Welsh and Denny 1980). Say and Whitton (1983) found that concentrations of Cd and Pb in tips of *Fontinalis antipyretica* correlated with concentrations in the sediment, but not in the water, whereas Cu and Zn correlated with both. Presence of Zn, Mn, and filterable reactive phosphate influenced the concentration of Pb in the moss, and when these were considered, the concentration of Pb in the water became a good predictor of the concentration in the moss. The pH is likewise influential, with absorption ratios in highly acidic waters being less than those in less acid waters (Wehr and Whitton 1983*a*).

Whereas uptake in higher plants tends to be very seasonal, bryophytes are little influenced by season, except as the concentrations of the metals themselves vary seasonally (Wehr and Whitton 1983*c*; Glime and Keen 1984).

Bryophytes absorb heavy metals through their surfaces. They differ from phanerogams in that they behave essentially as an ion exchange medium and lack the regulatory mechanisms of roots (Empain 1985). All leaf cells have at least two surfaces exposed to the medium, and the proportion of cell wall to cell content is high, as reflected by high dry weight/wet weight ratios (Dietz 1973). Therefore, the ratio of leaf weight and area to stem weight will greatly affect absorption values that are based on weight of moss. However, Buck and Brown (1978) feel that low loss of Zn and Mg from *Cratoneuron filicinum* is probably due to the large ratio of stem and branch tissue to leaf tissue, providing more cell wall and less cytoplasm than leaf tissue on a dry weight basis. Differences in metal uptake between sites will also depend upon the array of metals present and reflect the differences in adsorption affinities: Cu, Pb > Ni > Co > Zn, Mn (Rühling and Tyler 1970). Thus, high concentrations of Cu may actually block adsorption of Mn and Fe to such a degree that the moss can suffer deficiency. Unlike the mosses, algal blooms appear to regulate uptake of individual metals in the order of Fe > Zn > Pb > Cu > Ni (Trollope and Evans 1976).

Mouvet *et al.* (1986) found high concentrations of Cd, Cr, Cu, Fe, Ni, Pb, and Zn in three aquatic mosses living in contaminated water and reported that within two months after diversion of one of the principal sources of pollution the concentrations decreased significantly in the mosses. In other transplant studies (Benson-Evans and Williams 1976) the improvement in survival of transplanted mosses coincided with improvement in water quality. On the other hand, Baudo *et al.* (1981) found that *Fontinalis* sp. and *Potamogeton crispus* had the lowest concentrations of Mg among the aquatic plants in one Italian lake. Although many authors claim that bryophytes are superior to tracheophytes for metal enrichment, some tracheophytes can concentrate metals more than some bryophytes in the same stream (Fig. 13.4). Södergren (1976) found the highest concentrations of metals in *Elodea* and *Myriophyllum* compared to those in *Fontinalis*; cobalt, on the other hand, was much higher in *Fontinalis*. Although abilities and sensitivities differ, all the taxa of bryophytes are good accumulators of heavy metals.

Empain *et al.* (1980) ranked 22 aquatic and subaquatic bryophytes and

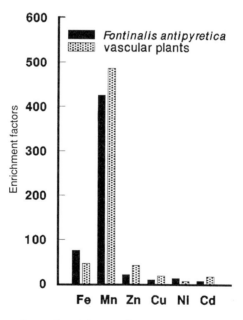

Fig. 13.4 Mean values of enrichment factors for *Fontinalis antipyretica* and the aquatic vascular plant with the highest mean in the same stream (redrawn from Ledl *et al.* 1981).

considered *Fissidens fontanus*, *Cinclidotus nigricans*, *C. fontinaloides*, and *F. crassipes* to be the most toxiphobic and *Eurhynchium riparioides* the least. When Mouvet *et al.* (1987) examined uptake of Cu, Cd, Zn, and Cr, they found that *Eurhynchium riparioides* and *Fontinalis antipyretica* had the greatest uptake, *Cinclidotus danubicus* the least. Under the same conditions, *Eurhynchium riparioides* usually has higher concentrations of Cu, Fe, Pb, Zn (Wehr *et al.* 1983; Wehr and Whitton 1983*b*; Mouvet 1985), Ca, Mn, Co, Ni, Cd (Wehr *et al.* 1983; Wehr and Whitton 1983*b*), Cr, and As (Mouvet 1985) than *Fontinalis antipyretica*. However, Mouvet (1985) found little difference in the Cd concentrations between these two species. McLean and Jones (1975) found that *Brachythecium rivulare* always had heavy metal levels similar to those of *Scapania undulata*, even though the former existed in areas with much lower ambient metal concentrations. *Fontinalis squamosa* generally had a higher concentration of Fe, Pb, Zn, and Mn than did *S. undulata* and McLean and Jones considered *S. undulata* to be a tolerant species. When transplanted to a site with only *S. undulata*, *F. squamosa* had significantly higher content of these metals. *Scapania undulata* remained healthy while the transplanted *F. squamosa* lost leaves, had disintegrating chloroplasts, and had cells that were completely empty. Say and Whitton (1981) likewise found *S. undulata* to be particularly tolerant of Zn; it will dominate in streams where heavy metal loading is combined with low levels of nutrients (McLean and Jones 1975; Wehr and Whitton 1983*a*). This contrasts with the resistant *Fontinalis antipyretica* in Spain, where *S. undulata* was the most sensitive (López and Carballeira, in press). McLean and Jones (1975) suggest that plants could tolerate the metals as long as these are sequestered on cell wall and internal building sites. Less permeability and a greater number of binding sites may permit some species to survive.

Concentrations of most metals in whole mosses are significantly higher (especially for Mn, Zn, and Cd) than in 2 cm tips, perhaps explaining why tips seem to be able to continue growing when the mosses are subjected to excessive loadings of heavy metals (Wehr *et al.* 1983). Whitton *et al.* (1982) likewise found a progressive increase in metal content from tip to base in *Scapania undulata*. Wehr and Whitton (1983*b*) found that 12 metals had higher concentrations in whole plants than in the tips of *Eurhynchium riparioides*. Na and K were exceptions to this spatial relationship (Wehr and Whitton 1983*b*). Soma *et al.* (1988) found higher concentrations of Al, Mn, Cu, Zn, and Pb 1–3 cm below the tip of *Pohlia ludwigii*, but Na, Al, P, K, Ca, and Fe differed little between the 1 cm tip and the lower parts.

The higher concentration of some minerals in older parts may be due to coatings on the leaves and stems of Fe and Mn oxides, which are known to increase the adsorption and coprecipitation of other metals (Robinson

1981), to greater exposure time of older leaves, or to greater permeability of older leaves and access to interior cell wall binding sites.

13.7.1 Zinc

Although Zn is an essential nutrient in small quantities, it soon becomes toxic in elevated amounts. At 0.1 mg l⁻¹ Pb or Zn, exposure of *Fontinalis antipyretica* for six days inhibits assimilation, but at 0.01 mg l⁻¹ Pb^{2+}, CO_2 assimilation is stimulated (Weise *et al.* 1985). When environmental conditions are optimum, e.g. 1200 lux, toxicity of both metals is reduced and positive photosynthesis can be maintained. Dark respiration likewise decreases as the concentration of Pb or Zn increases.

Levels of Zn in mosses are usually highly correlated with those in the water (Say *et al.* 1981; Whitton *et al.* 1982; Mouvet *et al.* 1986). When compared to Zn-rich sediment, mosses had nearly five times as much Zn as the sediment, while having less Cr than the sediment (Mouvet *et al.* 1987). Concentrations after removal to a metal-free medium dropped rapidly in the first 24 hours (from 13 mg g⁻¹ to 7.5 mg g⁻¹ for Zn, 26 mg g⁻¹ to 16 mg g⁻¹ for Pb) and continued more slowly after that.

Light influences the uptake of Zn in bryophytes (Pickering and Puia 1969). Bachmann and Odum (1960) found that benthic marine algae will take up Zn in the light but not in the dark, and that net uptake appears to be related to net O_2 production. Likewise, Mouvet (1979) found that uptake of Cr^{3+} in *Fontinalis antipyretica* increased with light intensity up to about 1500 lux. On the other hand, Broda (1965) found that anaerobiosis did not affect the uptake of Zn by the alga *Chlorella pyrenoidosa* or barley roots and concluded that uptake was a passive process based on ion exchange. Pickering and Puia (1969) found that in *Fontinalis antipyretica* Zn and Ca are absorbed competitively, which is consistent with ion exchange processes. They found that in *F. antipyretica* Zn seems to be absorbed in three stages:

1. The first stage of uptake was very short and was not influenced by light intensity or temperature, suggesting a passive uptake in the Donnan-free-space, primarily by exchange adsorption. The rate is dependent upon concentration, with about 50 per cent of the equilibrium amount absorbed in half an hour.

2. The second stage lasted no more than 90 minutes and was slightly affected by light intensity and temperature, with freshly killed plants absorbing more Zn than living ones. Zn entered the protoplast, with the cell membrane functioning as a diffusion barrier.

3. The third stage lasted several days, was very slow, and was influenced by both light and temperature. Active accumulation within the plant cell was dependent upon factors that affect cell metabolism (light, temperature).

The Zn is tightly bound within the cell, with 80–90 per cent in the cell wall (Burton and Peterson 1979), and less than 5 per cent is lost by subsequent washing (Whitton *et al.* 1982). Burton and Peterson (1979) reported that in *Fontinalis antipyretica* and *Solenostoma crenulatum* the Zn was all bound in its cationic state. No data are available on other aquatic plants, but in terrestrial vascular plants it can occur in the anionic state as well (Peterson 1969).

Whitton *et al.* (1982) found that Zn and Cd accumulations increase with a rise in pH and/or Ca, although the correlation with Ca may be due to the high correlation between pH and Ca.

Mouvet *et al.* (1986) found that *Eurhynchium riparioides* will take up Pb and Zn more quickly than it releases them, reaching asymptotic uptake in about 192 hours. Whitton and Diaz (1980) found *Eurhynchium riparioides* to be intermediate in its response to Zn, being superior to sensitive algae such as *Chamaesiphon confervicola* but inferior to the eutrophic algae *Hormidium rivulare* and *Euglena mutabilis*. In their study *E. riparioides* did not occur in water with more than 10 mg l^{-1} Zn. Young and Glime (unpublished data) found that at $ZnSO_4$ concentrations of 5 mg l^{-1} *Fontinalis duriaei* grown in Chlorophyta media (Prescott 1968) was unaffected, but at 56 mg l^{-1} it became slightly orange within two days, turning yellow in 12 days, when it had reduced concentrations of chlorophyll *a* and *b*. Hussey (1982) reported that *Brachythecium rivulare* was much more affected by Zn than was *Amblystegium riparium*.

In a downstream gradient study, Say and Whitton (1980) found that only algal taxa were restricted to regions where Zn levels exceeded 3 mg l^{-1}, whereas *Scapania undulata* occurred both where the mean was 9.2 mg l^{-1} and further downstream where the concentration was 0.064 mg l^{-1}.

Say and Whitton (1982) found the liverwort *Solenostoma crenulatum* growing among almost pure anglesite with high levels of Zn. That this species is highly resistant to Zn is not surprising because Brown and House (1978) had already found it growing near a lead mine and on the spoil of a copper mine.

13.7.2 Lead

Satake *et al.* (1989*b*) found Pb in the shoots of *Scapania undulata* ranged from 0.7 to 2.4 per cent of dry weight and had an enrichment ratio of 3.5×10^5–1.2×10^6. The Pb was localized in the cell wall and was not detected in the nucleus or other cell components, with no electron-dense particles in the cell. This differs from the electron-dense HgS in the cell wall of *Jungermannia vulcanicola* (Satake and Miyasaka 1984) and the nuclear electron-dense particles of Pb in the terrestrial moss *Rhytidiadelphus squarrosus*. The presence of PbS was undetectable, and insoluble sulphate, possibly as $PbSO_4$, was about 3 per cent of total Pb. Soma *et al.* (1988) used X-ray spectroscopy to examine Pb content and determined that the sulphate ion is not the sole binding ligand for Pb in the bryophytes.

13.7.3 Iron

Satake and co-workers (1989) found concentrations of 5–13 per cent dry weight of Fe in *Jungermannia vulcanicola* and *Scapania undulata* in the acid river Akagawa. The iron oxide was tightly bound to the leaves of both liverworts and sonification removed degraded leaves along with the Fe. Glime and Keen (1984) found a similar tight binding of Fe to the leaves of *Fontinalis duriaei* in a Michigan stream and considered that the photosynthetic moss was serving as a site of oxidation of the reduced Fe in solution. In solutions of ferric chloride and ferric sulphate, bryophytes become black (Field 1972).

At the Kootenay Paint Pots, British Columbia, Canada, the low pH (3.2–4.0), anaerobic conditions, and high levels of Fe (39–237 mg l^{-1}) and Zn (26.9–40.4 mg l^{-1}) result in an unusual assemblage of organisms, with the only non-algae being the liverwort *Cephalozia bicuspidata* and the moss *Dicranella heteromalla* (Wehr and Whitton 1983a). The *Dicranella* exists only as protonemata. Between 20 and 30 per cent of the dry weight of *Cephalozia bicuspidata* consists of Fe. Some cushions of the liverwort are entirely covered with a brittle iron oxide crust 1–2 mm thick, being formed there during photosynthetic release of oxygen. Liverworts are usually rare on acid mine drainage where Fe concentrations are high (Lackey 1938).

13.7.4 Copper

Plants require Cu and in aquatic systems this nutrient is often limiting, as demonstrated by Glime and Keen (1984) for *Fontinalis* grown in Lake Superior water. Shaw (1987) showed that unlike most plants, the copper moss *Scopelophila cataractae* was naturally Cu tolerant, even among populations that had not grown on Cu-enriched substrates.

On the other hand, most bryophytes are very sensitive to Cu. Sommer (1981) found that maximum damage to *Fontinalis antipyretica* by Cu occurred in 15 minutes, with a net photosynthetic reduction of 18 per cent at 0.01 mM and 30 per cent at 1.0 mM. Adding Ca to the solution reduced the photosynthetic difference between normal photosynthesis and Cu-damaged photosynthesis, especially at low concentrations of Cu (0.01 mM). It is known that Ca can control the cell permeability (Jacobson *et al.* 1961) and may therefore exclude the Cu. Membranes leak in the absence of Ca (Bates and Brown 1974). Furthermore, since Ca is a cation, it may offer some competition for binding sites. Uptake of Cu, Cd, and Cr^{3+} is likewise inversely related to concentration of the chelating agent EDTA (Mouvet 1987).

Fontinalis duriaei exhibited plasmolysis in 1.0 mg l^{-1} Cu; *F. duriaei, F. antipyretica* var. *gigantea*, and *F. dalecarlica* all exhibited deplasmolysis at 10 mg l^{-1}(Glime and Keen 1984). Branch tips are the first to show symptoms of Cu damage in *Fontinalis* (Glime and Keen 1984), whereas in terrestrial vascular plants the young shoots consistently have a low metal content (Baudo and Varini 1976).

In *Fontinalis antipyretica*, increasing the temperature increased the absorption from *c*. 4000 p.p.m. at 5°C to 10 000 p.p.m. at 20°C (Sommer 1981). There was a strong negative correlation between Cu absorbed and net photosynthesis, and there was a simultaneous increase in dark respiration. TEM demonstrated that the chloroplasts had been damaged: the lamellae were disorganized and grana were difficult to distinguish. There seemed to be no injury to the cell wall or other organelles. This damage to chloroplasts can explain why Glime and Keen (1984) found maximum chlorophyll in *Fontinalis duriaei* at 0.01 mg l^{-1}, with continuous decline as the Cu concentration rose. Of the four mosses studied, Glime and Keen found that *Fontinalis antipyretica* var. *gigantea* showed the greatest reduction of chlorophyll *a* and *Eurhynchium riparioides* the least. However, *E. riparioides* had the most visible colour change. They attributed this apparent discrepancy to the large leaf to stem ratio in *F. antipyretica* var. *gigantea* and small ratio in *E. riparioides*, causing the change on a per weight ratio to be more dramatic where there was a high proportion of leaf tissue.

Cu concentration in the moss is dependent upon concentration and exposure time (Fig. 13.5). Mouvet (1984) found that *Fontinalis antipyretica* transplanted into a contaminated stream accumulated enough Cr and Cu in nine days to show clear symptoms of impact. These elements are stored in electron-dense bodies in the cell wall and in vacuoles (Sommer 1981; Mouvet 1984).

Erdman and Modreski (1984) found that *Bryum* sp. and *Drepanocladus fluitans* concentrated up to 35 000 µg g^{-1} Cu, compared to 1700 µg g^{-1} in the sediment. *Fissidens rigidulus* from mining areas in New Zealand possessed approximately the same Zn concentration as its environment, but less Cu (Ward *et al.* 1977), suggesting a difference in ion exchange capacity of the moss for these two metals (Whitehead and Brooks 1969).

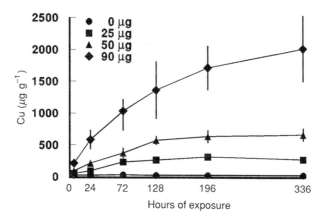

Fig. 13.5 Relationship of mean Cu concentrations (and error bars) in *Eurhyn-chium riparioides* to time of exposure at four concentrations (redrawn from Mouvet *et al.* 1988).

13.7.5 Cadmium

Although Cd is highly toxic for animals, plants tolerate large concentrations, wtih detrimental effects often requiring decades (Lepp 1981). In experiments by Glime and Keen (1984), chlorophylls *a* and *b* decreased in *Fontinalis duriaei* at Cd concentrations of 35 μg l^{-1} at 5°C. Sommer (1981) found that 1.0 mM Cd solution depressed net photosynthesis of *Fontinalis antipyretica* by 30 per cent at its optimum temperature of 10°C, and had a similar quantitative depression in gross photosynthesis, indicating that it was photosynthesis and not respiration that was primarily affected. Differences between Cd and Cd-free solutions had a slight negative effect (c. 7 per cent) upon dark respiration. Examination by transmission electron microscopy revealed that the mitochondria contained electron dense bodies, which could explain the slight depression in respiration. The cell wall and vacuoles bind much of the Cd, making it harmless.

Temperature has only a slight effect on Cd absorption, compared to Cu, increasing absorption from *c.* 3400 to 3700 p.p.m. with an increase in temperature from 5°C to 20°C (Sommer 1981). Exposure time seems to be a much more important factor than with Cu, with uptake increasing in the moss from *c.* 3800 p.p.m. at 15 min to *c.* 5400 p.p.m. after 25 min. As with Cu, adding Ca to the solution reduced the difference between normal photosynthesis and Cd-damaged photosynthesis, especially at low concentrations of Cd (0.01 mM).

Mouvet (1979) found that *Cratoneuron filicinum* had the highest accumulation of Cd and *Cinclidotus fontanus* the least among ten aquatic bryophytes studied. It is noteworthy that in some organisms Cd is more toxic than Cu, whereas in others, e.g. *Fontinalis antipyretica*, the reverse is true (Sommer 1981). When both Cd and Cu were added to the solution, the damage to *F. antipyretica* exceeded that of either alone. In addition to the damage to organelles, it is likely that these two metals will affect enzyme availability.

Ray and White (1979) considered the rhizome of the vascular plant *Equisetum arvense* to be a good accumulator of Cd, but *Fontinalis duriaei* can absorb up to approximately ten times as much (Glime and Keen 1984). McLean and Jones (1975) found that *Fontinalis* and *Scapania* accumulated Cd for more than 20 weeks, whereas Ornes and Wildman (1979) noted that aquatic vascular plants contained their greatest concentrations after one or two weeks, subsequently dying and releasing 97–99 per cent of their accumulation.

13.7.6 Mercury

Fontinalis hypnoides and *Ricciocarpus natans* concentrated three to five times as much Hg as did tracheophytes in the same streams in Finland (Suominen *et al.* 1977; Lodenius 1980). On the other hand, *Fontinalis antipyretica* had a concentration only slightly greater than that of the mean among aquatic vascular plants in a Finnish estuary (Cajander and Ihantola 1984). Satake *et al.* (1983) found the liverwort *Jungermannia vulcanicola* to have the highest known concentration of Hg in any aquatic plant (1.3 per cent dry weight); the Hg was localized primarily in the cell walls in electron-dense particles that appeared to be HgS, a form that is not toxic for organisms and is insoluble in water (Satake and Miyasaka 1984).

13.8 Salts

Most bryophytes are limited to freshwater, although their tolerance to metal salts might be inferred from the above discussion. *Fontinalis dalecarlica* is able to grow in brackish water (Söderlund *et al.* 1988) and several terrestrial taxa tolerate salt spray.

Fletcher (1973) found that the lichens *Verrucaria striatula* and *V. mucosa* were the lowest members of the tidal zone and were immersed for 52 per cent of the study year. At least 15 species of lichens in the British Isles live on the rocky coast where they are regularly inundated by sea water. If these lichens follow the pattern of some of the salt marsh vascular plants, we might find them eventually on emergent rocks of streams and rivers along highways where salt is used frequently to melt the ice.

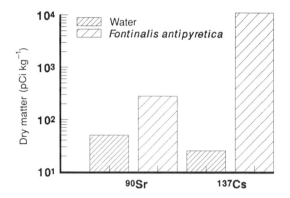

Fig. 13.6 Concentrations of the radionuclides ^{90}Sr and ^{137}Cs in water and *Fontinalis antipyretica* (redrawn from Kulikov *et al.* 1976).

13.9 Radionuclides

Associated with many natural heavy metal loadings in aquatic systems are high levels of such radioactive elements as uranium. Mosses incorporate radionuclides rapidly and maintain a higher concentration than the surrounding water (Hebrard *et al.* 1968; Whitehead and Brooks 1969; Foulquier and Hebrard 1976). Shacklette and Erdman (1982) attribute this in part to the ability of the mosses both to concentrate and to integrate the loadings in the water. For example, mosses had 1800 μg g^{-1} (ash) U whereas associated water had 6.5 μg l^{-1}. Lambinon *et al.* (1976) found that aquatic bryophytes concentrated radionuclides to about two to ten times as much as rooted macrophytes. *Fontinalis antipyretica* contained up to three orders of magnitude more ^{90}Sr and ^{137}Cs from airborne isotopes than did the surrounding water (Fig. 13.6) (Kulikov *et al.* 1976). Ge concentrated to 50 μg g^{-1} in mosses (Shacklette and Erdman 1982), whereas Hörman (1970) reported that Ge never exceeds 10 μg g^{-1} in plant ash and that it is toxic to plants at that concentration, causing death. Kirchmann and Lambinon (1973) considered mosses (*Cinclidotus danubicus*) to have superior ability compared to aquatic vascular plants for concentrating and retaining radionuclides. In *Eurhynchium riparioides*, an equilibrium between radionuclides (^{60}Co, ^{51}Cr, ^{137}Cs, [54]Mn, [22]Na) in the water and in the moss occurs within one to three days (Maurel-Kermarrec *et al.* 1983).

13.10 Fluctuating water level

Among the effects of dams for hydroelectric power are massive and sudden fluctuations in water level, sometimes on a diurnal cycle, but at other times on a longer cycle of days or even weeks. Aquatic bryophytes that are unable to tolerate such desiccation will disappear and tolerant riparian species such as *Brachythecium rivulare, Cinclidotus aquaticus*, and *C. fontinaloides* may become more abundant (Vrhovsek *et al.* 1985). On the other hand, the sudden onslaught of water when the dam is opened may dislodge the relatively quiet water *Brachythecium rivulare* and favour such species as the more flow-tolerant *Fontinalis dalecarlica* or *F. novae-angliae* (Glime 1970, 1980) provided the alternation of wetting and drying is infrequent (Glime 1971). Holmes and Whitton (1977) found that *Hygrohypnum ochraceum* and *Fontinalis antipyretica* increased visibly below a dam.

Peñuelas (1984*b*) found that loss of chlorophyll can occur in as little as one week in *Fontinalis antipyretica* and *Barbula ehrenbergii*, two weeks in *Hygroamblystegium tenax*, three weeks in *Eurhynchium riparioides*, four weeks in *Cinclidotus fontinaloides*, and five in *Fissidens crassipes* that are stranded above water. Phaeopigments increased during his experiment. However, *E. riparioides* and *C. fontinaloides* recovered to more than 50 per cent of the original chlorophyll *a* concentration later during the exposure. *Fontinalis antipyretica* was the least tolerant and *C. fontinaloides* was the most. In another experiment, *Fontinalis novae-angliae* and *F. dalecarlica* both had partial recovery when replaced into the stream after one year of exposure above stream level (Glime 1971).

Species such as *Fontinalis duriaei* may survive because of their rapid growth rate, permitting them to replace lost branches quickly (Glime 1987*a*). But if global warming trends force species farther north, new photoperiods may disrupt reproductive cycles (Glime 1984). Hussey (1982) found that when mosses dried on filter beds, especially *Amblystegium riparium*, they were easily detached, but small bits remained and were able to grow. Glime *et al.* (1979) suggested that periods of drying out may facilitate attachment of fragments of aquatic mosses.

James *et al.* (1977) feel that the frequency and extent of submersion are the most important controlling factors for aquatic lichens, and in fact, their zonation pattern has been used to determine river channel capacity (Gregory 1976). Thus, dams that alter flow and provide long periods of drought, followed by long periods of submersion, may be especially detrimental to these communities.

13.11 Filter beds

Moss growth, especially *Amblystegium riparium*, on filter beds in England was sometimes extensive, but did not cause any real problems (Hussey 1982). There was concern that more growth would cause ponding. Mosses tended to grow more easily on rough or pitted surfaces than on smooth ones, and when similar sizes of pebbles, stones, and granite occurred together, moss grew only on stones and granite. In my surveys of Appalachian Mountain streams, I found that smooth rocks lacked mosses, whereas rough-surfaced ones usually had them (unpublished observations).

13.12 Temperature

Numerous changes in our environment will influence the temperature of streams. Cooling water used in factories is returned to streams many degrees hotter. The projected global warming trend will render some streams warmer while others will become cooler.

Fontinalis growth is particularly sensitive to prolonged heat (three weeks above 20°C can cause cessation of growth) and may actually be favoured by cooling in some parts of the world (Glime and Acton 1979; Glime 1980, 1987*a*, *b*, *c*; Glime and Raeymaekers 1987). Sommer (1981) found that the optimum temperature for net photosynthesis in *Fontinalis antipyretica* was c. 10°C, and that the dark respiratory rate continued to climb as the temperature was elevated to 20°C. Glime (1987*a*, *b*, *c*) found that six species of *Fontinalis* had maximum growth at 10–15°C. Irmscher (1912) found that *Fontinalis* has the greatest cold resistance of 28 mosses studied.

Hygrohypnum ochraceum ceases growth above 26°C and experiences optimum growth at 18–21°C (Sanford *et al.* 1974). Construction of the Keswick and Shasta dams of the Sacramento River, California, USA, seems to have caused an increase in this moss by maintaining temperatures below the lethal levels. On the other hand, Florschütz *et al.* (1972) suggested that the increase of *Fissidens crassipes* in The Netherlands was due to a 2°C warming in the rivers from discharge of cooling waters of industries.

Aquatic mosses in general seem to be sensitive to temperature. Sanford (1979) found that *Amblystegium riparium* had optimum growth at 23°C but was killed at 33°C. Branch elongation and reproductive branch formation were altered by temperature, and high temperature stress resulted in a proliferation of branches. Heat resistance of the main axis apices can keep these plants alive through periods of stress, but growth is arrested until cooler temperatures return.

The northern lichen *Hydrotheria venosa* extends into the mountains of Tennessee, where it has an annual temperature range of 3–18°C (Dennis *et al.* 1981), and we can suppose that it would suffer from elevated temperatures. In the River Tweed in England, *Verrucaria* species occurred throughout the river; *Collema* and *Dermatocarpon* were absent in the uppermost and lowermost portions of the river, suggesting a possible relationship to flow rate or temperature (Holmes and Whitton 1975).

13.13 Summary

If the pH is high, bryophytes can successfully compete only in areas of high aeration (*e.g. Fissidens grandifrons*) or low nutrients or in shaded areas where light is insufficient for higher plants (Bain and Proctor 1980). Acidification can favour bryophytes by providing more CO_2 and reducing absorption of metals. Ability to sequester heavy metals and radionuclides in the cell wall, as electron-dense particles, and in vacuoles permits survival of bryophytes under conditions in which many vascular plants cannot survive. Increased temperatures, on the other hand, are likely to decimate both bryophytes and lichens in areas where cool temperatures once favoured their growth. Turbidity may favour bryophytes in shallow water by reducing the light intensity, but associated abrasives can damage buds and decrease reproduction.

References

Abaigar, J. M. and Diaz, M. S. (1987). *Efecto de la contaminacion organica sobre indices de feofitinizacion en trasplantes de briofitos acuaticos (Rio Iregua, La Rioja, Espnan)*, pp. 288–97. Actas del IV Congreso Espanol de Limnologia, Sevilla.

Allen, E. D. and Spence, D. H. N. (1981). The differential ability of aquatic plants to utilize the inorganic carbon supply in fresh waters. *New Phytologist*, **87**, 269–83.

Anschütz, I. and Gessner, F. (1954). Der Ionenaustausch bei Torfmoosen (*Sphagnum*). *Flora Jena*, **141**, 178–80.

Auerbach, S., Pruefer, P., and Weise, G. (1973). Gasstoffwechselphysiologische Schadigungskriterien bei submersen Makrophyten vom typ *Fontinalis antipyretica* L. unter Einwirkung von Schwermetallen oder Phenol. (Detrimental effects of toxical charge by heavy metals or phenol on submerged macrophytes (*Fontinalis antipyretica* L.)). *International Revue der Gesamten Hydrobiologie*, **58**, 19–32.

Bachmann, R. W. and Odum, E. P. (1960). Uptake of Zn^{65} and primary productivity in marine benthic algae. *Limnology and Oceanography*, **5**, 349–55.

Bain, J. T. and Proctor, M. C. F. (1980). The requirement of aquatic bryophytes for free CO_2 as an inorganic carbon source: Some experimental evidence. *New Phytologist*, **86**, 393–400.

Balch, R. F. (1970). *A biological water quality study of the Sacramento River in the vicinity of Anderson, California—1969*. Project 2328, Report 6: A progress report to Kimberly-Clark Corporation. The Institute of Paper Chemistry, Appleton, Wisconsin.

Bates, J. W. and Brown, D. H. (1974). Control of cation levels in seashore and inland mosses. *New Phytologist*, **73**, 483–95.

Baudo, R. and Varini, P. G. (1976). Copper, manganese and chromium concentrations in five macrophytes from the delta of River Toce (northern Italy). *Memorie dell'Instituto Italiano di Idrobiologia*, **33**, 305–24.

Baudo, R., Galanti, G., Guilizzone, P., and Varini, P.G. (1981). Relationships between heavy metals and aquatic organisms in Lake Mezzola hydrographic system (northern Italy). 4: Metal concentrations in six submersed aquatic macrophytes. *Memorie dell'Instituto Italiano di Idrobiologia*, **39**, 203–25.

Benson-Evans, K. and Williams, P. F. (1976). Transplanting aquatic bryophytes to assess river pollution. *Journal of Bryology*, **9**, 81–91.

Bolle, D., Wauthy, G., and Lebrun, P. (1977). Preliminary studies of watermites (Acari, Prostigmata) as bioindicators of pollution in streams. *Annale Societe Royale Zoologique Belgique*, **106**, (2–4), 201–9.

Broda, E. (1965). Mechanism of uptake of trace elements by plants (experiments with radio-zinc). In *Isotopes and radiation in soil-plant nutrition studies*, STI(PUB) 108, IAEA, Vienna.

Brown, D. H. and House, K. L. (1978). Evidence of a copper-tolerant ecotype of the hepatic *Solenostoma crenulatum*. *Annals of Botany*, **42**, 1383–92.

Buck, G. W. and Brown, D. H. (1978). Cation analysis of bryophytes; the significance of water content and ion location. *Bryophytorum Bibliotheca*, **13**, 735–46.

Burton, M. A. S. and Peterson, P. J. (1979). Studies on zinc localization in aquatic bryophytes. *The Bryologist*, **82**, 594–8.

Caines, L. A., Watt, A. W., and Wells, D. E. (1985). The uptake and release of some trace metals by aquatic bryophytes in acidified waters in Scotland. *Environmental Pollution (Ser. B)*, **10**, 1–18.

Cajander, V. R. and Ihantola, R. (1984). Mercury in some higher aquatic plants and plankton in the estuary of the River Kokemaenjoki, southern Finland. *Annales Botanici Fennici*, **21**, 151–6.

Cheniae, G. M. and Martin, I. F. (1969). Photoreactivation of manganese catalyst in photosynthetic oxygen evolution. *Plant Physiology*, **44**, 351–60.

Clymo, R. S. (1963). Ion exchange in *Sphagnum* and its relation to bog ecology. *Annals of Botany N. S.*, **27**, 309–24.

Dennis, W. M., Collier, P. A., Depriest, P., and Morgan, E. L. (1981). Habitat notes on the aquatic lichen *Hydrotheria venosa* Russell in Tennessee. *The Bryologist*, **84**, 402–3.

Dietz, F. (1973). The enrichment of heavy metals in submerged plants. In Proc. Internat. Conf. (Jerusalem), *Advances in Water Pollution Research*, **6**, 53–62.

Eidt, D. C., Sosick, A. J., and Mallet, V. N. (1984). Partitioning and short-term persistence of Fenitrothion in New Brunswick (Canada) headwater streams. *Archives of Environmental Contamination and Toxicology*, **13**, 43–52.

Empain, A. (1974). Relations quantitatives entre les bryophytes de la sambre Belge et leur frequence d'emersion: Distribution verticale et influence de la pollution. *Bulletin Societe Royale Botanique Belgique*, **107**, 361–74.

Empain, A. (1976). Les bryophytes aquatiques utilises comme traceurs de la contamination en metaux lourdes des eaux douces. *Memoirs Societe Royale Botanique Belgique*, **7**, 141–56.

Empain, A. (1985). Heavy metals in bryophytes from Shaba Province. In *The heavy metal-tolerant flora of southcentral Africa: A multidisciplinary approach*, (eds R. R.

Brooks and R. Malaisse), pp. 103–17. A. A. Balkema, Boston.

Empain, A., Lambinon, J., Mouvet, C., and Kirchmann, R. (1980). Utilisation des bryophytes aquatiques et subaquatiques comme indicateurs biologiques de la qualite des eaux courantes. In *La pollution des eaux continentales*, (2nd edn), (ed. P. Pesson), pp. 195–223. Gauthier-Villars, Paris.

Erdman, J. A. and Modreski, P. J. (1984). Copper and cobalt in aquatic mosses and stream sediments from the Idaho Cobalt Belt. *Journal of Geochemical Exploration*, **20**, 75–84.

Field, J. H. (1972). Blackening of various bryophytes in iron solutions. (Bryophyte Notes). *Proceedings of the Birmingham Natural History Society*, **22**, 134.

Fischer, H. (1948). Plasmolyseform und mineralsaltzgehalt in Alternden Blattern. I: Untersuchungen an *Helodea* und *Fontinalis*. *Planta*, **35**, 513–27.

Fletcher, A. (1973). The ecology of marine (littoral) lichens on some rocky shores of Anglesey. *Lichenologist*, **5**, 368–400.

Florschütz, P. A., Gradstein, S. R., and Rubers, W. V. (1972). The spreading of *Fissidens crassipes* Wils. (Musci) in the Netherlands. *Acta Botanici Neerlandica*, **21**, 174–9.

Foulquier, L. and Hebrard, J. P. (1976). Etude experimental de la fixation et de la decontamination du sodium 22 par une mousse Dulcaquicole: *Platyhypnidium riparioides* (Hedw.) Dix. *Oecologia Plantarum*, **11**, 267–76.

Frahm, J.-P. (1975). Toxitoleranzversuche an Wassermoosen. *Gewässer und Abwässer*, **57/58**, 59–66.

Frahm, J.-P. (1976). Weitere Toxitoleranzversuche an Wassermoosen. *Gewässer und Abwässer*, **60/61**, 113–23.

Glime, J. M. (1970). Zonation of bryophytes in the headwaters of a New Hampshire stream. *Rhodora*, **72**, 276–9.

Glime, J. M. (1971). Response of two species of *Fontinalis* to field isolation from stream water. *The Bryologist*, **74**, 383–6.

Glime, J. M. (1978). Insect utilization of bryophytes. *The Bryologist*, **81**, 186–7.

Glime, J. M. (1980). Effects of temperature and flow on rhizoid production in *Fontinalis*. *The Bryologist*, **83**, 477–85.

Glime, J. M. (1984). Physio-ecological factors relating to reproduction and phenology in *Fontinalis dalecarlica*. *The Bryologist*, **87**, 17–23.

Glime, J. M. (1987*a*). Growth model for *Fontinalis duriaei* based on temperature and flow conditions. *Journal of the Hattori Botanical Laboratory*, **62**, 101–9.

Glime, J. M. (1987*b*). Temperature optima of *Fontinalis novae-angliae*: implications for its distribution. *Symposia Biologica Hungarica*, **35**, 569–76.

Glime, J. M. (1987*c*). Phytogeographic implications of a *Fontinalis* (Bryopsida) growth model based on temperature and flow conditions for six species. *Memoirs of the New York Botanical Garden*, **45**, 154–70.

Glime, J. M. and Acton, D. W. (1979). Temperature effects on assimilation and respiration in the *Fontinalis duriaei*—periphyton association. *The Bryologist*, **82**, 382–92.

Glime, J. M. and Keen, R. E. (1984). The importance of bryophytes in a man-centered world. *Journal of the Hattori Botanical Laboratory*, **55**, 133–46.

Glime, J. M. and Raeymaekers, G. (1987). Temperature effects on branch and rhizoid production in six species of *Fontinalis*. *Journal of Bryology*, **14**, 779–90.

Glime, J. M., Nissila, P. D., Trynoski, S. E., and Fornwall, M. D. (1979). A model for attachment of aquatic mosses. *Journal of Bryology*, **10**, 313–20.

Grahn, O. (1976). *Macrophyte succession in Swedish lakes caused by deposition of airborne acid substances*, pp. 519–30. US Forest Service General Technical Report NE-23. (Unpublished report.)

Gregory, K. J. (1976). Lichens and the determination of river channel capacity. *Earth Surface Processes*, **1**, 273–85.

Hargreaves, J. W., Lloyd, E. J. H., and Whitton, B. A. (1975). Chemistry and vegetation of highly acidic streams. *Freshwater Biology*, **5**, 563–76.

Harrison, A. D. (1958). The effects of sulphuric acid pollution on the biology of streams in the Transvaal, South Africa. *Verhein Internationale Verein Limnologie*, **13**, 603.

Hasler, A. D. (1938). Fish biology and limnology of Crater Lake, Oregon. *Journal of Wildlife Management*, **2**, 94–103.

Hebrard, J. P., Foulquier, L., and Grauby, A. (1968). Apercu sur les modalites de la contamination d'une mousse dulcicole, *Platyhypnidium riparioides* (Hedw.) Dix. par le cesium-137 et le strontium-90. *Revue Bryologique et Lichénologique*, **36**, 219–42.

Holmes, N. T. H. and Whitton, B. A. (1975). Macrophytes of the River Tweed. *Transactions of the Botanical Society of Edinburgh*, **42**, 369–81.

Holmes, N. T. H. and Whitton, B. A. (1977). Macrophytic vegetation of River Tees in 1975—observed and predicted changes. *Freshwater Biology*, **7**, 43–60.

Hörman, P. K. (1970). Germanium. In *Handbook of Geochemistry*, Vol. II/2L., (ed. K. H. Wedepohl), pp. 32-L-1–32-L-3. Springer, New York.

Hussey, B. (1982). Moss growth on filter beds. *Water Research*, **16**, 391–8.

Irmscher, E. (1912). Über die Resistenz der Laubmoose gegen Austrocknung und Kalte. *Jahrbuch Wissenschaften Botanique*, **50**, 387–449.

Jacobson, L., Hannapel, R. J., Moore, D. P., and Schaedle, M. (1961). Influence of calcium on selectivity of ion absorption process. *Plant Physiology*, **36**, 58–61.

James, P. W., Hawksworth, D. L., and Rose, F. (1977). Lichen communities in the British Isles: A preliminary conspectus. In *Lichen ecology*, (ed. M. R. D. Seaward), pp. 295–413. Academic Press, New York.

Kirchmann, R. and Lambinon, J. (1973). Bioindicateurs végétaux de la contamination d'un cours d'eau par des effluents d'une centrale nucleaire à eau pressurisée: Evaluation des rejets de la centrale de la sena (Chooz, Ardennes Francaises) au moyen des végétaux aquatiques de la meuse. *Bulletin Société Royale Botanique Belgique*, **106**, 187–201.

Kulikov, N. V., Bochenina, N. V., and Molchanova, I. V. (1976). Peculiarities of ^{90}Sr and ^{137}Cs accumulation in some moss species. *Ekologiya, Sverdlovsk*, **1976**, (6), 82–5.

Lackey, J. B. (1938). The flora and fauna of surface waters polluted by acid mine drainage. *U. S. Public Health Report*, **53**, 1449–507.

Lambinon, J., Kirchmann, R., and Colard, J. (1976). Evolution recente de la contamination radioactive des écosystems aquatique et ripicole de la meuse par les effluents de la centrale nucleaire de la Sena (Chooz, Ardennes Francaises). *Memoirs Societe Royale Botanique Belgique*, **7**, 157–75.

Ledl, G., Janauer, G. A., and Horak, O. (1981). Die anreicherungvon Schwermetallen in Wasserpflanzen aus einigen oesterreichischen Fliebgewässern. *Acta Hydrochimica et Hydrobiologica*, **9**, 651–63.

Lepp, N. W. (1981). Copper. In *Effect of heavy metal pollution on plants. I: Effects of trace metals on plant function*, (ed. N. W. Lepp), pp. 111–43. Applied Science Publishers, London.

Lewis, K. R. (1973*a*). The effect of suspended coal particles on the lifeforms of the aquatic moss *Eurhynchium riparioides* (Hedw.) I: The gametophyte plant. *Freshwater Biology*, **3**, 251–7.

Lewis, K. R. (1973*b*). The effect of suspended coal particles on the life forms of the aquatic moss *Eurhynchium riparioides* (Hedw). II: The effect on spore germination and regeneration of apical tips. *Freshwater Biology*, **3**, 391–5

Lodenius, M. (1980). Aquatic plants and littoral sediments as indicators of mercury pollution in some areas of Finland. *Annales Botanici Fennici*, **17**, 336–40.

López, J. and Carballeira, A. (1989). A comparative study of pigment contents and response to stress in five species of aquatic bryophytes. *Lindbergia*, **15**, 188–94.

McLean, R. O. and Jones, A. K. (1975). Studies of tolerance to heavy metals in the flora of the Rivers Ystwyth and Clarach, Wales. *Freshwater Biology*, **5**, 431–44.

Malhotra, S. S. and Hocking, D. (1976). Biochemical and cytological effects of sulfur dioxide on plant metabolism. *New Phytologist*, **76**, 227–37.

Maurel-Kermarrec, A., Pally, M., Foulquier, L., and Hebrard, J. P. (1983). Cinetique de la fixation d'un melange de cesium 137, de chrome 51, de cobalt 60, de manganese 54 et de sodium 22 par *Platyhypnidium riparioides* (Hedw.) Dix. *Cryptogamie, Bryologique et Lichénologique*, **4**, 299–313.

Melzer, A. (1980). Ökophysiologische Aspekte der N-Ernährung submerser Wasserpflanzen. *Verhein Gesamten Ökologie*, **8**, 257–62.

Merlin, G. (1988). Contamination par le PCP décosystèmes aquatiques reconstitués—dégradation et effets sur les végétaux. Unpublished Thèse Etat. Université Joseph Fourier Grenoble I.

Morrison, B. R. S. and Wells, D. E. (1981). The fate of fenitrothion in a stream environment and its effect on the fauna, following aerial spraying of a Scottish forest. *Science of the Total Environment*, **19**, 233.

Mouvet, C. (1979). Utilisation des bryophytes aquatiques pour l'etude de la pollution des cours d'eau par les metaux lourds et les radionucleides. *Revue Biologie et Ecologie Mediterranean*, **6**, 193–204.

Mouvet, C. (1984). Accumulation of chromium and copper by the aquatic moss *Fontinalis antipyretica* L. ex Hedw. transplanted in a metal-contaminated river. *Environmental Technological Letters*, **5**, 541–8.

Mouvet, C. (1985). The use of aquatic bryophytes to monitor heavy metals pollution of freshwaters as illustrated by case studies. *Verhein Internationale Verein Limnologie*, **22**, 2420–5.

Mouvet, C. (1987). *Accumulation et relargage de plomb, zinc, cadmium, chrome et cuivre par des mousses aquatiques en milieu naturel et au laboratoire.* Laboratoire d'Ecologie, Université de Metz.

Mouvet, C. (1989). *Utilisation des mousses aquatiques pour la surveillance de la pollution des milieux aquatiques par les metaux lourds et les micropolluants organiques.* Laboratoire d'Ecologie, Université de Metz.

Mouvet, C., Galoux, M., and Bernes, A. (1985). Monitoring of polychlorinated biphenyls (PCBs) and hexachlorocyclohexanes (HCH) in freshwater using the aquatic moss *Cinclidotus danubicus. Science of the Total Environment*, **44**, 253–67.

Mouvet, C., Cordebar, P., Gallissot, B., and Roger, P. (1986) The use of aquatic mosses to monitor micropollutants such as cadmium and PCB's. Laboratory and field results. In *Environmental contamination*, 2nd International Conference, Amsterdam, September, 1986, pp. 115–19. CEP Consultants, Edinburgh.

Mouvet, C., Andre, B., and Lascombe, C. (1987). Aquatic mosses for the monitoring of heavy metals in running freshwaters comparison with sediments. Presented at the 6th International Conference, Heavy Metals in the Environment, New Orleans.

Mouvet, C., Cordebar, P., and Gallissot, B. (1988). *Evaluation de rejets de micropollutants mineraux (metaux lourds) et organiques (organochlores) par dosages dans les mousses aquatiques*, pp. III.5.1–III.5.8. Societe Hydrotechnique de France, Rapport No. 5.

Ornes, W. H. and Wildman, R. B. (1979). Effects of cadmium (II) on aquatic vascular plants. In *Trace substances in environmental health XIII. A*, (ed. D. D. Hemphill), pp. 304–12. University of Missouri, Columbia.

Peñuelas, J. (1984*a*). Pigments of aquatic mosses of the river Muga, NE Spain, and their response to water pollution. *Lindbergia*, **10**, 127–32.

Peñuelas, J. (1984*b*). Pigment and morphological response to emersion and immersion of some aquatic and terrestrial mosses in N.E. Spain. *Journal of Bryology*, **13**, 115–28.

Persson, H. (1956). Studies in 'copper mosses.' *Journal of the Hattori Botanical Laboratory*, **17**, 1–18.

Peterson, P. J. (1969). The distribution of Zinc-65 in *Agrostis tenuis* Sibth. and *A. stolonifera* L. tissues. *Journal of Experimental Botany*, **20**, 653–61.

Pickering, D. C. and Puia, I. L. (1969). Mechanism for the uptake of zinc by *Fontinalis antipyretica. Physiologia Plantarum*, **22**, 653–61.

Prescott, G. W. (1968). *The algae: a review.* Houghton Mifflin, Boston.

Ramaut, J. (1954). Modifications de pH apportaes par la tourbe et le *Sphagnum* secs aux solutions salines et à l'eau bidistilles. *Bulletin Cl. Science Academie Belgique (Ser 5)*, **40**, 305–19.

Ray, S. N. and White, W. J. (1979). *Equisetum arvense*—an aquatic vascular plant as a biological monitor for heavy metal pollution. *Chemosphere*, **3**, 125–8.

Roberts, D. A., Singer, R., and Boylen, C. W. (1985). The submersed macrophyte community of Adirondack Lakes (New York, USA) of varying degrees of acidity. *Aquatic Botany*, **21**, 219–35.

Robinson, G. D. (1981). Adsorption of Cu, Zn and Pb near sulphide deposits by hydrous manganese-iron oxide coatings on stream alluvium. *Chemical Geology*, **33**, 65–79.

Roelofs, J. G. M. (1983). Impact of acidification and eutrophication on macrophyte communities in soft waters in The Netherlands. I: Field observations. *Aquatic Botany*, **17**, 139–55.

Roelofs, J. G. M., Schuurkes, J. A. A. R., and Smits, A. J. M. (1984). Impact of acidification and eutrophication on macrophyte communities in soft waters. II: Experimental studies. *Aquatic Botany*, **18**, 389–411.

Rühling, A. and Tyler, G. (1970). Sorption and retention of heavy metals in the woodland moss *Hylocomium splendens* (Hedw.) Br. et Sch. *Oikos*, **21**, 92–7.

Samecka-Cymerman, A. (1983). The effect of phenol on aquatic mosses *Fontinalis antipyretica* L. and *Platyhypnidium rusciforme* (Neck.) Fleisch. in cultures. *Polskie Archiwum Hydrobiologii*, **30**, 141–7.

Sand-Jensen, K. and Rasmussen, L. (1978). Macrophytes and chemistry of acidic streams from lignite mining areas. *Botaniska Tidsskrift*, **72**, 105–11.

Sanford, G. R. (1979). Temperature related growth patterns in *Amblystegium riparium. The Bryologist*, **82**, 525–32.

Sanford, G. R., Bayer, D. E. and Knight, A. W. (1974). *An evaluation of environmental factors affecting the distribution of two aquatic mosses in the Sacramento River near Anderson, California.* Univ. Calif. Depts. Bot., Water Science Engineering. Davis, California.

Satake, K. (1983). Elemental composition of water and aquatic bryophytes collected from the central part of Kyushu (Mt. Kuju, Mt. Aso and the city of Kumamoto). *Proceedings of the Bryological Society of Japan*, **3**, (9), 137–40.

Satake, K. and Miyasaka, K. (1984). Evidence of high mercury accumulation in the cell wall of the liverwort *Jungermannia vulcanicola* Steph. to form particles of a mercury-sulphur compound. *Journal of Bryology*, **13**, 101–5.

Satake, K., Soma, M., Seyama, H., and Uehiro, T. (1983). Accumulation of mercury in the liverwort *Jungermannia vulcanicola* Steph. in an acid stream Kashiranashigawa in Japan. *Archiv für Hydrobiologie*, **99**, 80–92.

Satake, K., Shimizu, H., and Nishikawa, M. (1984). Elemental composition of the aquatic liverwort *Jungermannia vulcanicola* Steph. in acid streams. *Journal of the Hattori Botanical Laboratory*, **56**, 241–8.

Satake, K., Nishikawa, M., and Shibata, K. (1989a). Distribution of aquatic bryophytes in relation to water chemistry of the acid river Akagawa, Japan. *Archive für Hydrobiologie*, **116**, 299–311.

Satake, K., Takamatsu, T., Soma, M., Shibata, K., Nishikawa, M., Say, P. J., and Whitton, B. A. (1989b). Lead accumulation and location in the shoots of the liverwort *Scapania undulata* (L.) Dum. in stream water at Greenside Mine, England. *Aquatic Botany*, **33**, 111–22.

Say, P. J. and Whitton, B. A. (1980). Changes in flora down a stream showing a zinc gradient. *Hydrobiologia*, **76**, 255–62.

Say, P. J. and Whitton, B. A. (1981). Chemistry and plant ecology of zinc-rich streams in the northern Pennines. In *Heavy metals in northern England: environmental and biological aspects*, (eds P. J. Say and B. A. Whitton), pp. 55–63.

Say, P. J. and Whitton, B. A. (1982). Chemistry and plant ecology of zinc-rich streams in France. 2: The Pyrenees. *Annales de Limnologie*, **18**, 19–31.

Say, P. J. and Whitton, B. A. (1983). Accumulation of heavy metals by aquatic mosses. 1: *Fontinalis antipyretica* Hedw. *Hydrobiologia*, **100**, 245–60.

Say, P. J., Harding, J. P. C., and Whitton, B. A. (1981). Aquatic mosses as monitors of heavy metal contamination in the River Etherow, Great Britain. *Environmental Pollution (Ser. B)*, **2**, 295–307.

Schuurkes, J. A. A. R., Kok, C. J., and Den Hartog, C. (1986). Ammonium and nitrate uptake by aquatic plants from poorly buffered and acidified waters. *Aquatic Botany*, **24**, 131–46.

Schwoerbel, J. and Tillmanns, G. C. (1974). Assimilation of nitrogen from the medium and nitrate reductase activity in submerged macrophytes: *Fontinalis antipyretica* L. *Archiv für Hydrobiologie (Supplement 47)*, **2**, 282–94.

Schacklette, H. T. and Erdman, J. A. (1982). Uranium in spring water and bryophytes at Basin Creek in central Idaho. *Journal of Geochemical Exploration*, **17**, 221–36.

Shaw, J. (1987). Evolution of heavy metal tolerance in bryophytes. II: An ecological and experimental investigation of the 'copper moss,' *Scopelophila cataractae* (Pottiaceae). *American Journal of Botany*, **74**, 813–21.

Shaw, J. and Anderson, L. E. (1988). Factors affecting the distribution and abundance of the 'copper moss', *Scopelophila ligulata*, in North America. *Lindbergia*, **14**, 55–8.

Skene, M. (1915). The acidity of *Sphagnum* and its relation to chalk and mineral salts. *Annals of Botany*, **29**, 65–87.

Södergren, S. (1976). Ecological effects of heavy metal discharge in a salmon river. *Institute of Freshwater Research, Drottningholm Rep.* 55, 91–131.

Söderlund, S., Forsberg, A., and Pedersén, M. (1988). Concentrations of cadmium and other metals in *Fucus vesiculosus* L. and *Fontinalis dalecarlica* Br. Eur. from the northern Baltic Sea and the southern Bothnian Sea. *Environmental Pollution*, **51**, 197–212.

Soma, M., Seyama, H., and Satake, K. (1988). X-ray photoelectron spectroscopic analysis of lead accumulated in aquatic bryophytes. *Talanta*, **35**, 68–70.

Somers, I. I. and Shive, J. W. (1942). The iron-manganese relation in plant metabolism. *Plant Physiology*, **17**, 582–602.

Sommer, C. H. (1981). Reaktionen von *Fontinalis antipyretica* Hedw. nach Experimentellen Belastungen mit Schwermetallverbindungen. Unpublished Doctoral Thesis. University of Ulm, Germany.

Stokes, P. M. (1986). Ecological effects of acidification on primary producers in aquatic systems. *Water, Air and Soil Pollution*, **30**, 421–38.

Suominen, J., Häsänen, E., and Nuorteva, P. (1977). Vesikasvien elohopeapitoisuuksista Hämeenkyrössä. *Luonnon Tutkija*, **81**, 122–3.

Timofiejeva, S., Stom, D., Butorov, B., and Belych, L. I. (1977). In *First Soviet Union Conference on Higher Water and Littoral Plants Lecture Thesis*, pp. 145–7. Borok (in Russian).

Trollope, D. R. and Evans, B. (1976). Concentrations of copper, iron, lead, nickel and zinc in freshwater algal blooms. *Environmental Pollution*, **11**, 109–16.

Uotila, P. (1971). Distribution and ecological features of hydrophytes in the polluted Lake Vanajavesi, S. Finland. *Annales Botanici Fennici*, **8**, 257–95.

Vrhovsek, D., Martincic, A., and Kralj, M. (1981). Evaluation of the polluted River Savinja with the help of macrophytes. *Hydrobiologia*, **80**, 97–110.

Vrhovsek, D., Martincic, A., Krajl, M., and Stremfelj, M. (1985). Pollution degrees of the two alpine rivers evaluated with Bryophyta species. *Bioloski Vestnik*, **33**, (2), 95–106.

Ward, N. I., Brooks, R. R., and Roberts, E. (1977). Heavy metals in some New Zealand bryophytes. *The Bryologist*, **80**, 304–12.

Webster, H. J. (1985). Elemental analyses of *Pohlia nutans* growing on coal seeps in Pennsylvania. *Journal of the Hattori Botanical Laboratory*, **58**, 207–24.

Wehr, J. D. and Whitton, B. A. (1983a). Aquatic cryptogams of natural acid springs enriched with heavy metals: The Kootenay Paint Pots, British Columbia. *Hydrobiologia*, **98**, 97–105.

Wehr, J. D. and Whitton, B. A. (1983b). Accumulation of heavy metals by aquatic mosses. 2: *Rhynchostegium riparioides*. *Hydrobiologia*, **100**, 261–84.

Wehr, J. D. and Whitton, B. A. (1983c). Accumulation of heavy metals by aquatic mosses. 3: Seasonal changes. *Hydrobiologia*, **100**, 285–91.

Wehr, J. D., Empain, A., Mouvet, C., Say, P. J., and Whitton, B. A. (1983). Methods for processing aquatic mosses used as monitors of heavy metals. *Water Research*, **17**, 985–92.

Weise, G., Burger, G., Fuchs, S., and Schuermann, L. (1985). Zum Einfluss von Rueckstaenden von Zink—und Bleiverbindungen im Wasser auf die Assimilation von *Fontinalis antipyretica*. *Acta Hydrochimica et Hydrobiologica*, **13**, 25–34.

Welsh, R. P. H. and Denny, P. (1980). The uptake of lead and copper by submerged aquatic macrophytes in two English lakes. *Journal of Ecology*, **68**, 443–55.

Wenerick, W. R., Stevens, S. E., Jr, Webster, H. J., Stark, L. R., and DeVeau, E. (1989). Tolerance of three wetland plant species to acid mine drainage: a greenhouse study. In *Constructed wetlands for wastewater treatment*, (ed. D. A. Hammer), pp. 801–7. Lewis, Chelsea, MI.

Wetzel, R. G., Brammer, E. S., Lindstroem, K., and Forsberg, C. (1985). Photosynthesis of submersed macrophytes in acidified lakes. II: Carbon limitation and utilization of benthic CO_2 sources. *Aquatic Botany*, **22**, 107–20.

Whitehead, N. E. and Brooks, R. R. (1969). Aquatic bryophytes as indicators of uranium mineralization. *The Bryologist*, **72**, 501–7.

Whitton, B. A. and Diaz, B. M. (1980). Chemistry and plants of streams and rivers with elevated zinc. In *Trace substances in environmental health*, (ed. D. D. Hemphill), pp. 457–63. University of Missouri, Columbia

Whitton, B. A. and Diaz, B. M. (1981). Influence of environmental factors on photosynthetic species composition in highly acidic waters. *Verhein Internationale Verein Limnologie*, **21**, 1459–65.

Whitton, B. A., Say, P. J., and Jupp, B. P. (1982). Accumulation of zinc, cadmium, and lead by the aquatic liverwort *Scapania*. *Environmental Pollution (Ser. B)*, **3**, 299–316.

Yan, N. D., Miller, G. E., Wile, I., and Hitchin, G. G. (1985). Richness of aquatic macrophyte floras of soft water lakes of differing pH and trace metal content in Ontario, Canada. *Aquatic Botany*, **23**, 27–40.

14

The evolutionary capacity of bryophytes and lichens

A. Jonathan Shaw

14.1 Introduction

Most information about evolution in plants is derived from studies of flowering plants. The angiosperms are hardly typical of plants in general, however, having undergone an explosive adaptive radiation within the last 60–100 million years. It is the purpose of this chapter to summarize data on evolutionary change in natural populations of lichens and bryophytes. Although the bryophytes and lichens are often discussed together, they are not, of course, phylogenetically related.

Lichens are symbiotic organisms consisting of a fungus (the mycobiont, generally an ascomycete) and an alga (the phycobiont or photobiont, generally Chlorophyta). Weber (1977) discussed the lichen symbiosis in terms of the gene-for-gene hypothesis developed from studies of plant-host/fungal-pathogen genetic interactions. Although validity of this theory has been questioned (Barrett 1983), concepts of lichen evolution need to consider models of coevolution. A most pressing issue in lichen evolutionary biology is how closely coevolved the fungal and algal symbionts are. Genetic interactions between the mycobionts and phycobionts of lichens may be analogous to epistatic interactions between non-allelic genes in other organisms. That is, the fitness value of any gene in the mycobiont may depend upon its own genetic background as well as the genetic background provided by the phycobiont. The degree to which these epistatic interactions result in close genetic coadaptation between the fungus and alga depends in part upon the degree of taxonomic specificity between symbionts (Howe 1984).

An understanding of the factors leading to evolutionary change in the bryophytes is simpler in that at least we are here dealing with single rather than composite organisms. Nevertheless, the bryophytes are heterogeneous and many biologists consider them to include at least three divisions of plants that may not be monophyletic (e.g. Bold *et al.* 1987; Schofield 1985; Mishler and Churchill 1984). It has often been suggested that some kinds of evolutionary change occur more slowly in bryophytes than in flowering plants.

14.2 What is 'evolutionary capacity'?

Evolutionary capacity (evolutionary potential; Longton 1976) is an enigmatic concept that can be defined in different ways depending upon the context and the scientific questions being asked. In studies of phylogenetic history, evolutionary capacity may be thought of as the propensity for speciation. Biological and ecological factors that may promote or retard the rate at which genetic lineages branch have been hotly debated (e.g. Charlesworth *et al.* 1982; Longton 1976; Stanley 1979). Another measure of evolutionary capacity is the rate at which a population responds to natural selection. It is possible to distinguish between natural selection (measured as differences in fitness between individuals in a population), and the response to natural selection (measured as the difference in mean phenotype from one generation to the next) (Falconer 1981; Arnold and Wade 1984). Thus, natural selection can occur without a genetic response if differences in fitness are not heritable. The heritability of a trait is proportional to the amount of additive genetic variation for that trait in the population, and the potential response to selection is proportional to both the intensity of selection and the heritability (Fisher 1930; Falconer 1981).

This line of reasoning has led to the suggestion that measurements of intraspecific genetic diversity provide estimates of evolutionary capacity (Longton 1976; Cummins and Wyatt 1981; Wyatt *et al.* 1989*a*). Indeed, it is axiomatic that without genetic variation there can be no evolutionary change, either by genetic drift or by natural selection. Also, genetic diversity is readily estimated for natural populations making it an operational definition of evolutionary capacity (Cummins and Wyatt 1981). Nevertheless, although genetic diversity is the minimum requirement for evolutionary change, many other factors can constrain the rate of change in particular traits. For example, negative genetic correlations between traits can constrain rates of evolution in both traits even where genetic variability is ample (Antonovics 1976; Roach 1985; Falconer 1981). Thus, we need an approach to the question of evolutionary capacity that takes into account both the amount of genetic variability within species, and some way to infer if this genetic variation has or is likely to lead to phylogenetic diversification.

This information can be provided by detailed investigations of 'species genetic structure'. The genetic structure of a species is a description of both the amount and the spatial patterning of genetic variability among individuals of the species. Viewed in relation to the patterning of environmental variation throughout the range of the species, such genetic information provides insights into past history, present status, and future potential. Historical factors that contribute to species genetic structure include phylogenetic artefacts (characteristics that derive from ancestors irrespective of their present adaptive significance), as well as patterns of previous selection and genetic drift, founder events, population bottlenecks, and local extinctions. Species genetic structure provides clues about current processes, including mating behavior, dispersal patterns, mortality, and genetic adaptation to local ecological conditions. From all these clues, one can make predictions about the capacity of the species to respond to future environmental changes. Firstly, measures of genetic variability within populations provide general estimates of genetic diversity and establish the baseline potential for evolution. Secondly, measures of differentiation between populations in components of fitness provide insight into the extent to which species have actually undergone adaptive evolutionary change.

14.3 The genetic structure of lichen species

The first and undoubtedly the most crucial problem in the evolutionary biology of lichens is determining the basic unit of evolutionary change. Lichens are, of course, considered fundamentally to be fungi and there is good evidence that such characteristic features as secondary product chemistry and morphogenesis depend on the mycobiont rather than the phycobiont (Ahmadjian *et al.* 1980). Culberson and Ahmadjian (1980) showed that the mycobiont of *Cladonia cristatella* produced uniform chemistry when lichenized with several species of *Trebouxia* from twelve different lichens. More recently, Culberson *et al.* (1985) cultured spores from *Cladonia crytochlorophaea* and *C. grayi*, which differ in chemistry, with the alga *Trebouxia erici* which is the phycobiont of *Cladonia cristatella*, and found that the lichen progeny consistently displayed only the chemistry of the fungal parent.

The taxonomy of lichen phycobionts is not well understood because most species are rare in nature and laboratory culture is necessary in order to establish species limits (Tschermak-Woess 1988). It appears that there is much greater diversity among fungi than among their algal partners. Although there are estimated to be some 13 000 species of lichenized fungi, only about 100 species in 24 genera of Chlorophyta and 13 genera of cyanobacteria appear to be involved in these symbioses. Certain phycobionts are present in a number of distantly related lichens. For example, *Trentepohlia unbrina* occurs in species of *Chaenotheca, Graphis, and Opegra-*

pha (Hawksworth and Hill 1984). Ahmadjian and Jacobs (1981) tried lichenizing the fungus of *Cladonia cristatella* with about ten species of *Trebouxia* plus species of *Pseudotrebouxia* and *Friedmannia israeliensis*, and found that lichenization occurred with six different species of *Trebouxia* but none of the other algae. Similar results indicating relatively weak specificity of phycobiont/mycobiont symbioses were provided by Ahmadjian *et al.* (1980 and elsewhere). Some examples of more than one algal species occurring with a single fungus may reflect overly narrow algal species concepts (Tschermak-Woess 1988), but in certain cases the phycobionts of a single lichen are so different that the taxonomy is not in question. For example, *Chaenotheca brunneola* is lichenized with *Dictyochloropsis splendida* in Austria and elsewhere in Europe, but in Costa Rica, New Zealand, and Australia it is associated with a trebouxioid alga (Tschermak-Woess 1988).

'Diffuse coevolution' (Futuyma 1986) involving relatively low levels of symbiont specificity may significantly constrain the evolution of genetic coadaptation between partners. It is only in cases of very specialized interspecific interactions that close genic coadaptation is expected (Futuyma and Slatkin 1983). Additional information about phycobiont diversity, both within and between species, is needed in order to understand more clearly their effects on lichen ecology. It has frequently been observed, for example, that the detrimental effects of pollutants such as SO_2, NO_x, O_3, and flourides are exhibited first by the phycobiont and only much later by the mycobiont (Anderson and Treshow 1984; Galun and Ronen 1988). A wide range of variation in sensitivity occurs among species of lichens, and in many genera, related species differ significantly. In *Lecanora*, for example, *L. conizaeoides* is highly tolerant of SO_2 and has apparently expanded its range into polluted urban centres of Britain and Europe since the mid 1800s. *Lecanora dispersa* is also relatively tolerant, but *L. muralis* is considered to be quite sensitive (Marsh and Nash 1979). There is some evidence that intraspecific variation in tolerance of pollutants may occur in *L. muralis* (see Chapter 6). *Lecanora* might therefore be an appropriate genus for investigating the evolution of resistance to air pollution. Do these species share the same phycobiont? Are differences in resistance related to intra- or interspecific genetic differences between phycobionts? Is there a coevolved genetic interaction between phycobiont and mycobiont that determines different resistance levels? Do genetic differences between the mycobionts somehow affect resistance levels in the phycobionts, even with little or no genetic difference between phycobionts of the three lichens? Basic to any of these questions is the phylogenetic and genetic relationship between mycobiont and phycobiont.

An understanding of microevolutionary processes in lichens is further complicated by the observation that adjacent thalli can fuse in natural populations. Laundon (1978), for example, showed that two species of *Haematomma* growing on a single rock surface came into contact, fused, and formed composite thalli consisting of discrete regions differing in chemistry. Incomplete fusion was evidenced by the fact that yellow patches (with usnic acid) could be distinguished from grey patches (lacking usnic acid). Laundon speculated that soredia might simultaneously disperse both variants of the composite lichen.

Larson and Carey (1986) measured net CO_2 exchange rate, dark respiration, weight per unit size, and isozyme variability in *Umbilicaria mammulata* and *U. vellea*. Large thalli showed significant heterogeneity in all four parameters, but no such variation was observed in young thalli of either species, suggesting that the physiological and biochemical variation accumulates over time. Larson and Carey discussed a variety of mechanisms to explain this intra-thallus variation, including fusion of adjacent thalli, somatic mutation, heterokaryosis (not presently known in lichens; Bowler and Rundel 1975), parasexuality, and non-genetic effects such as developmental variation and subtle microenvironmental variation.

Larson and Carey (1986) thought that the most likely explanation for this remarkable intra-thallus variation was continual reinfection of the growing thallus by its own (genetically variable) ascospores, and/or by spores from adjacent thalli. Their results indeed suggest genetic heterogeneity within single thalli. It is especially noteworthy that the isozyme variability involved multi-enzyme phenotypes rather than single enzymes. Of nine enzyme systems they assayed, only two were monomorphic within thalli. On the other hand, it is apparent from their figures that there was little spatial congruence between patches revealed by each of the parameters they measured. If each of these types of measurements reflects genetic heterogeneity, then independence of patterns implies that the number of genetically distinct 'individuals' constituting these thalli must be extremely large.

Knowledge about the frequency of such intra-thallus heterogeneity and an understanding of its sources are critical to future evolutionary studies of lichen populations. Growth rate variations between different lobes of individual thalli have been documented in a wide variety of lichens (e.g. Armstrong 1988; Benedict and Nash 1990; earlier literature reviewed by Topham 1977). If these differences reflect intra-thallus genetic heterogeneity, the basic unit of evolution that might respond to natural selection is not the individual lichen, but the 'thallus patch'. Additional research designed to assess whether intra-thallus differences in growth rates and pollution resistance are maintained through successive cycles of asexual fragmentation and regeneration would almost certainly prove rewarding.

Ahmadjian (1964) found extensive variation in growth form, size, and pigmentation in single spore isolates of the mycobionts from twelve specimens of *Cladonia cristatella* collected in Massachusetts. Similarly, Fahselt (1987) showed that single spore isolates from one apothecium of *C. cristatella* showed variation in patterns of isozyme banding for esterase and alkaline phosphatase. She also found that the 12 single spore isolates differed in biomass production in liquid culture. Four fungal clones from *Cladonia cristatella* differed in the concentrations of secondary chemical compounds they produced under experimental conditions (Culberson *et al.* 1983). Furthermore, responses (in terms of chemistry) to light, temperature, and developmental stage also differed significantly among clones. More recently, sequence variation in rDNA has been demonstrated among single spore isolates from the same apothecium (see below).

A number of early workers observed conidia, apparently functioning as spermatia, attached to the trichogynes of various lichen mycobionts (Lawrey 1984). Ott (1987) made detailed observations on lichenization of *Xanthoria parietina* grown on pieces of bark inoculated with ascospores. The most common naturally occurring algae on the bark were of the *Pleurococcus* type, but *Pseudotrebouxia* (the normal phycobiont of *X. parietina*) also occurred in much lesser abundance. Ott observed that fungal hyphae from germinated ascospores were able to form a loose contact with foreign (non-trebouxioid) algae, possibly deriving temporary nutritional benefit from them, until contact was made with *Pseudotrebouxia*. When the latter occurred, tighter permanent contacts were made and thallus development commenced. Ott also observed that the fungal hyphae were able to extract cells of *Pseudotrebouxia* from dispersed soredia in nature.

Patterns of morphological variation in natural populations sometimes suggest genetic recombination. Eckman and Froberg (1988) described Swedish populations that contained numerous expressions intermediate between *Aspicilia contorta* and *A. hoffmannii*. They attributed intermediates to hybridization and introgression between these two normally distinct species. Anderson and Rudolf (1956) argued that morphological variation in a mixed population of *Cladonia* reflected hybridization and introgression. However, in the absence of experimental studies such interpretations remain speculative.

The most detailed information about sexual reproduction in nature comes from a study of intermingled chemotypes of the *Cladonia chlorophaea* complex in North Carolina (Culberson *et al.* 1988). Four of the fourteen known chemotypes of *C. chlorophaea* occur in North Carolina and were included in the study. Culberson *et al.* (1988) isolated three to six spores from each of 249 podetia growing within two centimetres of those belonging to other chemotypes. They innoculated the single spore mycobiont cultures with *Trebouxia erici* and obtained chemical information from 167 cultures.

In the Appalachian Mountains, *C. grayi* and *C. merochlorophaea* can interbreed, as can *C. cryptochlorophaea* and *C. perlomera* on the Atlantic coastal plain. Moreover, Culberson *et al.* demonstrated that reciprocal crosses occur within each pair of chemotypes. In contrast, no interbreeding was observed between *C. grayi* and *C. cryptochlorophaea*, which the authors consequently regarded as morphologically indistinguishable sibling species. These observations support earlier circumstantial evidence that microconidia function as spermatia in *Cladonia* (Jahns 1970; Honneggar 1984).

Recent work by DePriest (1990) has revealed extensive rDNA sequence variation in members of the *Cladonia chlorophaea* complex. Variation in rDNA genotypes occurred among sites in the southern Appalachian Mountains, within sites, within proximate mats, and among fungal progeny of individual podetia. Twelve rDNA restriction patterns were found among a sample of 50 individuals from one of the sites, and only five were represented by more than one individual. Eight genotypes for a variable region of the rDNA gene that DePriest amplified using the polymerase chain reaction occurred within just one small mat. Variation also occurred within the chemotypes defined by secondary product chemistry and the occurrence of shared genotypes between chemotypes demonstrated gene flow between them (past or present). Using single spore cultures of mycobionts, purified phycobiont cultures, and intact lichens, DePriest showed that exceptionally high levels of variation occur in the mycobiont, with much less in the phycobiont. Much of the variation was attributable to multiple insertion events and sequence divergence within the insertions.

Recent electrophoretic studies have also demonstrated remarkable levels of isozyme variation in natural lichen populations (e.g. Fahselt 1988, 1989; Hageman and Fahselt 1990). These workers have not attempted to apply standard genetic statistics such as genetic diversity or distance measures to their data because the allelic basis for banding patterns are not obvious in lichens, especially considering the unpredictable contributions of mycobionts and phycobionts (Fahselt 1985). Instead, they score gels for the presence/absence of bands and compute several measures of variability including the Shannon Diversity Index commonly used in ecological studies.

Scoring presence/absence of bands in this manner inflates levels of variation. Nevertheless, virtually every study to date has found intraspecific variability in isozyme patterns. Both within- and between-site variability have been demonstrated in a number of species (Fahselt and Hageman 1983; Hageman and Fahselt 1984; Fahselt 1988). Polymorphism occurred within morphotypes of *Cladonia cristatella*, although the morphotypes themselves could be distinguished using isozymes (Fahselt 1986). Morphological variants of *Umbilicaria muhlenbergii* could not be distinguished by their isozyme profiles (Hageman and Fahselt 1986*a*). Fahselt (1989) found evidence of higher levels of variability in *Umbilicaria virginicis*, an apotheciate species, than in *U. decussata* or *U. hyperborea*, which are sterile at least within the area of her study (Central Ellesmere Island). On the other hand, both sexual and asexual species contained significant variability.

Brown and Kershaw (1985) compared banding patterns for seven enzymes in three Canadian populations of *Peltigera rufescens*. These populations exhibited extensive isozyme differences with not even one enzyme system completely monomorphic across the three sites. These results, suggesting high levels of genetic differentiation, corroborated physiological observations from the same study which suggested significant differences in seasonal patterns of photosynthetic responses to temperature. This is in contrast to an earlier similar study of *Cladonia rangiferina*, in which morphological and physiological differences between sun and shade populations were not accompanied by significant isozyme differences (MacFarlane *et al.* 1983). Similarly, sun and shade forms of *Cladonia stellaris* showed low levels of isozyme polymorphism in general, and no consistent differences between the forms from contrasting habitats (Kershaw *et al.* 1983).

Studies of isozyme variation in lichens should be interpreted with caution. Hageman and Fahselt (1986*b*), for example, demonstrated differences in banding pattern between samples from the same stand of *Umbilicaria mammulata* gathered at five different times of the year. Variation was most evident in esterases and phosphatases. Thus, extensive intraspecific variation that has repeatedly been found may not reflect genetic polymorphism exclusively. In fact, the most variable enzyme systems in most studies by Fahselt and co-workers are precisely those found to vary seasonally in *U. mammulata*. Although other systems have also been shown to vary within lichen species, levels of genic polymorphism suggested by electrophoretic studies are significantly inflated.

Ecophysiological studies of photosynthetic rates and their relationships to light intensity, temperature, and degree of hydration have uncovered much phenotypic variation both within and between species of lichens. Not surprisingly, such variation frequently results from both genetic and environmental effects (e.g. Nash *et al.* 1990). Physiological studies have recently been thoroughly reviewed by Kershaw (1985), so only a brief summary of their significance in the present context is provided here.

It has long been known that many lichens are characterized by high levels of phenotypic plasticity, especially in morphological traits (Weber 1977). A common expression of intraspecific variation is the difference between sun and shade forms of many lichens (e.g. *Parmelia caperata* (Harris 1971); *Peltigera aphthosa* (Kershaw and MacFarlane 1980)). In general, morphological differences between sun and shade forms appear to be environmentally induced, and differences in photosynthetic responses to hydration level and light intensity between such forms may simply be a reflection of these induced features rather than an indication of genetic differentiation (Kershaw 1985). There appears to be a trade-off between water-holding capacity and protection for the phycobiont from excessive light intensities. High light induces the formation of a thickened upper cortex, which promotes water storage. This in turn may reduce the rate of gas exchange and result in the lower photosynthetic rates that are observed in sun forms at high levels of thallus hydration.

Although there is little evidence of true ecotypic differentiation (i.e. genetically based, habitat-specific variation) in lichens, there are differences, possibly adaptive, in the *patterns* of phenotypic plasticity expressed by different populations of some species. Brown and Kershaw (1984), for example, found that Arctic and temperate populations of *Peltigera rufescens* differ in the pattern of seasonal changes they display in net photosynthetic capacity. Temperate populations undergo a marked summer increase in photosynthetic capacity under high light and temperature conditions, whereas Arctic populations show no such seasonal variation.

The presence of intraspecific variation in patterns of phenotypic plasticity suggests that plasticity itself could be a trait subject to natural selection or genetic drift (Schlichtling and Levin 1986). Indeed, differences between related species of lichens in the patterns of physiological plasticity displayed in response to environmental variation (e.g. moisture, temperature, seasonality; Larson 1980) suggest that intraspecific variation may have been fixed in different phylogenetic lineages during the course of evolution. Work by Culberson *et al.* (1983) showed that variation in patterns of phenotypic plasticity do, in fact, occur within species. Whether particular patterns were fixed in different species by the direct forces of natural selection is more difficult to discern, but additional studies of intra- and interspecific variation in the patterns of phenotypic plasticity would be valuable.

14.4 The genetic structure of bryophyte species

In terms of basic aspects of genetics and life cycle, the bryophytes are unique among land plants. The free-living haploid gametophytes are with few exceptions perennial, and the sporophyte is small, unbranched, and almost invariably short-lived. Although the bryophytes, if circumscribed to include the mosses, liverworts, and hornworts, may be paraphyletic (Bower 1908; Bold *et al.* 1987; Mishler and Churchill 1984), similarities in life cycle may suggest similarities in microevolutionary processes (Shaw 1991*b*). On the other hand, it has been suggested that in general, evolution occurs more slowly in liverworts than mosses (Khanna 1964). There are only about 50–75 per cent as many species of liverworts as mosses (Schofield 1985). Liverworts are more uniform cytologically and appear to be characterized by lower levels of isozyme polymorphism in general than the mosses (Anderson 1980; Wyatt *et al.* 1989*a*).

Little is known about patterns of mating in natural bryophyte populations although the potential exists for complicated mixed systems that include intragametophytic self-fertilization, intergametophytic self-fertilization (the same as 'selfing' as applied to seed plants), various degrees of inbreeding, and outcrossing between unrelated gametophytes. The potential for asexual reproduction is also very high since virtually every gametophytic cell is totipotent.

In species with bisexual gametophytes there may be mechanisms to promote outcrossing, but experimental data are almost non-existent. Self-compatibility has been shown experimentally in *Desmatodon randii* (Lazarenko 1974), *Physcomitrella patens* (Cove 1983), and *Funaria hygrometrica* (Wettstein 1924; Weitz and Heyn 1981; Shaw 1990*b*). Individual gametophytes of the moss *Atrichum undulatum* produce male gametangia one season and female gametangia the next, making it functionally dioecious though genetically monoecious (Crum 1983). In the autoecious species, *Funaria hygrometrica*, most plants begin producing antheridia about 2–4 weeks before they form archegonia, although the period of full gametangial development may overlap substantially (Shaw 1991*a*). Anderson and Lemmon (1974) noted that only abut 10 per cent of the gametophytes in a population of the autoecious species *Weissia controversa* bear both antheridia and archegonia at the same time.

Since intragametophytic self-fertilization results in a completely homozygous sporophyte, meiotic spore progeny derived from such a sporophyte are genetically uniform and identical in genotype to the parental gametophyte. Consequently, isozyme variability among meiotic spore progeny from a single sporophyte provides evidence of outcrossing in the previous gametophyte generation. Zielinski (1986) used a peroxidase marker to estimate outcrossing rates in the monoecious liverwort, *Pellia borealis*, and found that only 5 per cent of 40 sporophytes were demonstrably outcrossed. Outcrossing appeared to be higher (25 per cent) in the related species, *Pellia epiphylla* (Zielinski 1984). In contrast, Shaw (1991*a*) found no evidence of outcrossing in two populations of the autoecious moss, *Funaria hygrometrica*. In that study, 13 isozyme loci were scored in 719 meiotic progeny derived from 132 sporophytes in the two populations. These results strongly suggest that intragametophytic selfing was the rule in both populations. Moreover, both populations, although fairly extensive in area and number of plants, may have been completely clonal as no variability was observed *among* families of progeny either. Additional studies of mating systems operating in natural populations of bryophytes are sorely needed.

The total phenotypic variance exhibited by a sample of gametophytes grown under uniform conditions can be partitioned into components attributable to differences among plants derived from the same sporophyte, and to differences among families of plants derived from different sporophytes ('haploid sib families'; Shaw 1991*b*). If all genetic variance for some quantitative trait of interest is additive, random mating will result in 50 per cent of the population variance occurring among haploid sib families and 50 per cent among sibs within families (de Jong, unpublished data; Shaw 1991*b*). Inbreeding increases the percentage of variance among families by decreasing the variance within families. Since gametophytes are haploid, dominance does not contribute to population variance. Under conditions of random mating, epistasis increases the variance within families, decreasing the variance between families (Shaw and Weir, unpublished data). Analyses of haploid sib families can therefore be used to infer the genetical architecture of gametophytic traits and mating behaviour in natural populations.

Most of the variance in protonemal growth rates and tolerance of metals among plants from two populations of *Funaria hygrometrica* grown in axenic culture occurs within families of related sibs (Shaw 1991*a*). This pattern of variability suggests high rates of outcrossing, a surprising result for a self-compatible species. Moreover, the growth rate data contradict isozyme evidence that suggests low or non-existing levels of genetic variability in the same populations (Shaw 1991*a*). Interpretation of these results is complicated by the observation that variance in growth rates and tolerance of copper also occurred both within and among experimentally produced homozygous families of *F. hygrometrica* grown in axenic culture (Shaw 1990*b*). These plants were known to be genetically uniform, yet the phenotypic differences between single spore sibs were consistent among replicate cultures grown in separate dishes (Shaw 1990*b*). The source of variability in growth rates and metal tolerance is not yet clear, but it is evident that significant variation can exist in the absence of meiotic segregation. The importance of subtle non-genetic sources of variation that can be maintained through successive cycles of asexual regeneration needs to be investigated further.

A haploid sib analysis of population structure in the dioecious moss *Ceratodon purpureus*, grown on soil rather than in axenic culture, also suggests high levels of growth rate variation both among and within families (unpublished data). Plants of *C. purpureus* in this study reached reproductive maturity and when the sex of individuals was determined it became evident that much of the within-population variation was attributable to growth rate differences between male and female gametophytes. Since male and female gametophytes segregate within families (sex determination is chromosomal in this species), more of the population variance occurs among sibs within families than among the families. In addition to the differences in growth rates between male and female plants, significant variation also occurred among gametophytes of the same sex, suggesting additional genetic variation.

Haploid sib analyses thus suggest significant variability in quantitative traits within populations of mosses. This is in agreement with accumulating data from isozyme studies (Wyatt *et al.* 1989*a*). Values for percentages of polymorphic loci per population, mean number of alleles per locus per population, and mean expected heterozygosity per locus per population from studies of mosses fall within the ranges published for flowering plants (Hamrick 1989). In some species, levels of electrophoretically detectable protein variation are comparable to those found in highly outcrossed, wind-pollinated trees. Values for G_{ST} from a study of 13 populations of *Plagiomnium ciliare*, a species that appears to contain exceptionally high levels of genetic variability, also suggest significant genetic differentiation between populations (Wyatt *et al.* 1989*b*). These authors attributed the high levels of electrophoretic variability in *P. ciliare* to multiple niche selection.

Levels of electrophoretic variability within species of liverworts are generally lower than in mosses. However, some variability occurs in almost every species, and cryptic genetic differentiation has been documented within morphologically stenotypic species (e.g. Krzakowa 1977; Dewey 1989). In the morphologically uniform thallose liverwort, *Concephalum conicum*, at least five 'races' can be distinguished by differences in isozyme patterns. Wyatt *et al.* (1989*a*) recently reviewed levels of isozyme variability in mosses and liverworts.

Isozymes and quantitative traits sometimes give conflicting information about the genetic structure of species (Hamrick 1989). Although a general view of 'evolutionary capacity' can be developed from isozyme surveys, the potential for a species to respond to natural selection for a particular characteristic depends upon the presence of variability in that trait. Moreover, the presence of genetic variability does not provide insights into the mechanisms by which organisms adapt to changing environments. Phenotypic plasticity, or physiological acclimatization, may be the best evolutionary solution to a particular pattern of environmental heterogeneity even when genetic variability for response exists.

A number of studies have compared responses of different populations of widespread mosses to such environmental factors as light intensity, temperature, day length, or habitat contamination by toxic metals, and have found surprisingly little evidence of ecotypic differentiation (see reviews by

Longton 1974, 1979, 1988; Shaw 1991*b*). Kallio and Saarnio (1986), for example, compared plants of *Pleurozium schreberi, Hylocomium splendens,* and *Racomitrium lanuginosum* from boreal, subarctic, and high Arctic sites, and found virtually no evidence of genetically based differences in optimum temperatures for photosynthesis, or response to day length. This and other studies have demonstrated apparent genetic differences in general vigour (measured either as maximum photosynthetic capacity—e.g. Kallio and Saarnio 1986, or vegetative growth rates—e.g., Longton 1981; Shaw and Albright 1989) between plants from different populations, but these generalized differences do not appear to be adaptations to specific environments. In summarizing the adaptation of bryophytes to polar environments, Longton (1988) noted that physiological acclimatization appears to be generally more important than genetic specialization. Kallio and Heinonen (1975) suggested that most widely distributed mosses appear to be 'pre-adapted' by broad physiological tolerances.

One of the most extensively studied species of mosses is *Funaria hygrometrica.* Dietert (1980) showed that populations from Alaska, Alberta, Massachusetts, and Texas exhibit differences in spore germination and growth under common garden conditions, although the differences could not be related to obvious climatic differences between sites of origin. Shaw *et al.* (1989) found that populations of *F. hygrometrica* vary in the degree of tolerance to copper and zinc, and that these differences could, in general, be related to habitat contamination by these metals. In a more detailed study, Shaw (1990*a*) found that levels of copper tolerance in populations of *F. hygrometrica* were closely related to substrate copper concentration, strongly suggesting differential responses to natural selection. However, differences among populations in tolerance of cadmium and zinc could not be so easily attributed to selection since levels of tolerance were not correlated with substrate metal concentration. The correlation between zinc tolerance and substrate zinc levels found in the first study (Shaw *et al.* 1989) may have been a result of the fact that contaminated substrates of plants used in that study contained high levels of both copper and zinc, and there appears to be a genetic correlation between tolerances to these metals in *F. hygrometrica* (Shaw 1990*a*). Thus, plants selected on the basis of copper tolerance in these habitats may also exhibit elevated levels of zinc tolerance as a correlated response.

In general, tolerances in *F. hygrometrica* appear to be less metal-specific than in most flowering plants that have been studied (Antonovics *et al.* 1971; Baker and Walker 1990). This is especially true in another species, *Bryum argenteum,* in which no evidence of genetically specialized metal tolerant ecotypes have been found in comparisons of populations growing in contaminated and normal environments (Shaw *et al.* 1989; Shaw and

Albright 1989). This lack of physiological specialization is in agreement with other studies on *B. argenteum* in which virtually no evidence of genetically based differences in response to climatic factors could be demonstrated between populations growing in polar, temperate, and even tropical latitudes (Longton 1981).

14.5 Summary and conclusions

The question of what constitutes an individual is the most pressing issue in the evolutionary biology of lichens. Specifically, we need to understand better the genetic and evolutionary relationships between phycobionts and mycobionts of lichen species. In addition, we need more information about genetic diversity *within* the mycobiont and phycobiont components of individual thalli. What are the phylogenetic relationships between phyco-bionts of different lichens? Do one, few, or many mycobiont genotypes contribute to the structure of lichen thalli? Is there as much genetic heterogeneity within the fungal symbionts of individual thalli as appears on the basis of the limited data now available? Lichen thalli might be better viewed as populations than as organisms. Does selection operate within lichen thalli? Is selection among thalli more comparable to classical group selection than to individual selection?

In apparent contradiction to the remarkable levels of variation that appear to characterize populations of lichen thalli, there is little evidence of ecotypic differentiation between populations with regard to important environmental characteristics. Instead, physiological acclimatization appears to be of great importance for survival of lichens in varying environments. This impression may well be attributable to a paucity of researchers and incomplete information. On the other hand, thorough studies of several species where genetic differentiation was expected have failed to yield such evidence.

In spite of predictions that populations of haploid bryophytes might contain low levels of genetic variability, they, too, have been shown in general to contain moderate or even high levels of electrophoretically detectable protein variation. Variation in quantitative traits also appears to be common within species and sometimes within populations. This level of variation suggests either that sexual reproduction occurs more frequently in bryophytes than has been thought, or that frequent sex is not necessary for the generation and maintenance of variability. The presence of ample variation makes it all the more remarkable that genetically specialized ecotypes appear not to be the most common response to environmental heterogeneity in this group of organisms. To the contrary, broad tolerances and the ability to acclimatize to environmental change characterize those species of bryophytes that have been studied in detail. In this important regard, bryophytes and lichens appear to bear significant similarity.

References

Ahmadjian, V. (1964). Further studies on lichenized fungi. *The Bryologist*, **67**, 87–98.

Ahmadjian, V. and Jacobs, J. B. (1981). Relationship between fungus and algae in the lichen *Cladonia cristella* Tuck. *Nature*, **289**, 169–72.

Ahmadjian, V., Russell, L. A., and Hidreth, K. C. (1980). Artificial reestablishment of lichens. I: Morphological interactions between the phycobionts of different lichens and the mycobionts of *Cladonia cristatella* and *Lecanora chrysoleuca*. *Mycologia*, **72**, 73–89

Anderson, E. and Rudolf, E. D. (1956). An analysis of variation in a variable population of *Cladonia*. *Evolution*, **10**, 147–56.

Anderson, F. K. and Treshow, M. (1984). Responses of lichens to atmospheric pollution. In *Air pollution and plant life*, (ed. M. Treshow), pp. 259–89. Wiley, Chichester.

Anderson, L. E. (1980). Cytology and reproductive biology of mosses. In *The mosses of North America*, (ed. R. J. Taylor and A. E. Leviton), pp. 37–76. Pacific Division, AAAS, San Francisco.

Anderson, L. E. and Lemmon, B. E. (1974). Gene flow distances in the moss, *Weissia controversa* Hedw. *Journal of the Hattori Botanical Laboratory*, **38**, 67–90.

Antonovics, J. (1976). The nature of limits to natural selection. *Annals of the Missouri Botanical Garden*, **63**, 224–47.

Antonovics, J., Bradshaw, A. D., and Turner, R. G. (1971). Heavy metal tolerance in plants. *Advances in Ecological Research*, **7**, 1–84.

Arnold, S. J. and Wade, M. J. (1984). On the measurement of natural and sexual selection: theory. *Evolution*, **38**, 709–19.

Armstrong, R. A. (1988). Substrate colonization, growth, and competition. In *CRC handbook of lichenology*, Vol. II, (ed. M. Galun), pp. 3–16. CRC Press, Boca Raton, Florida.

Baker, A. J. M. and Walker, P. L. (1990). Ecophysiology of metal uptake by tolerant plants. In *Heavy metal tolerance in plants: evolutionary aspects*, (ed. A. J. Shaw), pp. 155–93. CRC Press, Boca Raton, Florida.

Barrett, J. A. (1983). Plant-fungus symbioses. In *Coevolution*, (eds D. J. Futuyma and M. Slatkin), pp. 139–60. Sinauer Associates, Sunderland, Massachusetts.

Benedict, J. B. and Nash, T. H. III. (1990). Radial growth and habitat selection by morphologically similar chemotypes of *Xanthoparmelia*. *The Bryologist*, **93**, 319–27.

Bold, H. C., Alexopoulos, C. J., and Delevoryas, T. (1987). *Morphology of plants and fungi*, (5th edn). Harper & Row, New York.

Bower, F. O. (1908). *Origin of a land flora*. Macmillan, London.

Bowler, P. A. and Rundel, P. W. (1975). Reproductive strategies in lichens. *Botanical Journal of the Linnean Society*, **70**, 325–40.

Brown, D. and Kershaw, K. A. (1984). Photosynthetic capacity change in *Peltigera*. II: Contrasting season patterns of net photosynthesis in two populations of *Peltigera rufescens*. *New Phytologist*, **96**, 447–57.

Brown, D. and Kershaw, K. A. (1985). Electrophoretic and gas exchange patterns in two populations of *Peltigera rufescens*. In *Lichen physiology and cell biology*, (ed. D. H. Brown). pp. 111–28. Plenum, New York.

Charlesworth, B., Lande, R., and Slatkin, M. (1982). A neo-darwinian commentary on macroevolution. *Evolution*, **36**, 474–98.

Cove, D. J. (1983). Genetics of Bryophyta. In *New manual of bryology*, (ed. R. M. Schuster), pp. 222–31. Hattori Botanical Laboratory, Nichinan.

Crum, H. A. (1983). *Mosses of the great lakes forest*, 3rd edn. Contributions from the University of Michigan Herbarium. University of Michigan Herbarium, Ann

Culberson, C. F. and Ahmadjian, V. (1980). Artificial reestablishment of lichens. II: Secondary products of resynthesized *Cladonia cristatella* and *Lecanora chrysoleuca*. *Mycologia*, **72**, 90–109.

Culberson, C. F., Culberson, W. L., and Johnson, A. (1983). Genetic and environmental effects on growth and production of secondary compounds in *Cladonia cristatella*. *Biochemical Systematics and Ecology*, **11**, 77–84.

Culberson, C. F., Culberson, W. L., and Johnson, A. (1985). Does the symbiont alga determine chemotype in lichens? *Mycologia*, **77**, 657–60.

Culberson, C. F., Culberson, W. L., and Johnson, A. (1988). Gene flow in lichens. *American Journal of Botany*, **75**, 1135–9.

Cummins, H. and Wyatt, R. (1981). Genetic variability in natural populations of the moss *Atrichum angustatum*. *The Bryologist*, **84**, 30–8.

DePriest, P. (1990). Analysing gene flow in the *Cladonia chlorophaea* complex (lichen-forming Ascomycotina). Fourth International Mycological Congress Abstracts, p. 116.

Dewey, R. M. (1989). Genetic variation in the liverwort *Riccia dictyospora* (Ricciaceae, Hepaticopsida). *Systematic Botany*, **14**, 144–67.

Dietert, M. F. (1980). The effect of temperature and photoperiod on the development of geographically isolated populations of *Funaria hygrometrica* and *Weissia controversa*. *American Journal of Botany*, **67**, 369–80.

Eckman, S. and Froberg, L. (1988). Taxonomic problems in *Aspicilia contorta* and *Aspicilia hoffmannii*: an effect of hybridization? *International Journal of Mycology and Lichenology*, **3**, 215–26.

Fahselt, D. (1985). Multiple enzyme forms in lichens. In *Lichen physiology and cell biology*, (ed. D. H. Brown), pp. 129–43. Plenum, New York.

Fahselt, D. (1986). Multiple enzyme forms of morphotypes in a population of *Cladonia cristatella* Tuck. *The Bryologist*, **89**, 139–43.

Fahselt, D. (1987). Electrophoretic analyses of esterase and alkaline phosphatase enzyme forms in single spore cultures of *Cladonia cristatella*. *The Lichenologist*, **19**, 71–5.

Fahselt, D. (1988). Measurement of intrapopulational variation in five species of epiphytic lichens. *The Lichenologist*, **20**, 377–84.

Fahselt, D. and Hageman, C. (1983). Isozyme banding patterns in two stands of

Cetraria arenaria Karnef. *The Bryologist*, **86**, 129–34.

Falconer, D. S. (1981). *Introduction to quantitative genetics*, Longman, New York.

Fisher, R. A. (1930). *Genetical theory of natural selection*. Clarendon Press, Oxford.

Futuyma, D. J. (1986). *Evolutionary biology* (2nd edn). Sinauer Associates, Sunderland, Massachusetts.

Futuyma, D. J. and Slatkin, M. (eds) (1983). *Coevolution*. Sinauer Associates, Sunderland, Massachusetts.

Galun, M. (1988). Lichenization. In *CRC handbook of lichenology*, (ed. M. Galun), pp. 153–69. CRC Press, Boca Raton, Florida.

Galun, M. and R. Ronen. (1988). Interaction of lichens and pollutants. In *CRC handbook of lichenology*, (ed. M. Galun), pp. 55–72. CRC Press, Boca Raton, Florida.

Hageman, C. and Fahselt, D. (1983). Intraspecific variability of enzymes of the lichen *Umbilicaria mammulata*. *Canadian Journal of Botany*, **62**, 617–23.

Hageman, C. and Fahselt, D. (1986a). A comparison of isozyme patterns of morphological variants in the lichen *Umbilicaria muhlenbergii* (Ach.) Tuck. *The Bryologist*, **89**, 285–90.

Hageman, C. and Fahselt, D. (1986b). Constancy of enzyme electrofocusing patterns in a stand of the lichen *Umbilicaria mammulata*. *Canadian Journal of Botany*, **64**, 1928–34.

Hageman, C. and Fahselt, D. (1990). Multiple enzyme forms as indicators of functional sexuality in the lichen *Umbilicaria vellea*. *The Bryologist*, **93**, 389–94.

Hamrick, J. L. (1989). Isoenzymes and the analysis of genetic structure in plant populations. In *Isozymes in plant biology*, (eds D. E. Soltis and P. S. Soltis), pp. 87–105. Disoscorides Press, Portland.

Harris, G. P. (1971). The ecology of corticolous lichens. II: The relationship between physiology and the environment. *Journal of Ecology*, **59**, 441–56.

Hawksworth, D. L. and Hill, D. J. (1984). *The lichen-forming fungi*. Blackie, Glasgow.

Honneger, R. 1984. Scanning electron microscopy of the contact site conidia and trichogynes in *Cladonia furcata*. *Lichenologist*, **16**, 11–19.

Howe, H. F. (1984). Constraints on the evolution of mutualisms. *American Naturalist*, **123**, 764–77.

Jahns, H. M. (1970). Untersuchungen zur Entwichlungsgeschichte der Cladoniaceen unter besonderes Berücksichtigung des Potentienproblems. *Nova Hedwigia*, **20**, 1–177.

Kallio, P. and Heinonen, S. (1975). CO_2 exchange and growth of *Rhacomitrium lanuginosum* and *Dicranum undulatum*. *Ecological Studies*, **16**, 138–48.

Kallio, P. and Saarnio, N. (1986). The effect on mosses of transplantation to different latitudes. *Journal of Bryology*, **14**, 159–78.

Kappen, L. (1988). Ecophysiological relationships in different climatic regions. In *CRC handbook of lichenology*, Vol. II, (ed. M. Galun), pp. 37–100. CRC Press, Boca Raton, Florida.

Kershaw, K. A. (1985). *Physiological ecology of lichens*. Cambridge University Press, New York.

Kershaw, K. A. and MacFarlane, J. D. (1980). Physiological-environmental interactions in lichens. X: Light as an ecological factor. *New Phytologist*, **84**, 687–702.

Kershaw, K. A., MacFarlane, M. R., and Fovarque, A. (1983). Phenotypic differences in the seasonal pattern of net photosynthesis in *Cladonia stellaris*. *Canadian Journal of Botany*, **51**, 2169–80.

Khanna, K. R. (1964). Differential evolutionary activity in the bryophytes. *Evolution*, **18**, 642–70.

Krzakowa, M. (1977). Isozymes as markers of inter- and intraspecific differentiation in hepatics. *Bryophytorum Bibliotheca*, **13**, 427–34.

Larson, D. W. (1980). Seasonal change in the pattern of net CO_2 exchange in *Umbilicaria* lichens. *New Phytologist*, **84**, 349–69.

Larson, D. W. and Carey, C. K. (1986). Phenotypic variation within 'individual' lichen thalli. *American Journal of Botany*, **73**, 214–23.

Laundon, J. R. (1978). *Haematomma* chemotypes form fused thalli. *Lichenologist*, **10**, 221–5.

Lawrey, J. D. (1984). *Biology of lichenized fungi*. Praeger Scientific, New York.

Lazarenko, A. S. (1974). Some considerations on the nature and behavior of the relic moss, *Desmatodon randii*. *The Bryologist*, **77**, 474–7.

Longton, R. E. (1974). Genecological differentiation in bryophytes. *Journal of the Hattori Botanical Laboratory*, **38**, 49–65.

Longton, R. E. (1976). Reproductive biology and evolutionary potential in bryophytes. *Journal of the Hattori Botanical Laboratory*, **41**, 205–23.

Longton, R. E. (1979). Climatic adaptation of bryophytes in relation to systematics. In *Bryophyte systematics*, (eds G. C. S. Clarke and J. Duckette), pp. 511–31. Academic Press, London.

Longton, R. E. (1981). Inter-population variation in morphology and physiology in the cosmopolitan moss *Bryum argenteum* Hedw. *Journal of Bryology*, **11**, 501–20.

Longton, R. E. (1988). *The biology of polar bryophytes and lichens*. Cambridge University Press, Cambridge.

MacFarlane, J. D., Kershaw, K. A., and Webber, M. R. (1983). Physiological-environmental interactions in lichens. XVII: Phenotypic differences in the seasonal pattern of net photosynthesis in *Cladonia rangiferina*. *New Phytologist*, **94**, 217–33.

Marsh, J. E. and Nash, T. H., III. (1979). Lichens in relation to the Four-Corners powerplant in New Mexico. *The Bryologist*, **82**, 20–8.

Mishler, B. D. and Churchill, S. P. (1984). A cladistic approach to the phylogeny of the 'bryophytes.' *Brittonia*, **36**, 406–24.

Nash, T. H., Boucher, V. L., Gebauer, R., and Larson, D. W. (1990). Morphological and physiological plasticity in *Ramalina menziesii*: studies with reciprocal transplants between a coastal and inland site. *Bibliotheca Lichenologia*, **38**, 357–65.

Ott, S. (1987). Sexual reproduction and developmental adaptations in *Xanthoria parietina*. *Nordic Journal of Botany*, **7**, 219–28.

Roach, D. A. (1985). Genetic and environmental constraints to selection response to juvenile characters in *Geranium carolinianum*. In *Genetic differentiation and dispersal in plants*, (ed. P. Jacuard, G. Heim, and J. Antonovics), pp. 191–201. Springer, Berlin.

Schofield, W. B. (1985). *Introduction to bryology*. Macmillan, New York.

Schlichtling, C. D. and Levin, D. A. (1986). Phenotypic plasticity: an evolving character. *Biological Journal of the Linnean Society*, **29**, 37–47.

Shaw, A. J. (1990*a*). Metal tolerances and cotolerances in the moss, *Funaria hygrometrica. Canadian Journal of Botany*, **68**, 2275–82.

Shaw, A. J. (1990*b*). Intraclonal variation in morphology, growth rate, and copper tolerance in the moss, *Funaria hygrometrica. Evolution*, **44**, 441–7.

Shaw, A. J. (1991*a*). Genetic structure of sporophytic and gametophytic populations in the moss, *Funaria hygrometrica. Evolution*, **45**, 1260–74.

Shaw, A. J. (1991*b*). Ecological genetics, evolutionary constraints, and the systematics of bryophytes. *Advances in Bryology* (in press).

Shaw, A. J. and Albright, D. (1990). Potential for the evolution of heavy metal tolerance in *Bryum argenteum*, a moss. II: Generalized tolerances among diverse populations. *Bryologist*, **93**, 187–92.

Shaw, A. J., Antonovics, J., and Anderson, L. E. (1987). Intra- and interspecific variation in mosses in tolerance to copper and zinc. *Evolution*, **41**, 1312–25.

Shaw, A. J., Beer, S. C., and Lutz, J. (1989). Potential for the evolution of heavy metal tolerance in *Bryum argenteum*, a moss. I: Variation within and among populations. *The Bryologist*, **92**, 72–80.

Stanley, S. M. (1979). *Macroevolution: pattern and process*. Freeman, San Francisco.

Topham, P. B. (1977). Colonization, growth, succession and competition. In *Lichen ecology*, (ed. M. R. D. Seaward), pp. 31–68. Academic Press, London.

Tschermak-Woess, E. (1988). The algal partner. In *CRC handbook of lichenology*, (ed. M. Galun), pp. 39–92. CRC Press, Boca Raton, Florida.

Weber, W. A. (1977). Environmental modification and lichen taxonomy. In *Lichen ecology*, (ed. M. R. D. Seaward), pp. 9–29. Academic Press, London.

Weitz, S. and Heyn, C. C. (1981). Intra-specific differentiation within the cosmopolitan moss species, *Funaria hygrometrica* Hedw. *The Bryologist*, **84**, 315–34.

Wettstein, F. (1924). Morphologie und Physiologie des Formwechsels der Moose auf genetischer Grundlage. I: *Zietscrift fur Inducktive Abstammung- und Vererbungslehre*, **33**, 1–236.

Wyatt, R., Stoneburner, A., and Odrzykoski, I. J. (1989*a*). Bryophyte isozymes: systematic and evolutionary implications. In *Isozymes in plant biology*, (eds D. E. Soltis and P. L. Soltis), pp. 221–40. Dioscorides Press, Portland, Oregon.

Wyatt, R., Odryzykoski, I. J., and Stoneburner, A. (1989*b*). High levels of genetic variability in the haploid moss, *Plagiomnium ciliare. Evolution*, **43**, 1085–96.

Zielinski, R. (1984). Electrophoretic evidence of cross-fertilization in the monoecious *Pellia epiphylla*, $n = 9$. *Journal of the Hattori Botanical Laboratory*, **56**, 255–62.

Zielinski, R. (1986). Cross-fertilization in the monoecious *Pellia borealis*, $n = 18$, and spatial distribution of two peroxidase genotypes. *Heredity*, **56**, 299–304.

Taxonomic Index

Subject Index